The 25th Bomb Group (Rcn) in World War II

Norman Malayney

Schiffer Military History
Atglen, PA

ACKNOWLEDGEMENTS

I am grateful to the following people for providing information by correspondence, e-mails, personal interviews and telephone interviews regarding their wartime experiences, anecdotes, unit operations, and for sharing access to log books, photographs and personal diaries: Russel Abbey, Harold Aronson, Chris Ashcroft, Charles L. Austin, Robert M. Balser, Warren R. Barber, Larry Barbier, Dana Bell, Wendell A. Biggers, Bernard M. Blaum, Warren Bodie, Robert L. Boone, Warren Borges. Michael J.F. Bowyer, Roy Ellis-Brown, Roland H. Bushner, Joseph F. Capicotto, Frank Catalano, John M. Carter, Herbert J. Caudle, Edwin R.Cerruti, Vance J. Chipman, Keith A. Clinton, William A. Coleal, William G. Cannon, Roy C. Conyers, Ben Curry, Edward H. Darilek, Warren C. Davis, Peter D. Dustman, Marvin Edwards, Len A. Erickson, Aldon Ferguson, Thomas G. Fields, Robert W. Fish, George H. Fogarty OSS, Ian Ford, Harold L. Fordham, Richard O. Foster, Roger A. Freeman, Richard N. Geary, Charles F. Gieswein, Dewitt R. Goddard, Ken Godfrey, J. H. Gene Goodbread, Robert P. Gordon, William E. Grey, Albert J. Grosel, Clayton P. Hackman, James W. Handren, Donald F. Harris, Elbert F. Harris, Robert B. Hastie, Erik V. Hawkinson, John Hey, Dave Howley, William E. Huhtala, Morton M. Hunt, Michael J. Lorenz, Robert P. Howle, John K. Jackson, Harold L. James, Edward P. Jones, Coleman C. Kalick, Richard M. Kenny Kenneth H. Kenyon, Laurence J. Ketterer, Frank C. Klesch, Lydon F. Lakeman, Victor J. Layton OSS, Robert L. Lee, Joseph M. Levy, Alfred J. Lewis, Willis D. Locke, W. Eason Loden, Ken V. Lundberg, Milton R. Madsen, Benjamin C. Marshall, Florence McCarthy, Daniel A. McGovern, Frank A. McKee, Logan E. McKinley, James M. McNulty, James H. Meckes, Jr., Robert C. Menken, William G. Miskho, Claude C. Moore, Allan Morrow, John J. Morrison, Hugh V. Morrison, Earl L. Muchway, John E. Murphy, Jean Nater OSS, Roy Conyers Nesbit, Wiley Noble, Brunnon C. Normand OSS, John C. Nowak, Sam Odensky, Mark N. Ortega, Allen Ostrom, Thomas R. Owens, Andre E. Pecquet OSS, Alvin E. Podway, Robert K. Poe, William E. Poelstra. Byron H. Pollit, Ford D. Porter, Lionel A. Proulx, Albert D. Rasmussen, James B. Richmond, Louis A. Rogers, Dean H. Sanner, William L, Sawyer OSS, William E. Scharbach, Louis E. Schueler, Lewis J. Schumacher, Janet Scruggs, Jerry Scutts, Franz Selinger, Bill Setchell, Ben O. Shaver, Jack W. Sheen, William W. Siler. Steve H. Simpson Jr. OSS, Steve H, Simpson III, Edward F. Sofranko, Robert G. Spaight, Hinrik Steinsson, Allen R. Swerdlowe, John L. Swingen Jr., Thomas M. Tate, Richard C. Taylor, Ralph Thierry, Louis J. Toth, Anthony Turano OSS, Richard W. Tyhurst, Charles A. Veldon, Clayborne Vinyard, Robert P. Walker, Carl J. Wanka, Allan T. Wayne, Richard Wright, Richard L. Wulfkuhle, WRGI, William Young.

Book Design by Mark David Bowyer.

Library of Congress Control Number: 2011934908

Printed in China.
ISBN: 978-0-7643-3950-9

We are always looking for people to write books on new and related subjects.
If you have an idea for a book please contact us at the address below.

Published by Schiffer Publishing Ltd.
4880 Lower Valley Road
Atglen, PA 19310
Phone: (610) 593-1777
FAX: (610) 593-2002
E-mail: Info@schifferbooks.com.
Visit our web site at: www.schifferbooks.com
Please write for a free catalog.
This book may be purchased from the publisher.
Please include $5.00 postage.
Try your bookstore first.

In Europe, Schiffer books are distributed by:
Bushwood Books
6 Marksbury Avenue
Kew Gardens
Surrey TW9 4JF, England
Phone: 44 (0) 20 8392-8585
FAX: 44 (0) 20 8392-9876
E-mail: Info@bushwoodbooks.co.uk.
Visit our website at: www.bushwoodbooks.co.uk
Try your bookstore first.

Contents

Introduction

In the early-1980s, an England-based aviation researcher contacted me requesting information on USAAF Mosquito aircraft operations. A literature search revealed virtually nothing documented on American Mosquitos, nor on USAAF acquisition of these aircraft. In fact, it revealed poorly researched information on this subject that continues to this day. And so began my project to document the 25 BG (Rcn) history.

After attending a 25 BG (Rcn) Association reunion and meeting surviving veterans, I began in earnest to document the unit history. In the interim, many veterans had passed on and others could not be located, thus skewing the history to those surviving members. To overcome this handicap, missing information was obtained by researching official USAAF and OSS documents in an attempt to present a representative history of operations. Previously, Patricia Fussel-Keen detailed the 7 PG history in her book Eyes of the Eighth. The following presentation attempts to complement her study, and also place in perspective USAAF photo-reconnaissance development in the ETO.

The magnificent work of Eighth Air Force historian Roger A. Freeman served as a major reference source to assist in compiling the 25 BG (Rcn) history. His work is not without fault. Where conflict exists between his interpretation of events as compared to information compiled from interviews, logbooks, and USAFHRA files not referenced by him, I claim my documentation of events as more accurate.

Besides documenting the 25 BG (Rcn) operations, this study includes both 8 Combat Camera Unit and 50 Fighter Squadron histories. This required consulting microfilm unit diaries, locating lost souls, and OSS Joan-Eleanor project veterans, to document their recollections. Memories fade with time, but logbook entries do not.

All Watton-based aircraft flew individual missions, thus the lack of air-to-air photography images; therefore, flying photos of aircraft are indeed rare. Near the end of the war, one veteran documented the aerial Mosquito perspective during a Grapypea mission. Photo coverage is skewed towards the few veterans located and these images comprise mainly official accident photos.

The history of the OSS Joan-Eleanor Project (OSS/JE) first appeared in Joseph Persico's book Piercing the Reich. I have attempted to complement his account by detailing the role played by 25 BG (Rcn) personnel in these secret operations, following a particular format. It documents both OSS/JE Mosquito and A-26 aircraft missions flown from Watton, nicknamed Redstsocking. This follows with transfer of OSS/JE operations to Harrington under 492 BG control, where the Redstocking label ceased to exist. It continues to follow Mosquito communication flights to JE radio-equipped agents. For continuity, all A-26 missions are also detailed. B-24 insertion/supply missions to only JE-equipped agents behind enemy lines is noted. All Douglas A-20 flights are included since they are few in number. I hope this summarized presentation provides the researcher an acceptable perspective of OSS/JE operations with emphasis on Mosquito missions.

The last chapter briefly summarizes each JE-equipped agents' progress behind enemy lines. The appendix provides summarized data on JE mission flown, synopsis of OSS agents inserted behind enemy lines, and other pertinent 25th BG (Rcn) information. The final product of this research appears in the following pages as a best-case scenario representative history.

25th Bomb Group (Rcn) History

Norman Malayney

The United States involvement with the de Havilland Mosquito began in 1941. Army Air Corps (AAC) Commander Major General Henry H. Arnold traveled to England for an official fact-finding trip relating to British development of jet propulsion.[1] During the visit he viewed an RAF aerial exhibit on 20 April that included the demonstration of a de Havilland Mosquito.[2] As a result of this display, he returned to Washington with technical data pertaining to the Mosquito, Wellington bomber pressurized cabin, and Douglas DB-7 multiple nose gun installation.

The Mosquito data was dispatched and arrived at Wright Field on 4 June 1941. This included two rolls containing 28 sheets of drawings and one folder of descriptive material for study by the Experimental Engineering Division.[3] After examining the drawings, they replied on 27 August:

> "Although in some respect an interesting development (it) does not have a place in the defense program of this country … it would be extremely difficult to find anybody with the available time, personnel and facilities who could go in to production on this or any other British airplane … The famous Mosquito is yet to prove itself more than a hope. Its top speed has been less than 400 and the diving tests held up due to stresses, etc. The present industry in this country is loaded to the absolute limit. This office sees no practical way that foreign projects can be transported to this country and put in to production, especially when they are now in the stage of experimental development in England."

Despite this disclaimer, the technical material circulated among select contractors involved with Army Air Force (AAF) wood aircraft projects, i.e., Cessna, Curtis-Wright, Fairchild, Fleetwing, Beech, and Hughes, but none exhibited interest in the Mosquito-type construction. Only Beech and Fleetwings responded with negative assessments on the technical merits of wooden aircraft. Though they never replied, the Hughes Aircraft Company may have later incorporated this technology in the all-wood D-2/5 proposal and HK-1 Hercules construction.

If private industry displayed no interest, the US Navy certainly did when on 8 November 1941 they requested access to the Mosquito drawings. And on 12 December AAF contacted the British Air Commission (BAC) in Washington requesting loan of a flyable Mosquito, together with spares for flight tests and evaluation purposes. The BAC granted the request promising the first Mosquito bomber available from production.

The various air force departments had different photo requirements that no single aircraft could achieve. Charting required high-altitude photography for constant-scale navigation chart production. Mapping required low- and medium-altitude prints to create detailed maps for tactical and strategic use by both ground and air forces. Further, photo-reconnaissance (PR) provided low- and medium-level photos for tactical use (ground support), while medium- and high-level photography aided strategic targeting (destroy war production, morale, and economy).

Shortly after Japan attacked Pearl Harbor on 7 December 1941 an RAF official visited Washington to explore the AAF Intelligence requirements. The British Air Ministry (AM) agreed to supply the US with Intelligence concerning Axis forces and targets. Not only did AAF lack tactical and strategic Intelligence expertise, they had no aircraft specifically designed to collect this information. Air Transport Command (ATC) had diverted four B-24s to photography use for charting but not mapping as requested by the Corps of Engineers.

In June, four B-17s modified for mapping in the European Theater of Operations (ETO) proved unsuitable for tactical use. In late 1942 a number of B-25s underwent modification for mapping but again were not considered prudent operationally in a war theater. In combat areas, units converted conventional P-39s and P-40s to camera ships on an ad hoc basis for short-range, close-support role. By 1942, the conversion of Lockheed P-38 to F-4 status with camera nose finally provided AAF the only tactical PR aircraft then available.

The Engineering section at Wright Field initiated a program evaluating foreign-built aircraft technology for possible application to AAF requirements. The British shipped a Mosquito to the newly opened De Havilland Aircraft Company of Canada (DHC) factory in Toronto for their engineers to examine. Crated Mosquito DK287 arrived slightly damaged in Halifax, Nova Scotia, on 12 October 1942 and assembled in preparation for the flight to Toronto. The RAF provided one airframe fitter and two mechanics for engine maintenance.

Geoffrey de Havilland visited DHC, Toronto to inspect the new Mosquito manufacturing facility and test flew the first production model KB300. He borrowed the Mosquito for a U.S. demonstration tour and arrived on 4 December 1942 at Washington National Airport. An hour later he presented a convincing exhibition of Mosquito handling and maneuverability. The impressive thirty minute aerial display closed with a safe landing.

KB300 at Wright Field, Ohio. *Photo NARA.*

1 US Army Corps became the US Army Air Force on 20 May 1941.
2 C. Martin Sharp & Michael J F Bowyer, Mosquito, UK,:Crecey Books, 1995: 96.
3 Airplanes-Mosquito-British-General RD2612, NARA, College Park, MD. cited throughout text.

When Lockheed Aircraft Company learned KB300 was in Washington, they requested the Mosquito be ferried to California for performance testing with the P-38. The BAC agreed to the west coast visit for official AAF use only, but, "if it was (for) some manufacturer, out of idle curiosity, they didn't want to." Therefore, AAF stipulated the aircraft would not participate in any comparative performance test and only allow various manufacturers to inspect the aircraft. DHC pilot James R. Follett with DH engineer Francis Burell flew KB300 to Santa Monica arriving on 11 December with letters from BAC stating no demonstration flights be allowed. Northrop, Hughes Aircraft, and other manufacturers inspected the Mosquito under these restrictions.

Elliott Roosevelt, second son of the U.S. president, profoundly influenced AAF policy acquiring Mosquito aircraft. At age 22 he was in partnership with the New York advertising agency of Kelly, Nason and Roosevelt. A student pilot since 1928 he began working for publishing magnate William Randolph Hearst's newspaper chain in 1933 as aviation editor. He advocated expansion of U.S. air power and established contacts with aircraft manufacturer Donald W. Douglas and AAC Lt.Col. Henry "Hap" Arnold, then in charge of March Field.[4] In 1936 Hearst hired Elliott as vice-president of his radio enterprises. He commissioned him to lobby the reversal of U.S. administration cost-cutting policy of canceling government airmail contracts to commercial airlines, a task President Roosevelt assigned to AAC on 19 February 1934. He also became a regular radio commentator on the Mutual Broadcasting System.[5]

With war clouds hovering over Europe and Asia, Roosevelt enlisted in the AAC with rank of Captain because of age and aviation experience. A visual problem precluded his flying military aircraft. However, Republican Party furor over his commission and appointment to AAC procurement division forced him to submit his resignation with plans to enlist in the RCAF.[6] AAC rejected this request and reassigned him to 21 Reconnaissance Squadron, Gander Lake, Newfoundland, Canada. Here he assisted surveying potential sites along the northern Atlantic rim for development as staging areas to ferry lend-lease aircraft to England.[7,8,9] He later returned and in December 1941 graduated a navigator from Kelly Field.[10] In April 1942 he accompanied a B-17 to western Africa on an aerial photographic survey.[11]

In January 1942 the AAF had only four units experienced in aerial charting work, designated as mapping squadrons. Of these, the 5 Photographic Squadron (PS) formed 1 February 1942 with Lockheed F-4 aircraft and assigned to Eighth Air Force, Fort Dix, New Jersey on 25 May. By 13 July the entire 5 PS ground echelon arrived in England by ship. The aircrews remained at the Photo Reconnaissance Operational Training Unit (PROTU), Colorado Springs, Colorado, for flight training, arriving in England during September 1942. The 5 PS was then reassigned to Twelfth Air Force on 10 September and arrived in Oran, Algeria, in November.

On 1 July 1942, 3 Photo Group (PG) was activated at PROTU with one mapping (15th) and three PR (12th, 13th and 14th) squadrons. The Group (less 14th) transferred to England in September with Major Roosevelt as CO and received RAF PR training. On 18 November he took the unit to North Africa following TORCH, the Allied invasion, to join Twelfth Air Force which assigned 5 PS to his Group. Lacking aircraft, 13 PS remained in England as the nucleus around which the 7 Photo Reconnaissance Group (7 PRG) would later form. En route, Roosevelt and Major Harry T. Eidson stopped at Gibraltar to borrow an RAF Mosquito.[12] Upon arrival in North Africa both men flew the Group's first combat sorties over enemy held Tunisia.

At Oran, Algeria, 3 PG joined No.4 PRU RAF (including 682 Squadron with support and interpretation personnel) dispatched from Gibraltar during the first week of Allied occupation. The 3 PG (5, 12 and 15 PS) responsibilities included providing bombardment units with target information and bomb damage assessments; photo Intelligence to all Mediterranean Theater of Operations (MTO) commanders; Intelligence for Naval commands regarding movements of all Axis shipping in the Mediterranean.[13] The operational debut of the 15 PS Boeing B-17Fs as a camera platform proved disastrous. Losses forced withdrawing the Flying Fortress to operate as a courier aircraft and the aircrews reassigned to 5 and 12 Squadron with Lockheed F-4s.

F-4 on PR flying over MTO area. *Photo USAFHRA*

B-17 attached to 3 PG. *Photo USAFHRA*

The Mosquito became Lt.Col. Roosevelt's favorite mount allowing his participation in operational missions. On one sortie with newly promoted Lt.Col. Eidson at the controls, Roosevelt was forward taking pictures when Eidson happened to glance down and notice his disconnected oxygen supply hose on his lap. He reached down but lacked coordination to grasp it. He called on the interphone to Roosevelt, who came back and steadied the control column with one hand, while reconnecting the oxygen tube. Roosevelt continued maintaining flight control for five minutes until the pilot recovered. Eidson felt no other ill effects until after landing when his legs buckled upon exiting the aircraft.

By January 1943, 3 PG operated from Maison Blanche, Algiers. At this time, Prime Minister Winston Churchill and President Roosevelt, along with other Allied representatives, met at Casablanca, Morocco, 14-20 January 1943. Besides deciding the invasion of Sicily and Italy, they called for greater cooperation to further Intelligence gathering in the MTO. The AAF on the Axis western flank and RAF middle-east forces in Egypt, therefore amalgamated.

4 Elliot Roosevelt and James Brough, A Rendezvous With Destiny, New York: Putnam's & Sons, 1975: 101.
5 W. A. Swanberg, Citizen Hearst, New York: Charles Scribner & Sons, Ltd., 1996: 477.
6 Ibid: 272.
7 Ibid: 281.
8 Glenn B. Infield, Unarmed and Unafraid, New York: Macmillan, 1971: 71.
9 Elliott Roosevelt, If My Parents Could Hear Me, Parade Magazine (October 10, 1982): 18.
10 Journal of the 7 Photo Recon Group, 10:3, (February 1991): 5.
11 Roy Conyers Nesbit, Eyes of the RAF, UK: Sutton Publishing Ltd., 1996: 140.
12 Sharp & Bowyer, Mosquito, Two Mosquitos were assigned to Roosevelt by RAF: 372.
13 Infield, Unarmed and Unafraid, 71.

On 18 February, 3 PG and 4 PRU RAF consolidated to form the Northwest African Photographic Reconnaissance Wing (Prov) with Lt.Col. Elliott Roosevelt as CO and W/C Eric Fuller, Deputy Commander. As part of Lt.Gen. Carl A. Spaatz's Northwest African Air Force, NAPRW comprised a mixture of 2,800 American, British, Canadian, Australian, New Zealanders, Czechs, and Poles cooperating to avoid any duplication of efforts. It included a joint Central Interpretation Unit (CIU) and the Americans learned well from their more experienced British counterparts.[14]

At the Casablanca conference, Lt.Col. Roosevelt provided glowing accounts of the Mosquito's capabilities vis-a-vis the Spitfire and F-4. These favorable reports filtered back to HQ AAF in Washington, DC. His praise was not shared by other Twelfth Air Force officers: "The Mosquito with low- or medium-altitude engines is useless for our purposes. With the Merlin 61 engine its usability has yet to be proven."[15]

A detachment of 3 PG Lockheed F-4s transferred to Malta in March 1943 attached to 683 Squadron RAF under G/C Adrian Warburton. Their duty involved photographing Sicily for map production to facilitate the invasion campaign known as Operation Husky. Upon completion of this project, similar PR work began over mainland Italy for Operation Avalanche. On Wednesday, May 12, the Wehrmacht's *Afrikakorps* capitulated, ending all organized Axis resistance in Tunisia. The 3 PG then expanded PR work covering all Italy and southern Europe. Engaged in this endeavor, a Mosquito carrying Colonels Plothinger and Roosevelt made a crash landing at Mason Blanche aerodrome on 21 May. By June, 60 Squadron SAAF (Mosquito) joined NAPRW, along with 683 Squadron RAF.[16] Meanwhile, Lt.Col. Eidson went to Wright Field and joined PR authority Lt.Col. Karl Polifka as consultants to the AAF PR aircraft program.

USAAF and RAF airmen gather around Wing Commander Adrian Warburton. *Photo USAFHRA.*

14 Roy M. Stanley II, World War II Photo Intelligence, New York: Scribner, 1981: 58.
15 WG-90-SW, USAF Historical Records Agency, Maxwell AFB, Alabama. (Herein know as USAFHRA).

16 Nesbit, Eyes of the RAF, 145.

Capt. Arthur Power returns from a PR mission in an F-4. *Photo USAFHRA*

In preparation for Operation Husky and the invasion of mainland Italy, a B-25 underwent conversion to carry 12 photo flash bombs (PFB) to capture on film the night movements of enemy troops and material. Equipped with K-19A cameras and electric eye, the bomber released PFBs from 12,500 feet at 20-second intervals with a delayed fuse set to ignite at 3,000 feet.

U.S., British, Canadian, and French forces invaded Sicily 9/10 July. Late that evening 3 PG sent Major George Humbrecht on the first B-25 night photo mission with Colonel Roosevelt as copilot/navigator. They left La Marsa, Tunisia, and flew over Palermo. The photo run started one mile prior to the target. The successive PFB flashes alerted antiaircraft and searchlight defenses to mount a horrendous response. Unscathed, both men flew the next two night photo missions over Sicily. Actor Joe E. Brown flew as guest observer during a night B-25 mission over Italy.

Continued enemy activity presented problems. On one occasion, two German night-fighters were encountered without action. Heavy flak proved consistent on most flights, and except for one explosion blowing off the top-turret and slightly wounding the gunner, the B-25 night photo missions proved successful.[17]

L-R. Lt.Col. Elliott Roosevelt with Major Leon Gray, NAPRW. *Photo. USAFHRA*

Two seasoned NAPRW veterans, who would later command the 25 Bomb Group, earned their distinction with 3 PG. On 5 June 1943, one hour after B-17's bombed Spezia, Italy, Capt. Leon Gray flew his F-5A, and by PR covered Spezia Harbor for bomb damage and Leghorn Harbor for shipping information. Arriving over Leghorn first, he made two runs through heavy accurate flak before being chased by three enemy fighters. After flying out to sea and losing the fighters, he performed four runs over Spezia Harbor and one over an aerodrome in the vicinity, at which time five enemy fighters chased him. Returning again, he made high-speed photo runs over two ships off the Italian coast and over two points in Corsica.

On 27 July, Major Gray completed a dicing mission over the northern coast of east Sicily. As he approached on his first run at less than 300 feet, the shore guns began firing. For twenty minutes he continued taking oblique pictures of enemy activity from as low as fifty feet. Tracers and antiaircraft shells bracketed the aircraft, yet he maintained a straight photographic run at 300 feet. Undeterred, he continued obtaining pictures of beaches, defiles, strong points, barbed wire, and photos of antiaircraft batteries in action on the beaches. Later, while performing a low level run in Yugoslavia, ground fire severely damaged his F-5. He ditched the aircraft in the Adriatic Sea and was soon rescued. The other luminary, 1/Lt. John R. Hoover, on 18 July 1943, flew to the Milan area of northern Italy, located beyond the accepted F-5A range. He was cautioned to remain only twenty minutes at this extreme distance. Instead he spent an hour in the target area and fully covered 23 aerodromes and pinpoints. He spent six hours at 30,000 feet on oxygen during the seven-hour mission. Fuel consumption was so controlled, he returned with 100 gallons fuel remaining, setting a new standard on operational range for all future F-5A flights.

17 WG-90-SW, USAFHRA.

NAPRW at La Marsa, Tunisia, expanded in July.[18] Reconnaissance Group II/33, Free French AF arrived and became operational 3 August. Its flight crew included writer/poet Major Antoine de St. Exupery. And on 30 October, 682 Squadron RAF and 60 SAAF Squadron formed 336 Wing, RAF and were soon joined by 683 Squadron from Malta on 21 November. The US contingent also underwent reorganization when 5 PG joined 3 PG to form 90 Photographic Wing (Rcn) on 22 November, providing PR for both Twelfth and Fifteenth Air Forces.[19] At this time, NAPRW became Mediterranean Allied Photo Reconnaissance Command (MAPRC) having both 336 Wing RAF and 90 PW with Col. Roosevelt as CO. Note, the AAF Group is equivalent to RAF Wing, and an AAF Wing is equivalent to RAF Group.

The 90 PW created an AAF photo and Intelligence organization independent of the British, capable of transferring intact to another theater of operation. Behind the scenes, identical plans were formulated to develop a similar independent PR AAF organization (325 PW Rcn.) in the ETO.

The AAF decision to acquire Mosquito aircraft remained an unsettled issue. For many years, qualifications for PR aircraft was subject to debate between Chief of Engineering and HQ AAF. Aircraft to perform this role were never developed nor procured, thus mandating conversion of combat aircraft to perform this duty. The importance of the task was summarized in a 23 January 1943 letter by the Secretary of War:

> "In view of the urgent requirement for mapping photography, action will be taken to provide suitable equipment and trained personnel as early as practicable. The minimum requirements of photographic airplanes for such a mission include capability for sustained operations at altitudes in excess of 40,000 feet, heated and pressurized cabins accommodating a minimum crew of three, and requisite photographic and navigational equipment."

The RAF shot down a Ju 86 PR aircraft at 51,000 feet over Egypt, therefore, the above requirements were realistic. On this recommendation, General Arnold directed on 25 January, that action be initiated to develop and produce a photographic airplane suitable to meet this requirement.

Returning reconnaissance pilots from SWPA and North Africa indicated the Lockheed F-4 proved less than satisfactory in this role. It performed well but lacked range for deep penetration of enemy territory. Officers attending the Casablanca conference brought endorsements from Lt.Col. Roosevelt that the British Mosquito proved superior for this role in high speed, longer range, higher altitude performance, plus a crew of two increased precision work.

The Chief of Air Staff, General George E. Stratemeyer directed Assistant Secretary of War for Air, Robert A. Lovett to coordinate staff efforts to investigate experimental aircraft, e.g. McDonnell XP-67 to meet the medium-range PR requirements; develop an aircraft specifically to fulfill the PR role – the Aerial Mapping Committee (AMC) was formed and delegated this duty; in the interim, procure PR Mosquito aircraft.

The AMC met 9 April 1943 to discuss technical problems concerning prerequisites for a PR airplane to meet the mapping mission of the Corps of Engineers. The F-5 had one crewmember and its rigid camera installation resulted in excessive tilt and crab, deemed unsuitable for mapping photography. The heavier B-17, B-24 and B-25 lacked sufficient operational altitude over heavily defended areas. The minimal requirements for a "super" PR aircraft included: a minimum crew of two in a heated cockpit with maximum visibility for all members; a speed of 400 mph at 30,000 feet with a 3,000 miles range; a pressurized cabin capable of maintaining an interior pressure of 10,000 feet at a service ceiling of 60,000 feet; carry a minimum of three vertical-camera stations, plus two oblique cameras. For that time period, this proved a formidable technological task to achieve.

After further deliberation, AMC concluded the gestation of a super PR aircraft required two or three years, by which time fighter aircraft development would nullify its effectiveness. Reviewing this matter further, they confronted the problem of justifying development and production of a special PR airplane, because the small number required would never justify the diversion of developmental personnel and cost involved. Instead they suggested investigating current aircraft whose basic design allowed conversion to a mapping airplane. They previously expressed interest in the XP-71, XP-67 and a redesigned P-38; also suitable were the XB-35, XB-36 and the XB-39. AMC recommended meeting PR needs through conversion of production combat aircraft to the PR role; in the interim, procure the Mosquito for PR role above 30,000 feet and cancel the super PR airplane development. General Arnold ignored the later endorsement and on 26 May, issued to industry the AMC minimum requirements for a special photograph-mapping airplane.

Meanwhile, General Spaatz pleaded with Arnold his need for long-range PR aircraft to cover the vast MTO area. Since the lack of Rolls-Royce Merlin 61 series engines impeded UK Mosquito delivery, he suggested obtaining 24 airframes, minus engines, and dispatch by highest priority, 48 Packard-built Merlins to England instead. This would provide 12 aircraft to the MTO and 12 for PRU RAF Benson. In the interim, he requested four of the 12 Mosquitos at PRU Benson, England, be assigned to NAPRW.

General Stratemeyer contacted the AM and requested Mosquitos for AAF units in the ETO/MTO. The RAF experienced operational problems with the light-bomber Mosquito, and he suggested allocating 24 production examples, minus engines, which the AAF would supply. Unfortunately, the Packard Merlin engine proved critical for the P-51 program and unavailable from USA production. On 24 February, he proposed delivery of 200 Mosquitos on a quid-pro-quo basis for 1,000 P-51Cs delivered to UK under the lend-lease program. The British balked at this proposal stating their PR Mosquito output was only one-in-ten of total production. Instead, the AM allotted five Mosquitos (designated F-8) from DHC awaiting delivery while another 19 would receive consideration.

Generals Ira C. Eaker (ETO) and Carl Spaatz (MTO) disagreed over the Mosquito's merits. Spaatz coveted its high altitude, long range, and remarkable speed. But General Eaker had great reservations:

> "There have been numerous requests for Mosquito planes for special photo jobs. The Mosquito is not the answer. The F-5s outperforms them in almost every way. Only advantage is the Mosquito carries a navigator and has long-range ... The British have lost nine out of 13 Mosquito missions over the continent. The F-5 may not get the pictures wanted, but the Mosquito, even though they get the pictures, will probably be shot down before they get back ... "

On 1 April, AAF again requested 200 Mosquitos for 1943: 90 percent for high-altitude charting, mapping, and PR work; 10 percent for oblique PR at low altitude requiring defensive and offensive armament. The British considered supplying these from the UK-financed Canadian production to reequip five existing AAF squadrons with further provision for two new squadrons. This 1943 requirement totaled 235 aircraft, allotted to 5, 12, 23 (Twelfth Air Force), 13, 14, 22 and 27 Photo Squadrons (Eighth Air Force) with 13 each, three in reserve, and the remaining 127 for attrition; and 2 PG, PROTU at Peterson Field, Colorado, with 13 training aircraft and three for attrition. Unfortunately, DHC production would not peak until April 1944; therefore, the British offered 111 DHC aircraft against the 134 minimum AAF requirement for 1943.

General Arnold informed Spaatz the AAF planned procuring 35 Mosquitos from DHC: four to Peterson Field and one to Wright Field, with 30 more promised for 3 PG, MTO. In the meantime, AAF continued supplying Lockheed F-5s to ETO squadrons. This proved costly with three P-38s lost to combat squadrons for every production-line aircraft modified to F-5 standards.

The British on 15 July allocated 40 DHC and 80 UK airframes for 1943. The 40 DHC bombers designated F-8 would undergo PR conversion for 3 PG, MTO. The 80 UK Mosquitos were assigned as follows: 60 Mk.VI reserved for PR conversion earmarked as Far East replacements for F-7s (B-24): one Combat Mapping Squadron (CMS) for SWPA; one CMS for India; one CMS for China and 12 aircraft for training in USA. Conversion of 20 Mk.XIII to PR role was impractical; these were planned for AAF night-fighter units.

During a Washington visit, General Eaker again defended the F-5. "There is an immediate need for accurate mapping in certain parts of Europe. The British are trying to do it with the Mosquito ... is unsuited for the work and actually is not as good as our own F-5." This debate left an element of uncertainly at AAF HQ in acquiring the Mosquito.

18 Nesbit, Op. Cit., 164.
19 Nesbit, Op. Cit., 170.

The AAF Mosquito testing program began when UK-built Mosquito IV DK287 arrived at Wright Field from Toronto on 14 February 1943. After initial acceptance and fifty-hour safety inspection it was weighed and instrumented for tests. These began 25 February carrying four 500-pound bombs on a series of sawtooth climbs between 18,000 and 20,000 feet at maximum continuous power.

Mosquito DK287 undergoing evaluation at Wright Field. *Photo. NARA*

DK287 crash scene on 14 March 1943. *Photo N. Rosengarten*

Test pilot Col. Osmund J. Ritland readied DK287 for another test on 14 March. Taxing and takeoff were routine, and all engine temperatures and pressures remained normal throughout the climb to 19,500 feet. A shallow dive was initiated and a power calibration run begun at full throttle. The left engine temperature began to rise, and after four minutes reached maximum as smoke entered the cockpit. Ritland terminated the test and banked towards Wright Field in a descending turn.

Shortly after, the left engine began missing as flames spread from the inboard nacelle and wing leading edge. He feathered the prop and turned off the magneto, but the engine would not stop. The fire was out of control with flames extending past the wing behind the engine. Ritland ordered his navigator, 1/Lt. Nathan Rosengarten to jump. After considerable difficulty, he finally lifted the lower inner door and jettisoned the outer hatch. This immediately cleared the cockpit of smoke, and Rosengarten jumped from 16,000 feet. Upon exiting safely, the inner door slammed shut.

The fire now raged and all attempts to stop the windmilling failed. Fearing an imminent explosion, Ritland climbed through the bottom hatch and safely exited. Taking height loss in freefall, his parachute later deployed abruptly, snapping several stringers, resulting in a rapid descent and heavy landing. Accident investigators attributed the cause to a blown exhaust manifold on the left engine.

A conference held at Wright Field on 17 March requested the loan of a DHC-built Mosquito to continue the tests. Mosquito B Mk.VII, KB306 arrived at Wright Field on 25 March. After special equipment installation the aircraft was subjected to a test flight program from 1 April to 26 May. The evaluation report concluded:

> "The Mosquito ... longitudinal stability ... unstable in ascent at speed-of-best climb. It was tail heavy and unstable longitudinally during landing approach, especially with full fuselage tanks and center of gravity located near the aft limit, and rather precarious for inexperienced pilots to land in this condition. The DHC Mosquito was more unstable longitudinally at landing speed than the first UK Mosquito."

The U.S. Navy experienced a dire requirement for night-fighter and PR aircraft. Previously, the Joint Chiefs of Staff on 3 November 1942 rejected a request for 18 P-38s and 18 B-25s for PR work in SWPA. In late March 1943 the Navy displayed interest when Mosquito KB300 arrived at NAS Anacostia, DC, for evaluation. During a full power climb to 18,000 feet on 30 March, the test pilot observed severe blistering on the starboard top cowl. He throttled back, pressed the fire extinguisher button and commenced descent for an emergency landing. On final approach, he restarted the engine to align the aircraft for a perfect landing. The fire was attributed to a complete gasket disintegration at the rear three cylinders. The U.S. Navy inquired about securing 100 night-fighters and 50 PR Mosquitos from DHC.[20]

Damaged cowling and exhaust shroud of KB300 after overheating during 31 March test flight. *Photo NARA.*

20 Chiefs of Staff, Memorandum, 5 April 1943, RG24 Vol. 5187, File S-15-9-59, (Vol. 5), Ottawa, Archives Canada.

Colonel Carl Polifka on left talks with de Havilland, England, test pilot Pat Fillingham at Downsview.
Photo DCHC-TAM

Production line at DHC with F-8s. *Photo J. Griffin*

The pilot's perspective of the Mosquito cockpit. *Photo USAF Museum*

Navigator's perspective of a Mosquito cockpit. *Photo USAF Museum*.

Ground-level perspective of DHC F-8 production line.
Photo J. Griffin

Lt.Cols. Karl Polifka and Harry Eidson prepared a report detailing various camera installations and airframe modifications required for the PR role. Both men were engaged as consultants to discuss camera installation in a DHC Mosquito as a mock-up for prototyping.

On 15 April, DHC stated they would perform factory modifications, only if AAF accepted the plant's total output. Any allocation less than this was counterproductive and recommended contracting modifications in the USA. Bell Aircraft Co., Niagara Falls, New York, eventually performed the modifications. The initial F-8s were assigned serial numbers 43-34924 to 43-35123 inclusive including the projected 200 aircraft from Canadian production.

General Eaker, CO Eighth Air Force, sent a cable to General Arnold on 23 July requesting the latest and fastest PR airplane be sent to the ETO:

> "... Present F-5s do not have necessary speed advantage at altitude to accomplish successful photo missions ... Six sorties dispatched to northwest Germany during June 11 to July 17 were each time intercepted by enemy fighters ... three did not return ... Operations have been restricted to northern France and low countries in effort to conserve remaining operational planes ... for unarmed penetration ... of northwest Germany it is imperative ... the very best aircraft be made available. Request future aircraft be PR version of the P-38J or other single- or twin-engine types of superior performance."

Again, Eaker made no reference to Mosquito aircraft meeting this requirement. By 25 June 1943 protracted discussions between AAF and British AM became finalized with General Arnold approving the Mosquito PR program. Efforts were initiated to forecast requirements for the first and second halves of 1944.

General Spaatz requested Mosquitos for post-Operation Husky program producing maps that benefited future ground operations. This involved surveying mainland Italy, southern France, and Albania. The British COS assigned four RAF Mosquitos having rigid-mounted cameras, but they were unsuitable for this purpose. Only American-built, gimbal-mounted cameras could achieve the task. Therefore, General Arnold directed that four suitably equipped F-8s with both air and groundcrews be dispatched by 5 July.[21]

The British previously allotted five DHC Mosquito B Mk.VII of 24 requested by General Arnold which AAF designated F-8. The first F-8 43-34928 arrived 1 June at Wright Field as a mock-up for camera installations. Both AAF and DHC personnel examined the modifications on 9 July. Installation drawings together with changes were supplied to DHC and the F-8 mock-up prototype returned to Toronto for evaluation. After examining the proposed installation, DHC engineers decided to conduct static fuselage tests involving cutouts for camera ports thus delaying production until October.

The other four F-8s, 43-34924 to 927 arrived at 2 PG, Peterson Field, Colorado, and formed a training squadron. DHC assigned a pilot and mechanic to instruct and transition AAF personnel to the Mosquito. In July, 12 enlisted men attended No.6 Repair Depot, RCAF Station Trenton, Ontario, for a four-week, Rolls-Royce engine-training program, one week at the DHC factory, and one week at No. 30 OTU Greenwood, Nova Scotia, for additional familiarization.

Training accidents took a toll. On 1 July, 2/Lt. Brian K. Moyers took F-8 43-34927 on his first solo flight and ground looped the aircraft, wiping out the undercarriage. He achieved two hours flight time with DH test pilot Patrick Fillingham who felt Moyers competent to fly the F-8. AAF officials increased the dual-training hours required before pilots went solo. A week later on 7 July, F-8 43-34925 piloted by Capt. Paul M. Thorngren with navigator 2/Lt. William G. Miskho, started off for a flight to Salinas, California. After rolling several hundred feet the aircraft swerved left and ground looped.

The delayed DHC delivery schedule and proviso the AAF accept the entire production, resulted in cancellation to reequip all seven ETO/MTO squadrons with Mosquitos. Therefore the two remaining F-8s at Peterson Field transferred to the MTO to meet Spaatz's request. In August, F-8s 43-34924 and 43-34926 went to Wright Field for radio, radio compass, and pilot compass installation, while replacement F-8s 43-34929 and 43-34930 arrived 18 and 31 August, respectively for similar preparation. Mosquito KB306 at Wright Field transferred to AAF as 43-34931, the sixth B VII so allotted. Meanwhile the Corps of Engineers expressed approval with photography performed by F-5s. As pilots gained experience and better models became available, photography results improved.

Previous plans called for delivering 120 DHC Mosquitos in 1943. The British revised this in August with 40 from Canada, and 80 from UK production comprising 60 Mk.VI fighter-bombers and 20 Mk.XIII night-fighters. Because photo equipment could not be installed in the fighter version, AAF requested more PR models instead. Drawings of DHC Mosquito camera equipment installations were dispatched for adaptation to UK production aircraft.

Further analysis required further changes. After studying Mosquito procurements, Material Command requested DHC delay deliveries until the Packard V1650-3 (Merlin 61) engine became available. After further review, General Arnold requested a change to the two-stage super-charger, Merlin powered PR Mk.IX instead of the Mk.VI. The AM reported on 9 October 1943, this could be achieved by postponing delivery to 1944. Arnold accepted the decelerated rate and the AM amended this proposal to 40 F-8s and UK delivery of 125 Mk.XVI, and 43 Mk.VI fighter-bombers. In the interim, AAF engaged Boeing to study the feasibility of a jet-powered PR aircraft as a possible candidate.

During summer 1943, inclement weather both in England and at European targets grounded Eighth Air Force bombers for long periods. In September, General Eaker requested immediate delivery of 30 Mk.VI Mosquitos to form an ETO weather reconnaissance squadron. He tasked 7 PRG, Mount Farm with gathering data to improve target weather predication and forecasting using the Mosquito aircraft. Groundcrews from 14 Squadron prepared to attend airframe and Merlin engine school. However, the plans were short-lived. Instead, AAF formulated a strategy establishing a Photo Wing within Eighth Air Force with a PR Group flying Mosquitos to accomplish this assignment.[22] The proposal was one facet in a general scheme establishing an independent ETO AAF Intelligence command.

Col. Elliott Roosevelt arrived in Washington, DC, from the MTO and on 3 September met with General Arnold. Following debriefing on the super reconnaissance aircraft requirements, Roosevelt recommended the wooden Hughes D-5 as the only design suitable for photographic purposes. Arnold gave instructions to issue Hughes Aircraft a contract for 100 planes. Air Service Command opposition to wood aircraft structures resulted in Hughes modifying the proposal to effect the all-metal XF-11.

Col. Elliot Roosevelt approved the Hughes XF-11 as best proposal for super PR aircraft. *Photo H. McCormick.*

21 CCS, 2 July 1943, File 486-4, USAFHRA

22 Patricia Fussel-Keene, Eyes of the Eighth, Sun City, Arizona: CAVU Publishers, 1996: 75.

Mosquito evaluations continued at Wright Field. On 14 September, Capt. Gustav Lundquist landed KB306 after a test flight. As he taxied to the parking apron, the left engine began billowing white smoke from the rear engine nacelle. Fearing fire, he feathered both engines and cut the switches, while actuating the fire extinguisher. The F-8 came to a full stop and both engines ceased firing. The navigator, Lt. Martin Smith attempted unsuccessfully to open the bottom exit hatch, as Lundquist released the top canopy cover and proceeded to climb out. At this point Smith released the idle cutoff levers and prepared to follow him. Upon release, the levers slid back into position and both engines began firing, causing the aircraft to roll forward. Lundquist immediately returned inside the cockpit to apply the brakes while Smith reengaged the idle cutoff levers. During this brief period the left wing struck the rudder of a parked P-38G as the whirling right prop sliced the wingtip off a nearby PT-23.

Another incident occurred when F-8 43-34930 was delivered to Wright Field in error. On 24 September, groundcrew taxied it to a storage area. At idling speed, the engines failed to recharge the pneumatic air cylinder for braking. While swinging the aircraft around a parked B-25, the brakes failed and the Mosquito crashed into the tail section of an adjacent parked A-26 41-39100. The collision wrecked the Mosquito forward fuselage and damaged both props and spinners. The A-26 suffered severe tail damage.

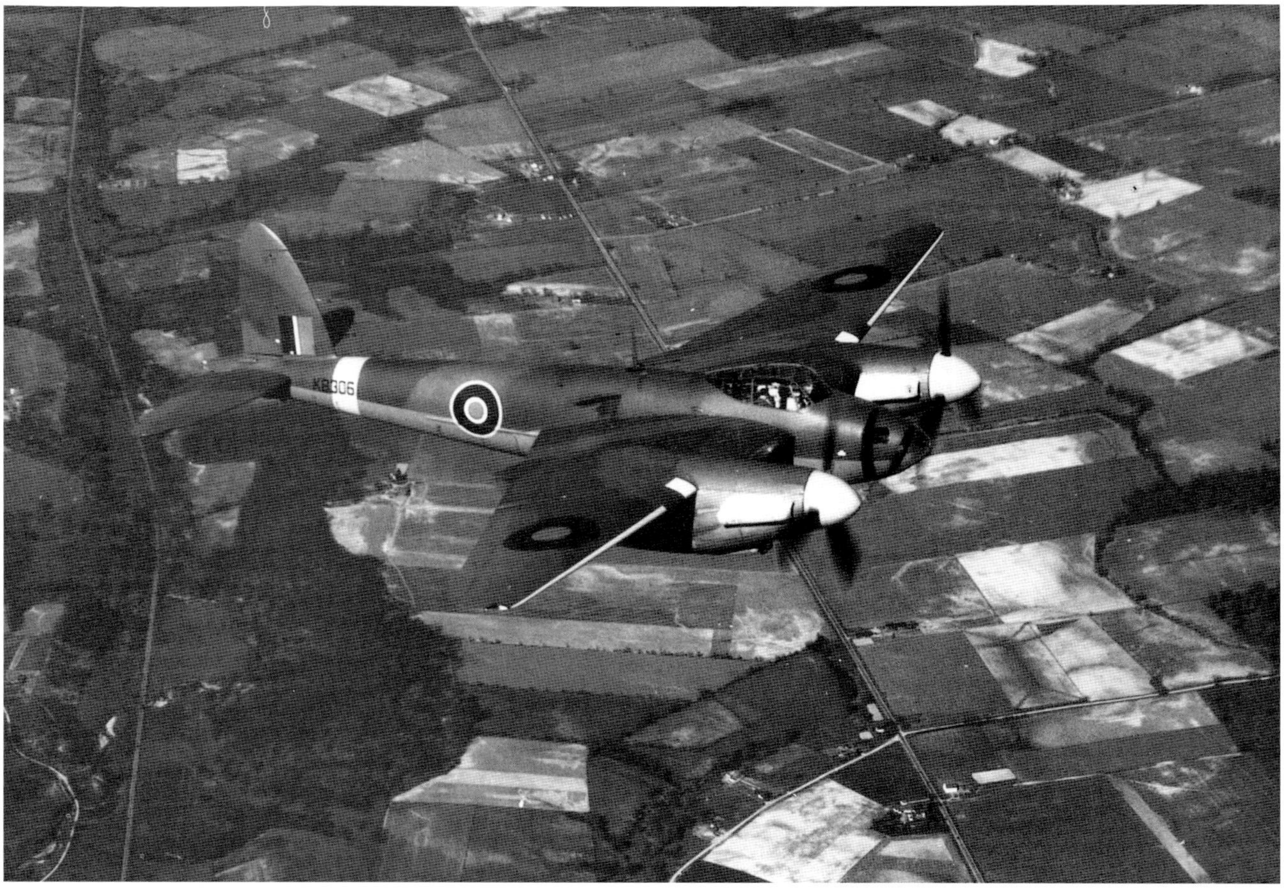

Flying over Ohio, KB306 eventually became F-8 43-34932. *Photo NARA.*

On 9 October 1943, Lt.Col. Karl Polifka ferried F-8 43-34926 from DHC, Toronto to Wright Field for CSR-294 transmitter installation. While removing a camera, groundcrew found highly flammable oil and gasoline soaked rags located in the fuselage, aft of the rear camera position. They were tightly placed and would not have been discovered except for the decision to reinstall the camera in another Mosquito.

Colonels Polifka and Roosevelt flew this F-8 to Bolling Field, DC, on the 13th. The landing gear failed to deploy, requiring Roosevelt to hand pump the undercarriage down. After a safe touchdown, an inspection revealed no hydraulic fluid in the system. A mechanic's drill had perforated a hydraulic line. An amateurish packing covered the puncture and the repair tightly clamped to the bomb bay ceiling. These events alarmed Roosevelt who instigated an investigation to determine if sabotage was involved.

Another Mosquito KB306 accident, now serial 43-34931, occurred 28 October, when Col. Thomas L. Thurlow landed at Bluethenthal Field, Wilmington, North Carolina. Instead of using the taxi strip, he cut across the field to reach the ramp. En route the right wheel dropped into a ditch damaging both propellers, the right engine, landing gear, nacelle and cowling.

F-8 43-34930 at wintry AAB Wright Field, Ohio, was involved in taxi accident with A-26 41-39100 on 24 September 1943. *Photo NARA*

On 13 January 1944, RCAF KB131 directly off the DHC production line has USAAF serial 43-34933 temporarily applied with water-based paint to the rear fuselage. It is ready for ferry flight to Niagara Falls, NY, for camera modification and permanent application of USAAF markings. It was involved in a fatal accident on 7 August 1944 at Rome, NY. *Photo DHC/TAM*

In November 1943, three F-8s were dispatched to North Africa over the South Atlantic air ferry route. The first, piloted by Lt.Col. Polifka successfully reached its destination in Africa, while a second Mosquito lost an engine en route to Belem, Brazil, and later safely returned to the USA. Lt.Col. Harry Eidson ferried the third F-8 43-34929 from Miami to Natal, Brazil, then on to Ascension Island by himself. On 27 November, he left Ascension alone at 0940 for Dakar, Africa. About 45 minutes later the left engine oil temperature went to maximum, so he feathered the Merlin and turned back. The island had an uphill sloping runway, and the local downdraft on approach did not help conditions any. The right engine started losing power once the undercarriage extended, causing rapid descent. The F-8 hit hard 250 feet prior to the runway, bounced and impacted an embankment, crushing the nose, ripping apart both engines and snapping off the tail. Eidson received a medical discharge from injuries incurred during the accident, which he claimed involved sabotage.

Tinker AAB, Oklahoma City, Oklahoma, prior to ferry flight to North Africa. It crashed 27 November 1943 at Ascension Island during an emergency landing. *Photo J. Campbell.*

Meanwhile, Major James Setchell and navigator Capt. Jerome C. Alexander successfully ferried F-8 43-34926 *The Spook* over the North Atlantic route to North Africa. The F-8 served with 32 Squadron, 5 PG, Twelfth Air Force, MTO.[23] After transfer to 3 PG, Major Setchell received the DFC for flying an F-8 on 16 and 17 January 1944 to successfully complete a PR mission, "at low altitude through accurate, intense flak with repeated hits on his aircraft."

L-R: Major James F. Setchell, pilot and Capt. Jerome Alexander navigator at DHC, Toronto to pickup F-8 Mosquito "The Spook." *Photo J. Griffin*

F-8 43-34926, "The Spook" preparing for flight to North Africa. *Photo J. Griffin*

In southern Germany, Fifteenth Air Force attacks encountered accurate and intense enemy radar-controlled Flak. The B-17F "Ferret" aircraft of 16 Rcn Squadron met with limited success. The formations' jamming activities interfered with obtaining accurate information. Ideally, a ferret aircraft should penetrate enemy defense alone to collect electronic information in advance of the formation. An appeal was made for three Mk.IX or Mk.XVI Mosquitos modified as ferret aircraft to fly unescorted deep in to enemy territory prior to actual bombing missions. None were available.

Mosquito aircraft allocations fluctuated. In October 1943 the AM assigned PR Mk.XVI Mosquitos to RAF MTO units, and on 23 November announced 43 Mk.VI Mosquitos allotted to AAF weather recon units. General Arnold objected. He questioned why the AM consigned inferior aircraft unsatisfactory for this role, and why they sent superior Mk.XVI models to MTO when enemy opposition proved more severe in the ETO? He requested suspending all Mk.VI deliveries. To satisfy Arnold, the AM in November allocated 40 DHC and 168 Mk.XVIs from 1944 UK production. In January 1944 a delivery agreement provided for 102 Mk.XVI scheduled at a rate of 10 per month for February to April, and 15 for May and June. An estimated attrition rate of 20 percent, or seven aircraft per month on reserve, was included in the agreement. By 23 May, deliveries for July to December 1944 was reduced to 42 aircraft, while the 1945 January to June allocation, scheduled only 36 aircraft. All were for the proposed 8 Reconnaissance Group (S) (P), Eighth Air Force.

And later on 28 March 1944 AAF contacted BAC in Washington requesting 50 Mk.XVI Mosquitos for weather reconnaissance in the Far East Theater. The BAC countered diverting aircraft at the expense of operational RAF units as unacceptable. They suggested AAF utilize F-8s for this purpose.

F-8 43-34949 at AAB Wright Field. *Photo NARA.*

Underside bomb bay view and entrance hatch to cockpit of F-8. *Photo NARA.*

23 Joe Halliday, *Mosquito!*, Toronto: Doubleday, Canada Ltd.: 195.

Directorate of Intelligence (DI)

General Spaatz visited Washington, DC, in November 1943 and received reassignment to England to establish the United States Strategic Air Force (USSTAF) on 1 January 1944 to command all AAF operational assets in ETO/MTO. The reorganization aimed to further develop American air power.

An AAF staff study dated 24 October 1943 began, "This, the most powerful air striking force in history, would be utterly blind without Intelligence … could not select targets, determine the proper composition of the attacking force, plan its routes or even assess the damage inflicted on the enemy." To achieve the objective dictated establishing an Intelligence community totally under AAF control. This required a fully staffed, equipped, and functional organization to meet all future AAF planning and operational needs. The Directorate of Intelligence (DI), HQ USSTAF, was constituted 6 January 1944. Personnel from A-2, HQ Eighth Air Force formed the nucleus of this organization. Col. George C. McDonald who served as A-2 to General Spaatz, received appointment as Director of Intelligence (DoI).

The DI had four subsections to carry out its mission. 1. Targets. 2. Operational Intelligence and "Y." 3. Secret Intelligence. 4. Photo Reconnaissance and Intelligence. The development of the later was Col. McDonald's first concern. The bulk of PR and interpretation came under British control. At Mount Farm, 7 PRG flew PR missions in cooperation with No. 106 PR Wing, RAF. Although under operational and administrative control of Eighth Air Force, its missions were dictated by the AM section concerned with photo Intelligence. Similarly, American photo interpreters were under operational direction of the British at CIU, Medmenham. In brief, ETO AAF Intelligence came under total British control.

The AAF personnel at CIU achieved sufficient technical expertise and sophistication to function independently in this role. The planned invasion of Europe required transfer of Eighth and Ninth Air Force units to the continent, thus obligating technical and interpretation support. General Spaatz felt they should be "self contained in all phases of photography, photo interpretation, preparation of target folders, and weather reconnaissance." He previously built such an organization in Africa, the 90 PW Rcn. Also, he wanted an independent American Intelligence organization in the event elements of Eighth Air Force were sent to another theater, the Pacific.[24]

To achieve this goal, General Spaatz summoned Col. Elliott Roosevelt, CO MAPRC. He arrived in England 19 January 1944 accompanied by several PW officers. They received orders to survey the AAF ETO PR and photo Intelligence competence, and determine the feasibility of establishing an American organization independent from the British. He submitted his report 28 January recommending development of a Reconnaissance Wing within Eighth Air Force.

On 2 February, General Spaatz outlined this objective during a meeting of RAF and USAAF officers. The British were not pleased with this revelation. Several more meetings discussed ramifications of the proposed implementation. While this took place, expansion and reorganization proceeded independently with establishment of 8 Reconnaissance Wing Provisional [8 RW (P)] on 8 February 1944 by General Order No. 69, HQ Eighth Air Force. This provisional organization existed on paper only without personnel. Col. Roosevelt served both as photographic staff officer to General Spaatz and Wing CO.

One month later, HQ USSTAF directed Eighth Air Force to bring the 8 RW (P) to a degree of efficiency as rapidly as possible. It also directed close RAF cooperation to eliminate any duplication and correlate all efforts to guarantee maximum collaboration with the commanders of 106 Wing RAF and CIU. A Reconnaissance office was established in the DI to coordinate efforts between USAAF/AM/RAF, Eighth and Ninth Air Force. The CIU became Allied CIU (ACIU) with the AAF section at 8 RW (P), High Wycombe, and the British section at Medmenham, several miles east of RAF Benson. The main British PR contingent was located at RAF Benson, Oxfordshire; five miles away 7 PRG stationed at Mount Farm, while the future 8 Reconnaissance Group (Special) (Provisional) would locate at Watton. The USAAF now instituted an initiative to control all photo Intelligence directly through its own organization.[25] Later, on 15 May 1944, both RAF 106 Wing and ACIU combined to form 106 Photo Reconnaissance Group (106 PRG) RAF.

The proposed second element under 8 RW (P), the 8 Reconnaissance Group (Spec) (P) [8 RG (Spec)(P)] would complement and not duplicate PR functions already assigned to 7 PRG. Development began 4 March when two provisional units were established: 8 Reconnaissance Courier Squadron (P) for the purpose of providing B-17 dispatch service in the ETO, herein called Courier Squadron; 8 Reconnaissance Weather Squadron (Light)(P) flying Mosquitos in the ETO gathering essential meteorological information, herein called Light Weather Squadron. Personnel from 50 Fighter Squadron formed the Mosquito unit. Major George O. Doherty assumed command 14 March, attached from 50 FS that arrived from Iceland.

The 8 RG (Spec)(P) became instituted on 22 March, and HQ, less personnel and equipment, established at Cheddington, Buckshire, operating under a Medium Bombardment Group T/O. Both Courier and Light Weather Squadrons were incorporated into the Group. Next day 23 March, the 8 Reconnaissance (Special) Squadron (P), herein called Special Squadron, formed to secure special PR photographs and radar maps with H2X-equipped Mosquitos, and assigned to Group. Capt. George F. Felger became CO and continued as such except for two short periods when Capt. Walter D. Gernand and Major Marshall Wayne were appointed to command. On 24 March, Major George Doherty became CO of the provisional Group but remained in 50 FS on detached service to Group. Major Hugh Bozarth took command of Light Weather Squadron.

In accordance with General Order 8, HQ 8 RW (P) dated 25 March, the following organizations were assigned to Group: HQ & HS, 6 CCRC Group; 5 Replacement and Training Squadron (Bomb); and 50 Fighter Squadron (TE) on paper for administrative purposes. These organizations provided personnel for all provisional units comprising the Group. Major Doherty and his officers on 26 March began formulating administrative and operational polices. These men included Major Marshall Wayne (Deputy Commander) from 7 PRG and Capt. Levan A. Cottingham (Executive Officer). It was their responsibility to bring the organizations to an operational level.

Personnel were obtained from various Eighth Air Force units, transferred to 6 CCRC Group and placed on detached service with the provisional organizations. Many individuals held an MOS inconsistent with their qualifications, and a shortage of qualified men to fill vacancies required special training programs. To correct this, a number of men were selected to attend schools to prepare them for specialized assignments. A reclassification program was instituted to correct the misclassification and reassignment of personnel within the Group.

History of 50 Fighter Squadron (TE)

Movement to North Africa from California:

29 June 1942	Hamilton Field to March Field, California for lunch
29 June	March Field to Albuquerque, New Mexico, RON
30 June	Albuquerque to Midland, Texas, RON
1 July	Midland to Kansas City, Missouri, RON
2 July	Kansas City to Cleveland, Ohio, RON
3 July	Cleveland to Bradley Field, Connecticut
28 July	Bradley Field to Westover Field, Massachusetts, RON
29 July	Westover Field to Presque Isle, Maine, RON
30 July	Bad weather delayed the flight
31 July	Presque Isle to Goose Bay, Labrador, Canada

The 8 RG received seasoned pilots drawn from 50 FS. The unit formed at Hamilton Field, California, part of 14 FG. When the USA entered Word War II, both 1 and 14 FG were assigned to ferry their aircraft to England across the Atlantic Ocean under Operation Bolero. During the ferry operation, 1 FG left the 27 FS in Iceland to provide protection for later 14 FG units. Both 48 and 49 FS of 14 FG flew P-38s over the northern Atlantic route to England. The 50 FS followed on 29 June 1942, ferrying their P-38s cross-country [see box] arriving at Bradley Field, Connecticut, on 2 July for further training. The first major accident occurred here.

24 John F. Kreis, general editor, The Piercing Fog, Washington, DC, Air Force History and Museums Program, GPO, 1996: 89.

25 History, Directorate of Intelligence, USSAF Europe, January 1944-May 1945, File 519.600, USADHRA..

50 FS airmen, 7 May 1943. L-R. 1st Row F.R. Vierling, H.J. Havre, W.A. Guilfolyle, D.A. Lawhead, M.F. Scales, W.W. Korges, J.L. Roy, H.H. Bozarth, T.E. Hughes, P.J. Peltier. 2nd Row R.P. Howle, F.C. Jones, G.I. Stanford, W.P. Cumiskey, W.C. Locke, H.C. Burton, J.A. Baker, J.S. Gray, G.E. Felger, W.A. Clair, R.M. Holly, W.G. O'Brien, C.P. Hackman, J.M. McNulty, C.O. Vinyard. 3d Row C.R. Bottorff, R.D. Freyermuth, R. Aldrich, H.R. Stengle, G.T. Eshelman, R.C. Grimes, T. Bowling, W.D. Gernand, R.C. Peeke, E.J. Maslow. Missing H.E. Caldwell, U.L. Meeter, W.E. Bethea, E.B. Rodriguez, C.F. Stubblefield, G.O. Doherty, E. Carpenter, E.L. Scott, W. Sneddon, J.H. Walter, H. Pflughoaft. *Photo G. Scruggs*

On 17 July, 2/Lt. Albert R. Dawson in P-38F 42-12588 began a two plane formation takeoff at 0945, with 2/Lt. Temple Bowling Jr. Bowling was formation leader and the flight echeloned to the right with Dawson as wingman. They had no contact with the tower due to a difference of radio frequencies between the aircraft and air traffic control. After covering 900 feet on the takeoff run, a truck pulling a fuel-trailer unit, crossed the active runway from the right-hand side without obtaining a green light from the tower. Upon sighting the approaching aircraft, the driver attempted to turn off parallel to the runway when the outer right wing of Dawson's aircraft struck the tanker unit. Upon contact the P-38 wheeled into and collided with trailer resulting in a massive fireball engulfing both vehicle and aircraft. The lead P-38 managed to takeoff, but Dawson, boxed-in between his flight leader and the tanker unit, had insufficient speed to become airborne.

The force of collision interlocked both P-38 and fuel-trailer unit together. The impact momentum forced the entangled wreckage 200 yards down the runway, leaving a fiery trail of burning fuel. After the combined wreckage came to a halt, Dawson released his hatch, removed himself from the cockpit and walked through the flames with clothing ablaze. He rolled himself on the ground to smother his burning garments. Emergency equipment dispatched to the scene transported him to the hospital where he died the following day from massive burn injuries.

The P-38s departed Bradley Field on 28 July, and staged through Westover AB, Massachusetts, and Presque Isle AB, Maine, before delayed by bad weather. Key mechanics and radio personnel in C-47s preceded the fighters enabling servicing en route. The P-38s departed on the 31st for Goose Bay, Canada, and remained overnight before flying to BW-1, Greenland, an airfield at Narsarssuak located 60 miles up the Tunugdliarfik Fjord from the sea. The facility afforded good anchorage near sizable areas of flatland with an airstrip.

On 1 August, the 50 FS left Goose Bay accompanied by four B-17 aircraft providing navigation support. Each B-17 escorted four P-38s except the fourth flight that had five Lightnings. They flew in four, line-astern formations. En route, the last group attempted to climb through a light overcast to clearer weather and join the other three. This formation consisted of an escort B-17 with P-38s on the right wing flown by Capt. Scales and 2/Lt. Clayborne Vinyard, while off the Forts left wing were 2/Lt. Willis Locke followed at intervals by 2/Lt. James McNulty and 2/Lt. Herbert Goodrich in P-38F 42-12589.

In preparation for ascent the escort B-17 ordered the five P-38s to close on his wing. McNulty flew closer to Loche with Goodrich trailing. After 45 seconds in overcast conditions the formation broke out on top with Goodrich nowhere in sight. The Fortress sent him radio messages with no reply. McNulty achieved radio contact and Goodrich answered he was still in the overcast. Several minutes later he failed to appear, nor replied to radio calls. The B-17 from the third formation turned back to perform a search while the fourth Fortress picked up the additional P-38s and continued on course to BW1. After descending, the search B-17 broke from the undercast at 600 feet over rough waves and whitecaps. He continued surveying the ocean below a further 25 minutes before climbing for altitude and heading for Greenland. AAF organized a search and rescue operation, but no trace of Goodrich or his aircraft was ever found and the search abandoned 31 August.

Bad weather en route to Iceland delayed the P-38 ferry flight. Capt. George O. Doherty became CO of 50 FS on 18 August as they departed BW-1 for Patterson Field, Iceland. On arrival in Iceland, the ferry flight halted. The 50 FS would replace the 27 FS, 1 FG, and assigned the task to cover for the next group to come through to England. But none arrived due to the high attrition rates from lost pilots or force landings on the Greenland ice cap, thus canceling Operation Bolero. The AAF reassigned 50 FS from Eighth Air Force to Iceland Base Command to strengthen the island's air defense. Increased U-boat activity and Luftwaffe reconnaissance flights in and around Iceland forced this decision. The ground echelon from 50 FS that already arrived in England, along with the remainder of the 14 FG, formed a cadre for the newly created 4 FG and reassigned to North Africa.[26]

On 11 September 1942, 342 Composite Group, Iceland, became activated to which 33 FS and 50 FS became attached. The 50th began operations and numerous aircraft were assigned to isolated airfields around the island on alert status. If an unidentified aircraft appeared on the island's defense radar, they dispatched two P-38s to investigate. Other missions included flying patrols covering surrounding waters for U-boat activity, local fishing ships in distress, missing aircraft, and provide air protection to passing convoys.

The first combat action occurred 22 October when at 0930, two P-38Fs piloted by Lts. Robert C. Grimes and Richard. M. Holly were dispatched to investigate an unidentified aircraft. Having confirmed a Fw 200K, both airmen fired at the aircraft which immediately took refuge in heavy overcast clouds.

When the runways at Patterson Field, Keflavik, required resurfacing, 50 FS transferred to RAF Kaldadarnes on 23 October, where the British operated Lockheed Hudson aircraft on antisubmarine patrol. The first Lightning accident occurred here on 7 December. 2/Lt. Robert C. Grimes in P-38F 42-12590 returned to base and on landing, his aircraft rolled several hundred feet when the right tire blew, causing a right swerve. Momentum sheared off the right undercarriage causing a ground loop and collapsing the remaining landing gear.

Five days later 2/Lts. William E. Bethea and Walter D. Gernand returned from patrol on 12 December. Gernand landed first and Bethea in 42-12577 followed close behind. Thirty feet over the runway, he encountered a violent prop wash forcing his left wing and nose to strike the runway first. The ship bounced and the ensuing momentum sheared away the landing gear. The aircraft burst into flames as it slid straight down the runway. Bethea was not injured.

With the fall of France in July 1940, *Kampfgeschwader (KG) 40* transferred to Bordeaux-Merignac airfield, France. Equipped initially with long-range Fw 200C-1, they attacked Allied shipping in the English Channel. Later, they dispatched missions west of Ireland towards Iceland to attack convoys. RAF attempts to intercept the Fw 200s proved futile because their Bristol Blenheims lacked speed, range, and armament for the task. Later, *KG 40* Condors flew shuttle missions skirting west of Ireland to land in Norway. After a brief stopover, they retraced their route to Bordeaux, attacking shipping en route.[27]

The Kreigsmarine attempted to destroy Allied convoys with surface raiders such as the *Bismarck* and *Prince Eugen*, but this effort failed when the *Bismarck* was sunk 27 May 1941. The Allies routed convoys further north, allowing smaller escort vessels to refuel in Iceland and provide convoy protection to UK. The Luftwaffe then began stationing reconnaissance Fw 200 and Ju 88 aircraft in Norway to assist U-Boat packs with the current position of Allied convoys.

The peak operations for *KG 40* occurred in December 1942 when nearly daily, one or two Fw 200 reconnoitered the Arctic seas between Greenland, Iceland, and Jan Mayen Islands. This included daily missions of weather reconnaissance Heinkel He 111 and Junkers Ju 88 of the *Wettererkundugsstaffel 5* from Trondheim-Vaernes, Norway. By this period, the Fw 200 C-1 version was replaced by the C-3 and C-4 with improved armament and range.

Junker Ju 88 reconnaissance aircraft of *Aufkarungsgruppe 120* from Stavanger patrolled the sea between Iceland and Scotland. The same areas received weather reconnaissance coverage by *Wettererkundungsstaffel 1 Ob.D.L.*, (First Weather Reconnaissance Squadron), and Squadron No. 5 of Trondheim under regional command of *Luftlotte 5* in Norway. The missions between Iceland and Scotland were flown from Aalborg, Stavanger, or the main base of the Squadron at Zwischenahn in northwest Germany.

The Allies countered this threat with AAF fighters protecting Iceland and its environs. They intercepted both Luftwaffe units and shot down several bombers. Long-range patrol bombers from Iceland, North Ireland, and Scotland, also deployed against marauding U-Boat wolf packs.

Daily 50 FS patrols continued when on 11 January 1943, 2/Lt. William A. Clair flying near the Keflavik area in P-38F 42-12613 lost an engine and attempted an emergency landing at Patterson Field. A pranged P-40 obstructed the NE/SW intersecting runway, so he circled to the EW runway. He forgot to switch fueltanks and on approach, the good engine sputtered, but with sufficient flying speed, he landed wheels-up to a safe stop.

Not long after on 23 January, 2/Lt. Rodney C. Peeke taxied P-38F 42-12592 from dispersal at RAF Kaldadarnes for air alert scramble. He proceeded between two parked P-38s towards the runway, when his ship began sliding to the right on thick ice covering the taxi area. Attempts using brakes and right engine, then cutting throttles and applying both brakes failed to halt the slide towards P-38 42-12587 located 75 feet away. His right wing struck the fighter's nose, swinging his ship directly towards it. He shut off both engines, but the right prop slashed the nose section of the parked aircraft.

26 SQ-FI-49-HI. Reel No. A0742. USAFHRA.

27 Kenneth Poolman, Scourge of the Atlantic, London: Book Club Associates, 1979: 14-55.

"Skeeks" in hangar undergoing overhaul and repair. *Photo G. Scruggs*

23 January 1943 accident during icy taxiway conditions. *Photo Flypast*

Climatic conditions caused problems. On 1 February, 2/Lt. George I Standford returning in P-38F 42-12594 from patrol to Kaldadarnes received runway landing instructions. Upon arrival, the wind shifted 180 degrees and unknowingly he approached down wind. The ice covered runway had sand dispersed two-thirds of its length. A stranded RAF Hudson sat at the approach. Stanford landed 100 yards further and applied brakes with little if any traction. His aircraft left the runway end where the nose wheel collapsed.

An early spring thaw in 1943 caused increased melting, creating an ice jam on the Thjorsa River. This caused heavy flooding of Kaldadarnes with small ice flows scattered over the field. This event forced transferring the unit to Reykjavik on 10 March.

The first successful combat engagement occurred 24 April, when ground radar detected an unidentified aircraft over Iceland that afternoon. Lts. James M. McNulty and Harry R. Stengle scrambled to investigate. Both P-38s climbed up through a puffy strato cumulus cloud layer at 4,000 feet, and immediately spotted a Ju 88 flying ahead and away from them. The Junker Ju 88D-5, 430087, registration *A6+CH* from *l.(F)/120* (the First Squadron of the Long-Range Reconnaissance Wing 120) had left Stavanger-Sola for photographic reconnaissance of ships located in the ports of Iceland. It carried three cameras. The crew consisted of four airmen: pilot *Feldwebel* Werner V.G. Bullerjahn, navigator/observer *Leutnant.* Karl M. Bruck, radio operator *Unteroffizier* Anton M. Mynarek, and mechanic *Feldwebel* Theodor Scholtyssek.

Both P-38s gave pursuit. McNulty fired first and struck the Ju 88 port engine setting it on fire. The bullets cut the radio operator's earphone cable and disabled the radio. The attack caused a cabin fire and wounded the German pilot who ordered the radio operator to jettison the roof hatchcover and prepare for bailout. Both engines continued running smoothly, so the radio operator remained inside, unstrapped near the roof hatch cover. Streaming smoke, the Ju 88 took shelter in clouds with the P-38s in pursuit.

The P-38s continued flying through the cumulus clouds. When McNulty emerged, he found the Ju 88 off to his right with Stengle close behind. Stengle began firing, striking the starboard engine and fuselage. The Ju 88 continued flying and entered more clouds. Both P-38 pilots followed but lost the intruder in overcast and returned home.

The German aircraft sustained mortal wounds. After the second attack, the Ju 88 suddenly flipped over on its back and *Unteroffizier* Mynarek fell from the aircraft, deployed his parachute and landed safely on the ground. Elements of the British Army took him in custody. According to an official Luftwaffe report, the Junker crashed at 1405, six miles from Kleifarvatn Lake, killing all remaining crew members.

Unteroffizier Mynarek was brought to Camp Pershing for interrogation, then later taken to the 50 FS officers mess at Reykjavik to await transfer to USA. Born 22 February 1920, he was appointed to *l.(F)/120* in April 1942 and had flown 40 missions. Through an interpreter, Mynarek said he believed the second engagement killed the remaining crew, causing the Ju 88 to flip on its back.

Camp Pershing, Iceland. *Unteroffizier* Anton M. Mynarek sits during meal at U.S. Army Intelligence HQ. He is the first German recovered alive after Ju88 shoot down on 24 April by 50 FS pilots. *Photo. H. Steinsson*

The first major fatality occurred 25 April, when P-38F 42-12575 on a search mission crashed near Sandskeio Dum, killing the pilot Lt. Frederick E. Eichman. The circumstances surrounding the fatality is undetermined.

The 50 FS relocated from Reykjavik to Patterson Field, Keflavik, on 16 June. Lts. Holly, Bethea, Locke and Gernand took turns performing duty at Kassos Field, Melgerdi, near the town of Akureyri. Two P-38s remained stationed here with pilots and two mechanics, on a two-week rotation. They refueled the aircraft from barrels. The first two pilots remained for two weeks, flying routine patrols every day, but no enemy aircraft were seen. Then Lt. Richard M. Holly received this assignment for the following two weeks as flight leader with Lt. William E. Bethea. According to Holly:

> "Several days of routine air patrols proved uneventful. But on 5 August 1943, Bethea and I were on routine patrol flying from Akureyri at 1,000 feet altitude, some 10 miles from Grimsey Island, located north of Iceland. While flying over a thin overcast, I called Bethea by radio at 1230, notifying him I was going under the cloud layer to look around. I descended through the light overcast, and upon reaching clear air, I observed 1,000 yards ahead, a large bomber flying towards me that I thought was a B-17. My first thought on seeing the aircraft, 'What was a B-17 doing in this area?' The bomber was heading east and flying towards me at approximately 45 degrees from my left side.
>
> "As we flew closer, I realized it was ... a Focke-Wulf Fw 200 Condor reconnaissance patrol aircraft. I had never seen an enemy plane before, and it was a great shock to me. I flew over the bomber and circled in behind him. I charged my cannon, switched on the machine guns, maneuvered in to position, and commence firing, causing the outer left engine to burst in flames. After I started shooting, the German gunners responded, firing back.
>
> "The Fw 200 climbed up towards the overcast trailing smoke. I was 100 yards behind and radioed for Bethea to watch for the bomber to emerge on top. The Fw 200 reached the overcast layer, but then started descending. I radioed to Bethea and he came down through the overcast, 100 yards directly in front of the bomber. The Fw 200 forward gunners opened fire on him, but scored no hits. Meanwhile, the side and rear-top gunners started firing at me.
>
> "The German pilot placed the bomber in a dive and headed for the ocean with flames and black smoke trailing from the left wing and engine. I knew he was going to ditch – there was no place to go! He executed a perfect landing on the calm sea, and almost immediately a yellow life raft came out the door on the left side, and seven crewmembers hastily exited the floating bomber. One crewman suffered an injured foot when the plane impacted the water.
>
> "Fortunately, the U.S. Navy had a small contingent in that fjord, so I sent my wingman Bethea back to refuel and have a ship dispatched to pickup the survivors. The bomber floated for a long time, then slowly disappeared beneath the waves. Two life rafts carrying seven crewmembers bobbed in the water nearby. I orbited the rafts for 45 minutes to an hour. Bethea returned at 1530 and remained on the scene until a rescue ship arrived and transported the survivors to Akureyri where the Navy detained them overnight. The following day, the captured German airmen were brought to the field. One of the Germans spoke broken English, and I had the opportunity to talk with them. Pictures were taken of both Bethea and myself with the German crew members."

A Douglas B-23 arrived and transported the Germans to Reykjavik, They arrived at the officers' mess for a meal. The seven-man crew came from *3. Staffel, l. Gruppe* (Squadron) *Kampfgeschwader 40* stationed at Trondheim, Norway flying Fw 200C4, Nr. 0200, tactical code *F8+FL*. They were: *Oberfeldwebel* (Master Sergeant) Karl Holtrup (pilot), *Unteroffizier* (Buck Sergeant) Gunter Karte (copilot), *Feldwebel* (Technical Sergeant) Josef Teufel (radioman), *Unteroffizier* (Buck Sergeant) Herbert Richter (radioman), *Oberfeldwebel* (Master Sergeant) Emil Brand (flight mechanic), *Obergefreiter* (Senior Corporal) Siegfried Klinkman (gunner), *Gefreiter* (Corporal) Wilhelm Lehn (gunner). All expressed relief to be safely out of the war. Eventually a C-47 transported them to the USA for internment.

Meanwhile, the previous day 27 August, Lt. Stanley J. Burkot, 33nd FS, while taxing P-38 42-12594, accidentally collided with a tow tug as it turned from a taxiway on to a parking apron, resulting in minor aircraft damage.

2/Lts. Holly and Bethea with Luftwaffe crew of downed Fw 200 prior to leaving in a Douglas B-18 for Patterson Field. L-R. Brand, Holtrup, 2/Lt. Bethea, Teufel, Lehn, 2/Lt. Richard Holly, Klinkmann, Richter, and Karte. The German pilot gave Holly his flying jacket and helmet that were later donated to the Bomber Command Museum, Nanton, Alberta, Canada. *Photo R. Holly*

50 FS personnel in lounge. Front Paul Peltier, engineering officer. 1st Row, L-R. Clayton Hackman, Willis Locke and Rodney Peeke, all pilots. Rear row William O'Brien, armament officer and Edward Maslow, pilot. Note Ju 88 machine gun on fireplace mantel taken from a German bomber shot down over Iceland. *Photo G. Scruggs*

Orders received on 6 October, reassigned 50 FS to England. While plans for the transfer were formulated, a mysterious tragedy occurred 14 October, when 2/Lts. Alfred A. Arnold and Douglas A. Lawhead flew a search mission to Kleifarvatn Lake area seeking a missing PBY. Visibility was six miles in scattered showers and low ceiling. The radar controller ordered the men to split and search different areas. Lawhead turned and headed for hills located northeast of the lake. That was the last he saw of Arnold. The wreckage of P-38F 42-12586 was later discovered having struck a mountainside and a search party recovered his body.

2/Lt. Alfred A. Arnold killed 14 October 1943 in a flying accident near Kleifarvatn Lake, Iceland. *Photo W. Locke*

On 18 October, 1/Lts. George E. Felger and Harry R. "Casey" Stengle performed a routine patrol between the mouth of the Thjorsa River and Gullfoss (waterfall). Stengle, flying P-38F 42-12576, discovered smoke belching from the left engine. He banked towards Kaldadarnes trailing smoke that quickly became a solid sheet of flames. Both P-38s descended and banked to follow the river towards the ocean. The fire and smoke gradually receded when Stengle decided to bailout from 1,000 ft. After release of the canopy hood the aircraft slowly went in to a right spiral and crashed. The impact created a deep crater and Stengle's body found 50 feet west from the wreckage with parachute pack intact. After leaving the ship, he struck the tail section, then fell to earth unconscious and killed upon impact. Stengle, from Pittsburgh, Pennsylvania, previously received credit in sharing the destruction of a Ju 88 on 24 April.

Unexpected events occurred during patrol missions. On the morning of 30 October, Lts. Robert C. Howle Jr. and Walter D. Gernand flew patrol in their P-38s five miles northwest of Reykjavik. They viewed four Swordfish aircraft peel off for landing on board a British aircraft-carrier while nearby, three 33 FS P-40s flew in formation. Howle glanced around and observed a P-40 leave formation and dive straight in to the sea. Only an oil film remained on the ocean surface as he flew over the scene.

Mishaps further plague the Iceland operations. On 11 January 1944, four P-38s returned to Kaldadarnes from a patrol. Lt. William A. Clair lowered the landing gear of 42-12613. The tower operator at first preoccupied, turned his attention to the P-38 now 50 feet above the runway and noted the undercarriage failed to fully extend. He fired a red warning flare, but too late. The aircraft slid down the runway on its belly to a safe stop.

On 17 January, Lt. Clayton Hackman taxied P-38 42-12595 to takeoff position when the nose wheel collapsed due to mechanical failure. A few weeks later on 1 February, he took off in the same aircraft, when the oil cooler shutters failed to operate. Declaring an emergency, he performed a wheels-up landing on the runway. An external fueltank ruptured and the ensuing blaze destroyed the aircraft.

Finally plans arrived for transferring 50 FS to Eighth Air Force. February 4 marked their initial departure for England. Each section comprising three P-38s following behind a navigation aircraft. The multi-formation of 24 P-38s escorted by B-17s and B-25s departed Meeks Field at 1000 hours. The majority flew direct to Prestwick, but nine aircraft landed at Stornaway, Scotland, followed by 23 groundcrew members transported in four B-24s. The following day all continued to Prestwick. Back in Iceland the squadron's remaining personnel embarked 5 February on the HMS Empress of Russia for England.

Both aircraft and men remained at Prestwick until 8 February, then flew to Nuthampstead, Hertfordshire, and officially rejoined Eighth Air Force. Squadron personnel formed the nucleus around which the new 8 Reconnaissance Group (P) would develop. Ferry crews flew the Lightnings away and conversion training began on Mosquito aircraft. Two RCAF-trained Americans, F/Os Richard Wulfkuhle and Oliver Emmel arrived to instruct the conversion program. Both previously flew night-fighter Mosquito aircraft with 410 Squadron RCAF. Lt. Robert Howle flew a dual-control Mk.III trainer with instructor Wulfkuhle. After flying the next sortie solo, he then acted as instructor assisting other pilots transition to the Mosquito.

Lt. Willis Locke remembers his first flight on 24 February. As he waited his turn, RAF Mosquito LR343 from 487 Squadron flew over the field at 2,000 ft. Two miles north of the runway the aircraft disintegrated in midair and crashed to the ground killing the crew.[28] The incident made Locke apprehensive of his impending excursion in the Mosquito. However, his flight with instructor F/O Emmel proved uneventful.

"Skeeks" P-38 42-2595 with collapsed nose gear on 17 January 1944.
Photo G. Scruggs.

28 Croman, G.P. And Smith D.J. DH Mosquito Crash Log, Wigan, UK: Clarington Press Ltd.:16.

P-38 42-12595 after fiery crashlanding on 1 February 1944.
Photo G. Scruggs

Major George Doherty and F/O Emmel in Mosquito LR516 performed a training flight the 27th. Both ground and aircrew failed to notice the tail pitot tube cover still in place. During the takeoff run when no air speed registered, Emmel took control but failed to slow the Mosquito on the slick, moisture covered runway. He overshot the end where the undercarriage collapsed damaging the aircraft beyond repair. Several days later on 3 March, Lt. Edward J. Maslow with instructor F/O Emmel (MM337) veered right on landing, went off the runway and hit a drainage ditch, wiping out the landing gear.

On 8 March, 50 FS officially integrated into the newly designated 8 Weather Recon Squadron (Light) (P) and moved from Nuthampstead to Cheddington on 16 March. Here further training continued in Airspeed Oxford aircraft learning the peculiarities of British aircraft control systems, followed by further Mosquito familiarization flights. The conversion to bomber-type aircraft was not well received, therefore, the pilots had the option of remaining on Mosquitos, or transferring to operational fighter units. Many pilots chose the latter and were reassigned within Eighth Air Force.

MM337 accident 3 March 1944 with 1/Lts Edward J. Maslow and Oliver Emmel. *Photo R. Abbey*

LR516 takeoff accident 27 February 1944 with Maj. George Doherty and 1/Lt. Oliver Emmel. *Photo R. Abbey*

USAAF Mosquito in RAF markings at Nuthempstead February 1944.
Photo A. Abbey

New Mosquitos arrived at 1 Base Air Depot Area (BADA), Burtonwood for modification with American equipment. The factory-fresh aircraft exhibited poor workmanship and lacked proper lubrication, while dirt accumulation, corrosion, and metal filings pervaded various parts of the airframe. In fact the aircraft arrived faster than modifications could be performed resulting in damage from neglect while in storage. F/O Wulfkuhle was assigned to BADA as liaison adviser.

In preparation for night PR Joker missions, Colonel Roosevelt on 17 March had B-25 42-53357 from 7 PRG at Mount Farm converted for this purpose.

An RAF Burtonwood Association photo shows five USAAF Mosquitoes in RAF markings among aircraft on the parking ramp.

An increased training accident rate created the need for dual-control Mosquitos. Unable to obtain these, several factory Mosquitos were drawn from 1 BADA for both Nuthampstead and Cheddington. Training accidents continued: 18 March, Major Charles W. Himes (MM340) on a check flight from Warton ground looped on takeoff. The left undercarriage collapsed causing extensive damage to the left wing and engine nacelle; 22 March, Lt. George C. Hand (MM344) went solo, lost control and ground looped the aircraft; 24 March, Capt. Thomas Hughes (MM385) began his run and veered left off the runway and ground looped, shearing off the right undercarriage. By month's end, extensive repairs were required to Oxfords AB720, V3780, V3628, T1066 and R6034 leaving only P8917. Until repaired, Oxfords, DF528 and DF417 were assigned to Cheddington as training aircraft for 20 days.

On 30 March, while at Cheddington, the 8 RG (P) became the 802 Reconnaissance Group (Sp)(P). Eighth Air Force decreed all provisional Groups and squadrons be delegated the 800 series numbering, i.e. 801, 802, and 803. The 802 RG (S)(P) [herein listed as 802nd or 802 RG] was assigned permanent facilities at Watton, Norfolk.

B-25 42-53357 obtained by 7 PRG, Mount Farm, in August 1943 as an executive transport. It was later converted to carry photo flash bombs for nighttime Joker PR missions and transfered to Watton. *Photo R. Astrella*

MM385 accident 24 March 1944 by Capt. Thomas E. Hughes. *Photo G. Scruggs*

Situated in each of three parishes, Watton, Carbonate and Griston, construction of RAF Watton started in the mid-1930s, and the station officially opened in January 1939, having only a grass runway. RAF 34 Squadron Blenheims arrived in February, joined a few days later by 21 Squadron, also with Blenheims. In August 1939, No.34 Squadron transferred to the Far East and replaced by 82 Squadron with Blenheims Mk.IV. In March 1940 a grass strip satellite airfield at Rodney located five miles west of Watton came in to use to disperse aircraft and later became home to the AAF 352 Fighter Group. On 7 May 1941, 90 Squadron reformed at Watton with B-17 Fortress 1s, but flew from Rodney, not the parent station.

In July 1943 units of 31 and 46 Air Depot Groups occupied RAF Watton. Their responsibility involved operation of 3 Strategic Air Depot (SAD) providing third and fourth echelon maintenance, plus all supply function for the 14 numbered B-24 bombardment Groups located in East Anglia. Duties also included salvaging crashed B-24s in the UK and the continent. If the downed B-24 could be made flyable it was flown to Watton. If not, all components were salvaged and returned to 3 SAD supplystock.[29]

In January 1944 the newly constituted 801 BG (P) formed at Glastonbury (later to become 492 BG) and assigned to Watton. Poor PSP runway conditions caused this organization to maintain operations from Glastonbury, before eventual relocation to Harrington.

On 4 April, the 802 RG dispatched an advance party with enlisted men to prepare Watton for occupancy. Between 6 to 10 April, the 8 Rcn Weather Squadron (H)(P) with B-17s from Bovingdon, Buckshire, transferred to Watton. They were referred to as Heavy Weather flight. On Easter Sunday, 12 April, 3 SAD relocated to a temporary living area called the Griston Site located directly across the Watton runway. That same day, 802nd elements moved by convoy from Cheddington to the permanent Watton base facilities. Pilots and key personnel of Light Weather Squadron moved baggage and equipment as much as possible by air, as did the majority of Courier Squadron. The remainder of the two units moved by road convoy.

The 3 SAD personnel continued to manage Watton functions such as control tower, flight line, bomb dump, four large hangars and telephone service. The 802nd performed its own administrative duties and operational tasks from buildings on the permanent side. Thus Watton contained two different Eighth Air Force bases using the same runway, each performing different functions under different commands, and each unit reported to separate base commanders. The 802 RG operated as AAF Station 376 on the Carbrooke side, and 3 SAD as AAF Station 505 from the Griston side.

After serving one and one-half years in Iceland and traveling to Scotland, the 375 Service Squadron arrived at Cheddington 16 March and assigned to Eighth Air Force Composite Command. They were reassigned to 8 RW(P) on 5 April under command of Capt. Leo A. Karawsky, responsible for third and fourth echelon work on 802nd aircraft. They also joined the transfer to Watton.

Accommodations at Watton boosted morale, supplying spacious lawns and comfortable brick barracks. The squadrons were assigned their specific areas and Group HQ established in a centrally located brick building. Shortly, the Group began operating as an independent organization with newly attached 1070 Quartermaster Company, the 375 Service Squadron, the Detachment A, 1274 Military Police Company (Avn), the 594 Army Postal Unit and two batteries of the 386 Anti-Aircraft Battalion.

Heavy Weather Flight

Weather affects all military operation in wartime. The RAF realized this as pertinent to air operations and conducted weather flights from UK bases. Accurate forecasts were also essential to Eighth Air Force daylight bombing campaign over Europe. Thus developed the need for weather reconnaissance (WR).

The RAF operated twin-engine Handley Page Hampdens gathering weather data. The aircraft lacked range and proved difficult to fly during emergency single-engine flight. Many times the missions were limited by adverse weather or attack by Luftwaffe aircraft operating over the Atlantic area. The four-engine Halifax aircraft replacements would not arrive until December 1943, and an interim aircraft was required.

On 31 August 1943, four crews from 814 BS, 482 BG transferred to the 2900 CCRC Group (Bomb) (Prov) on temporary duty operating as a WR flight for not more than six weeks. The pilots were 2/Lts. Norman F. Herget, Samuel E. Harris, William E. Poelstra and Ernest H. Patterson. Under command of l/Lt Leo S. McIntire with a complement of 40 groundcrew members and four B-17E aircraft, they served attached to 517 Weather Squadron, RAF Coastal Command, St. Eval, on the southwest Cornwall coast. The 517 Squadron formed 11 August 1943 at St. Eval by expanding No. 1404 Met. Flight already on base.

The RAF looked upon the AAF unit with great admiration to collect better meteorological data for long-range weather forecasting. They provided the flight with experienced RAF meteorological observers (met/obs) familiar in this work.

Lt. McIntire had no idea the six week duty would continue functioning two years later. The men arrived at St. Eval on 8 September and commenced local training flights. Operational sorties began the 18th. The B-17s were equipped with both regular and Tokyo wing tanks. A weather flight was scheduled each day but none at night. Soon the bombardier, waist gunner and tail gunner were removed from each crew and reassigned to other Eighth Air Force bomber units, leaving seven of the original ten man crew.

The mission profile of the WR flights required extreme aircrew proficiency. Each mission lasted 14 to 18 hours duration with at least 11 hours flying over the endless expanse of Atlantic Ocean waters. Frequently due to severe frontal weather conditions, the pilot and copilot were required to fly an entire mission on instruments. When mechanical failure resulted in feathering an engine hundreds of miles from the coast, the crew unloaded all excess equipment to lighten the load as the Fort limped back through turbulent and icing cloud systems. Fear of ditching in the cold, inhospitable Atlantic waters was foremost on each crewmember's mind. Flying low above an ocean at night in rough weather with 50 to 100 foot swells was fraught with danger.

On takeoff, a Fortress loaded with maximum fuel capacity required full power all four engines could furnish and skill the pilot could muster to lift the aircraft from the runway. There was no allowance for error. One small flaw could result in a fiery disaster.

The navigator's job required utilizing all known means of navigation then available with no single method solely relied upon. Dead Reckoning (DR) depended on compass deviation and wind, both determined frequently by the navigator. Atmospheric and enemy radio interference with radio beams and stations caused radio bearings to be useless. Often the enemy sent out a signal with a false bearing. The final course check using celestial navigation proved both accurate and difficult to measure. Even when employing all these methods to determine position, there were times when none proved truly effective.

Many missions were flown through solid cloud formations extending from the sea surface to as high as 24,000 feet. In such foul conditions, it proved impossible to determine wind because drift could not be read, and the lack of visible stars hampered celestial measurement. Invariably, radio interference during such conditions became problematic. If 60 miles south of course, the flight would pass directly over the Brest Peninsula and enemy defenses. To avoid such pitfalls required a grueling 14 hours of constant calculation and computation by the navigator.

The initial WR flights were labeled Epicure. The B-17 took off and made a low level pass over St. Eval at 50 feet to check the accuracy of settings on both radio and barometric pressure altimeter. It then flew over the Atlantic on a track of 260 degrees from true north, 270 degrees magnetic. At 25 miles from Land's End, the aircraft descended to 50 feet above the ocean to recheck both barometric and radio altimeters.

Weather observations were then made every 50 nautical miles along the track at 14 numbered positions at a pressure level of 950 millibars or roughly 1,800 feet. At each position, the base and tops of clouds passed through were recorded by pressure level using the aneroid barometer. All cloud-types visible were recorded giving tenths-of-sky covered by each layer, plus base and top of each layer were measured when practicable. Otherwise the necessary information was estimated by eye. Visibility, weather changes, turbulence and airframe icing were also recorded. Provision also made for recording and reporting changes of weather encountered between standard observing positions.

At position 3, 6, 9, 12 and 14 the Fort, using the radio altimeter, descended to measure sea level pressure 50 feet above the ocean, then climbed quickly to 950 millibars for further observations. Both 50 and 1,800 feet readings of dry and wet bulb temperatures were taken and dew point calculated. In calm weather the radio altimeter reading proved accurate, but in stormy conditions the Fort remained higher over the ocean because of varying radio readings with each wave.

29 Wiley Noble, ed. 3 SAD Newsletter (1 July 1989): 3.

Photo of 482nd BG members and aircraft assigned to Weather Flight at St. Eval. *Photo W. Poelstra*

Winds were estimated by triangulation. Inside the aft Fort, a compass-rose etched on the floor encircled the ball turret. The turret's Perry Computing gunsight was replaced with a rifle-type sight and the turret-sight marked zero with the compass-rose. A second man dropped a flare or smoke bomb through a floor chute and the ball turret operator aimed his gunsight at the smoke base on the water during day and the burning flare at night. He notified the second man by intercom, who then called off the degrees on the floor compass-rose to the navigator. The navigator computed wind velocity using the double-drift measurement by having the Fort changed heading twice before returning to the original track while dropping additional smoke bombs.

Observation were logged on a six-foot long by one-foot wide scroll, then transformed in to five-figure groups and ciphered for radio transmission at every third position. Once received on land and deciphered, the code groups were transmitted to Meteorological Office HQ, Dunstable, by teleprinter and then relayed to all interested meteorological offices allowing data plotting on appropriate weather charts.

At position 14, some 700 miles distant, observation were taken from sea level all the way up to 24,000 feet recording measurements every 50 millibars along with cloud bases and tops passed through and any temperature inversion encountered. Climbing a box pattern, the aircraft leveled off every 1,000 feet for two minutes allowing thermometers to settle for each observation. Heights of pressure levels were calculated on a graph called a tephigram using temperatures and dew points. This data was quickly encoded and ciphered for radio transmission.

From 24,000 feet the Fort took a reciprocal heading while slowly descending to pressure level of 500 millibars, roughly 18,000 feet and flew this height until position 7 where a box descent was made, the exact reverse of position 14, to sea level. Once the information was encoded, ciphered, and transmitted, the met/obs duties were completed and the Fort headed for St. Eval.

On 28 September, Lt. Norman F. Herget lost an engine immediately after lift off and feathered the motor. He circled the field and came in for a three-engine landing. Unfortunately, a strong crosswind created a difficult approach. The added weight of the heavily loaded bomber completely washed out the landing gear relieving the flight of B-17F 42-29811. Ground crew cannibalized the airframe for much needed parts to maintain the other three Forts airworthy.

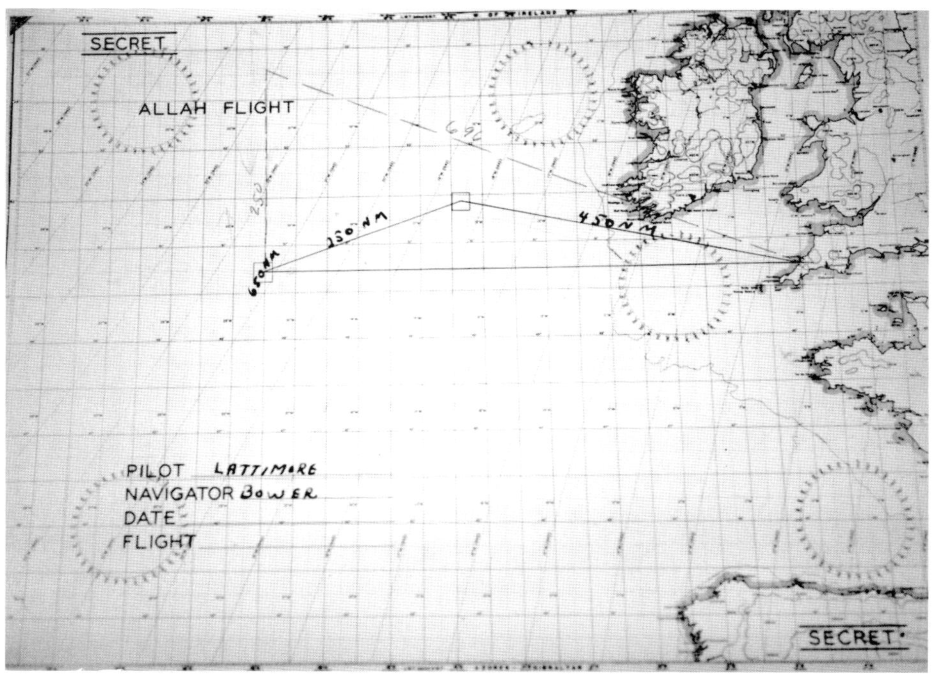

Solid line indicates new Allah route and broken line the old route. *Photo O. Bower*

St. Eval records note on 23 October, "the Met Flight in trouble – no aircraft available." It proved a difficult task maintaining three flying B-17s from an RAF base. Replacement parts became difficult to obtain, and facilities for overhaul and repair were far from adequate. Runway resurfacing at RAF St. Eval in November required transferring weather flight operations to St. Mawgan, two miles away that had a large American contingent.

Atlantic weather data improved long-range forecasting for bomber operations. RAF St. Eval experienced problems procuring Halifax aircraft, further increasing the importance of AAF WR missions. Therefore, the weather flight became a permanent unit, and on 25 October formalized at St. Mawgan as Detachment A of a newly formed Combat Weather Detachment, 1 CCRC, Bovingdon. The men were assigned to 1 CCRC for rations and pay purposes while orders came from HQ Eighth Air Force. On 5 November the flight expanded with four additional aircraft and crews. Continuing problems with repairs required transfer to the l Replacement & Training Squadron, l CCRC at Bovingdon on 23 November.

New pilots, navigators and radio operators were given the option to volunteer for this dangerous assignment. Thus 1/Lts. Ernest W. Stephenson, Robert H. Grafton, LaVerne P. Rissinger, John W. Perdue, Sam Nudelman and their men were the first five crews to willingly submit themselves to this work. When Lt. McIntyre returned to USA, Captain Alvin E. Podwojski transferred from lst R&T Squadron and appointed CO. He previously completed an operational tour with the 92 BG.

In December, Lt. Herget's crew returned to 482 BG and transferred to an operational bomber unit. Two more volunteer crews were added bringing strength to ten. This allowed each crew to fly every fourth day, thus relieving the strain of flying every third day. Night WR flights were now scheduled.

Two 500-gallon tanks were installed in the bomb bay to increase range but made the Fort heavier. Bovingdon had a short runway requiring an "aircraft-carrier type" takeoff procedure. Brakes were applied with all four engines advanced to full throttle. When the tail was felt to rise, the brakes were released and the Fort lumbered down the runway. The WR Forts commuted 200 miles from Bovingdon to St. Eval to set the radio altimeter, then commenced the WR mission.

All new crews required checkout flights. Veteran pilot Lt. William E. Poelstra took Lt. LaVerne Rissinger as copilot on a mission to fly the track, pattern and procedures required during a WR flight:

> "He then flew his first mission with his own crew. I was on duty as operations officer. After briefing, the senior RAF weather observer commented the flight should be canceled because the weather was terrible, saying, 'I just don't think you should send us.' But, I thought they could takeoff with no problem."

On 9 December at 1856 hours, Rissinger in B-17G 42-37744 filled with 3,600 gallons of fuel took off from Bovingdon. Visibility was 1,000 yards and ceiling 200 feet in light rain. According to Poelstra:

> "They started off. It was dark and all you could see was the blue glow from the turbo superchargers underneath the aircraft. They used the full 1,600-yard runway and became airborne. At an altitude of 100 feet the bomber went down and plunged in to a grove of trees 600 yards from the runway exploding in a fiery pyre of 100-octane fuel and exploding .50 cal. machine-gun bullets. No one heard any engine sputter or quit – they were going full blast.
>
> "I had to investigate the wreckage. When I arrived the aircraft was upside-down and burning fiercely. The B-17 has a 100-foot wingspan. When one wing hit the trees, the impact turned the aircraft completely over and it crashed inverted. I'll never forget experiencing that terrible feeling."

Listed KIA were 2/Lts. LaVerne P. Rissinger, William H. Holcomb, Otto H. Ahlers, M/Sgt. John P. Buchanan, T/Sgt. Peter W. Costello, S/Sgt. Bernard Cohen, Sgt. Ralph W. Harding, Cpl. Mathew Ekes, and two RAF meteorology officers, S/L Jack H. Osborne and F/L John Leigh-Clarke. The cause of the accident was not determined but suspicion focused on the possibility the carrier-type takeoff preparation was too brief.

First crew to complete 500 hrs on WR mission. L-R. Top row Henry L. Wilcox, Willie C. Harrell, Charles C. Balady, Ernest V. Shearerl. Front row 1.Lts. William E. Poelstra, Edward R. Heeking, and Alexander H. Chris. *Photo W. Poelstra*

On 20 December a tour of duty became 500 hours; 20 hours of WR flight equaled one sortie, but the completion of five did not merit the award of an air medal as in Bomb Groups. Also on this date, the weather flight transferred to Cheddington while runways at Bovingdon were resurfaced.

Five days later on Christmas eve another accident occurred. Fortress 42-37869 returned from a meteorological survey. Because adverse weather forced closing the airfield at Cheddington the Fort landed at St. Eval in Cornwall on completion of the survey. When the all-clear signal arrived from Cheddington, the aircraft took off for its flight home. The weather in the west of England reported a cloud base of 2,500 feet and visibility of two to four miles. Near Bridestow, the aircraft entered mist in the hilly area of Okehaptom Moor, skimmed over the first hill which was 1,430 ft. above sea level and then entered an area of heavy mist. The pilot went on to instruments and started to climb, but the Fort struck the next hill 1,750 ft. high and bounced along for 3/4 mile, split open, then caught fire. The pilot and copilot managed to escape the wrecked Fort when it came to rest; two other bodies were thrown clear.[30]

Five crew members were killed in the accident: 2/Lt Richard T. Neary, navigator; T/Sgts. Sherwood R. Rennar and Samuel L. Craig, S/Sgts. Mario A. Panetti and Albany J. Blanchard, and RAF met/obs Sgt. Basil Brown. Both Lts. Ernest H. Patterson and his copilot, Raymond A. Coats suffered severe injuries and burns. At the end of March 1944, they were released from hospital and resumed flying duties.

30 Pat Carty, Secret Squadrons of the Eighth, London: Ian Allen, 1990: 29.

In January 1944, the flight transferred back to Bovingdon where General Doolittle arrived for a photo session posing with Lt. Poelstra's crew who on 9 January flew the longest Eighth Air Force operational mission of 16 hours and 10 minutes.

On 2 February, a third and most mysterious disaster occurred. Poelstra recalls:

> "We were sent on a day weather mission. It was as rough as I've ever seen it over the Atlantic. We entered a front when all of sudden the aircraft dropped and my copilot, Lt. Chis and I hit our head on the ceiling. The men in the rear fuselage were having a lunch break when they started floating in the air along with the coffee and food. The aircraft was continuously thrown about throughout the flight by the horrendous weather conditions.
>
> "Finally we were returning to Bovingdon when somewhere over England we met Lt. Robert H. Grafton and his crew beginning their WR flight. We talked by radio and I briefed him on how atrocious it was out there, detailing our experiences. That was the last we heard from him. They disappeared into nowhere and never returned."

The Grafton crew left Bovingdon at 1819 hours in B-17F 42-30131. Listed as MIA were: 2/Lts. Robert H. Grafton, Ralph E. Williams and Casper M. Durgin, T/Sgt. Louis Feller, S/Sgts. Angelo M. Nicoletti and Tillman O. Law, Sgt. Charles Green, and RAF met/obs Sgt. Edward J.B. Pye.

Soon after this event, Poelstra took a colonel from HQ Eighth Air Force on a night WR mission:

Heavy Weather crew stand in front of B-17G 42-39777 on 10 February 1944 at Bovingdon. L-R. standing are Lts. A.M. Sobelman, copilot; E.F. Sofranko, navigator; E.W. Stephens, pilot, and Sgt. David E. Lake, engineer. Kneeling Sgt. Irvin Stilwell, radio operator; T/Sgt. D.F. Harris, engineer, and Sgt. Paul Goble. *Photo F. Sofranko.*

"The colonel flew as my copilot. We taxied to the runway, applied brakes, lowered flaps and advanced full power. The brakes were released and we bore down the runway. As we approached the end, the colonel appeared to be sitting on 'pins and needles' as to whether we would get off or not. I lifted the aircraft and slowly climbed to begin our weather flight.

"At the first position, we descended in darkness to within 50 feet over the ocean with landing lights on. The rough weather heavily bounced the aircraft around. The colonel again appeared very worried. I think he wanted to know if I knew what I was doing? We obtained our measurements and advanced to the next position. The colonel appeared pale and sighed with relief as we started to climb. This behavior continued throughout the flight when sea level measurements were required. We later completed the mission and returned to Bovingdon. After that flight the colonel changed the policy. We were instructed not to descend lower than 300 feet at night."

Having only seven crews, the weather flight experienced operational trouble. In February, Capt. Podwojski requisitioned three more crews who were not given the option to volunteer for this type of duty.

Each mission brought new experiences. During the early morning of 1 April, the crew of Lt. Harold Aronson prepared to depart Bovingdon in B-17G 42-39871. Customary procedure required the radio operator secure signal codes for the day from Operations, and also emergency packets containing foreign currency and maps in case of being forced down. During this process, the crew became engaged in conversation with staff members, then boarded a vehicle for transport to the Fort. After warm up and preflight check the aircraft took off.

It was beginning to crack dawn as they approached St. Eval when the radio operator asked Aronson for the radio codes, passwords, etc. Aronson replied, "I never pickup the codes, you've got them!" Apparently while conversing at operations, the radio operator forgot to procure the material. It was pointless continuing the mission because weather data could not be transmitted without the codes.

Lt. General James Doolittle with Poelstra's crew. L-R. Unknown, 1/Lt. Edward R. Heeking, RAF met/obs, Unknown, 1/Lts. Alexander H. Chis, and William E. Poelstra next to Doolittle. *Photo NARA.*

Normally, the flight checked-in with St. Eval prior to performing a low-level pass across the field to calibrate the radio altimeter. Aronson radioed the tower but received no acknowledgment. After a lengthy period he called a second time and circled within view so the RAF could visually identify and confirm he was an American B-17. Finally they responded and Aronson explained the current predicament. He maintained an orbit below the overcast while the RAF contacted Bovingdon for further instructions. A message arrived ordering Aronson to land and pickup the required codes.

The heavily laden Fort approached the runway at 0700 in light rain. Upon crossing the threshold, it lost lift and dropped heavily from 50 feet. The impact broke the right landing strut, collapsing the undercarrriage. The aircraft skidded and when it slowed, the right wing dug into the ground causing an 180 degree swing, damaging propellers, nacelles, and wingtip. Gasoline poured from the aircraft as the men hastily exited.

Combat Camera Unit History

The 8 CCU was activated 21 November 1942, without personnel and equipment at Bolling Field, Washington, DC. On 10 January 1943, the unit was assigned to the First Motion Picture Unit (FMPU), Culver City, California. FMPU supplied personnel and equipment to accomplish the actual activation. Major Ted. D. Tetzlaff, former director of photography for Paramount Studios, assumed command. FMPU was responsible for both still and motion picture documentation of AAF war activities.

Landing accident at St. Eval on 1 April 1944 by Lt. H. Aronson.
Photo C. Ashcroft

1/Lt Harold Aronson poses for photo. *Photo H. Aronson*

A three-week training program was held at Bonham's Gulch, a former a Boy Scout camp in Laurel Canyon, Hollywood. In order to prepare for combat photography, physical training consisted of crawling through underbrush with rifle in one hand and a hand-held camera in the other. Classroom work included nomenclature, loading and operation of all cameras. Then commenced two months further training at Hal Roach Studios including photography classes, hikes, drills, cross-country marches and low-altitude flights in AT-11 aircraft.

The unit left Culver City on 7 May 1943, and arrived in New York city the 11th. On the 22nd, the men embarked on the SS Mariposa and docked at Liverpool, England, on 1 June. They traveled to Hurst Park near Hampton Court, Middlesex, their new home. Quartered in an old racetrack near the Thames River, the men enjoyed a three-week hiatus before their equipment arrived. Classes in photography began once the equipment was unpacked. On 10 June the enlisted men received gunnery training at Bovingdon, and on 2 July Major Tetzlaff presented them wings, followed by a celebration party.

Detachment D formed on 6 June, and one month later left for Marks Hall to fly operations in B-26 Marauders. Two weeks later the detachment moved to Andrews Field near Great Sailing. By the end of July, Detachment A at Alconbury, B at Chelveston, and C at Snetterton Heath were now combat ready.

The first operational mission was flown by T/Sgt Daniel A. McGovern to Villacoubley, France, on 14 July. He manned one of the Fort's guns during fighter attacks and claimed a Me 109, but similar declarations by numerous gunners voided his kill. Bill Wood made the unit's second mission, to Haroya, Norway, on 24 July and McGovern flew his second mission on 28 July when the Mighty Eighth bombed Kassel, Germany. And thus started the combat operation of this unit.

On 12 August, Sgt. George Gamble received a flak wound above the right wrist while filming a mission to Gilsenkirken in the Ruhr Valley. The first shuttle raid to Regensburg on 17 August with cameramen Bartholomew, Howell, Hawn, Mullen and McFarling met fierce resistance from flak and enemy fighters. S/Sgt. Hawn flying in tail-end-Charlie position ended up as the lead bomber after leaving the target billowing smoke. Hawn, Howell and McFarling filmed bombing of Bordeaux, France, on the return shuttle from North Africa, while Mullen's group hit the secondary target of Villacoubley. Bartholomew's Fort aborted over the Mediterranean, headed for North Africa, and he returned to England two weeks later. Sgt. Bill Wood flew his third mission on 27 August to St. Omer, France, and became MIA

During September, Detachment D completed 27 sorties in Marauders. On 8 September, Lt. Robert H. Edwards flew his first mission to Boulogne, France. His formation missed the target and the following day repeated the mission when Edwards went MIA.

All detachments were recalled to a permanent 3 Bomb Division (BD) station at Bury St. Edmunds on 18 October. Only 12 sorties were flown in October and 14 in November. Sgt. Dave Barker and T/Sgt. Irven G. Slate flew in B-24s to Bremen on 13 November when Slate went MIA. Bill Wood, listed MIA in August, returned December 17 after spending four months dodging the Gestapo in France before reaching Spain.

Twelve sorties were flown in January. On the 11th to Brunswick, Germany, Lt. George Bartholomew returned minus his right little finger when a flak burst struck his aircraft. During this mission, S/Sgt. David A. Barker was KIA flying in a 410 BS, 94 BG B-17F 42-30248. Rockets fired by attacking Fw 190s damaged the Fort and wounded the tail gunner. The pilot ordered a bailout. Just as tail gunner Sgt. Bernard B. Brown staggered from his position and collapsed on the rear floor from serious wounds, both waist gunners jumped. Instead of following, Barker went to aid Brown. The bomber then entered a spiral dive and centrifugal forces prevented any chance of parachuting to safety. The bodies of both Barker and Brown were found beside the wreckage near a road between Abbeile to Eltze, Germany.

Assigned to 8 Rcn Wing (P) on 27 February, CCU remained attached to 3 BD for duty, having operated from Bury St. Edmunds since October. March saw 21 sorties and on 22 April, CCU transferred to High Wycombe. On 14 May 1944, all still photographers transferred to 3 CCU and motion picture men to 8 CCU. Finally on this day, 8 CCU arrived at Watton, their permanent home.

Joint Photo Reconnaissance Committee

The DI was tasked with maintaining RAF cooperation to eliminate duplicating PR efforts. The AAF now had an independent 8 RW requiring sharing mission decisions. To achieve this goal at ACIU, the Joint Photo Reconnaissance Committee (JPRC) formed on 14 May by an AM directive reorganizing the combined RAF-AAF photo-reconnaissance establishment.

JPRC consisted of four voting members: one RAF, one Royal Navy, one British Army and one AAF representative. Each had an alternate who joined the meeting but had no voting power unless their principal was absent. The meetings were held in a house bordering the Benson aerodrome, twice a day, seven days a week with 100 percent attendance.[31] Col. Carrolton G. Ketchum became the first AAF committee member:

> "At our morning and afternoon meetings there were laid before us requests for flying and picture-taking all over enemy-held territory in Europe originating in all the major headquarters of the British and American forces. We also were called upon frequently for intensive coverage for mapping the areas between our lines and those of the enemy. Much of the time, the request for reconnaissance services exceeded the availability of our flying forces, even making allowance for rejection for some requests because current weather conditions in the area specified did not permit picture taking. This situation necessitated very careful consideration as to priorities, but we had substantial guidance for most of those decisions from General Orders establishing some priority and others directing us for a specified period to give first choice to the calls from one of the agencies requesting actions. Priorities varied from day-to-day, and JPRC had to screen out duplicate and unnecessary requests ... Our two daily sessions averaged a total of four or five hours, and we all had to study relevant information for most of the day."[32]

Problems arose in developing Mosquito maintenance functions at Watton. Operating a wooden aircraft required personnel changes. It became necessary to reclassify men from an MOS number held under a bombardment squadron to an MOS number required to maintain a wooden-type aircraft, i.e. armorers became dope and fabric workers. Those having a technical expertise unessential to wooden aircraft transferred to another unit in exchange for woodworkers and airframe specialist to maintain the wooden Mosquito. Assembling sufficient skilled craftsmen to carry on an operational unit proved a colossal task.

The *Stars and Stripes* newspaper published weekly a Mosquito photograph seeking veterans to fly the aircraft. Many pilots who completed a combat tour on bombers, volunteered for a second tour with the 802nd. Rather than return to USA for reassignment to the Far East, most preferred to remain in ETO. Many submitted an official request for transfer, and the Mosquito squadrons soon became staffed almost totally with volunteers.

With the influx of bomber personnel, a training program was devised using P-38-Js to transition B-17 and B-24 pilots to twin-engine aircraft. This became revised utilizing an A-20 Havoc, and Oxfords to familiarize pilots with mechanical features of British-type aircraft.

Experienced AAF Mosquito pilots were almost non-existent. Those few available became important for instructional purposes: Lts. James Luma, Royal Collard, Oliver Emmel, Victor Doroski, Richard Geary, Vance Chipman, along with F/Os Richard Wulfkuhle and Russell Whitmer. All served in the RAF/RCAF with Mosquito backgrounds. They were delegated responsibility for training pilots the Mosquito's intricacies. In addition, aircrew received lectures in communication, weather, Intelligence, flight characteristics, and emergency abandoning procedures. The Rolls-Royce factory, Derby, provided Merlin engine familiarization classes.

---

31 Carlton G. Ketchum. The Recollections of Colonel Retread, Pittsburgh, PA: Hart Books, 1976: 235.

32 Carlton G. Ketchum, Ibid, 235-236.

Aronson crewmembers. L-R. S/Sgt. Francis J. Walsh, Sgt. Melvin R. Booze, Sgt. Leo E. McAvoy, and S/Sgt. Okey E. McCroskey. *Photo H. Aronson*

Douglas A-20 briefly at Watton to transition four-engine bomber pilots to twin-engine aircraft. *Photo. L. Poulx*

18th Weather Squadron men release balloon for optical tracking to determine winds aloft. *Photo J. Nowak*

Mosquito NS510 parked in storage next to another still in RAF markings. Cockpit is covered with canvas and no tail letter assigned. *Photo. G. Scruggs*

Mosquito navigators required proficiency in both meteorology and UK navigation aids. The majority completed a combat bomber tour that facilitated transitioning to Mosquito-type missions. Their prior combat experience benefited from Gee class instructions and actual operational flights. New navigators arriving from the USA received prior meteorological lectures and instruction at various American universities.

A training accident occurred 26 April when Lt. Otto E. Kaellner transitioned Lt. Richard E. Petersen (MM384). During takeoff the aircraft vaned in to a strong crosswind. Petersen attempted recovery using rudder, brakes and throttle but failed, and the left undercarriage sheared off from side-strain after skidding 150 yards. The 482 BG planned modifying Mosquitos to carry H2X radar equipment and their pilots required aircraft flight training. On the 27th, Lts. Carl Satterlund and Dennis Scanlan, flew from Alconbury to 94 BG at Bury St. Edmunds in Mk.III LR534. They performed a normal landing approach. After rolling 600 yards down the runway, the left landing gear locking device failed and the undercarriage slowly folded, causing the aircraft to ground loop off the runway.

MM384 comes to rest on 26 April 1944. *Photo J. Sheen*

802 RG Mosquito in RAF markings visiting Mount Farm. *Photo R. Astrella*

May 1944

The danger of low flying became vivid on 5 May, killing Capt. Richard B. Thompson and severely injuring Lt. David P. Hicks on a training flight in Oxford T1066. Watton received a report the aircraft crashed at East Dereham, Kings Lynn Railroad, some 18 miles from base. According to evidence, the Oxford struck power lines paralleling the railroad track while diving on a train, lost control, and crashed 150 yards further. As a mandatory reminder to consequences of unauthorized low flying, HQ posted a photo of the wreckage in the operations room with a sheet of paper for all pilots to initial.

On 8 May, Lt. Clarence Stephenson returned from an Atlantic WR in B-17 231766 when attacked by several Ju 88s. The Luftwaffe aircraft from a shipping surveillance unit in France departed home base in formation, flying to a point south of Ireland; then each Ju 88 flew in different directions to perform recce duties. They all met at a predetermined time for the return formation flight to France. The Fortress accidentally joined the Ju 88s at their rendezvous point.

The Fortress was 100 miles south of Ireland with the crew relaxing after the long overwater flight. Sgt. Keith Clinton exited the ball turret to stretch his long legs. Sgt. Eddie E. Walden came from the tail gun position and both men began conversing at the waist position when suddenly a blast shook the aircraft. There was a terrible, metal-ripping sound as shells ripped through the plane. Clinton received wounds to his head and back, while fragments struck Walden in the back.

Evidently four Ju 88s approached from behind and beneath the Fortress. Meeting no resistance, they immediately began firing 20 mm rounds at the aircraft. Exploding shells knocked out number one engine and damaged number three. The enemy rounds completely shot away the rudder cables and all manual control.

While the Ju 88s continued their attack, Clinton crawled to the radio compartment where Ward bandaged his head. He then assisted Clinton back into the ball turret. The RAF met/obs already manned a waist gun. The tail guns were useless without a gunner, but the other men quickly began return fire.

The pilot, wounded in the leg by bullet fragments, lost all manual flight control from the tail section. He switched to automatic pilot and the aircraft responded immediately. Enemy rounds did not damage the autopilot cables that connect to the regular control cables, attached at points past the cable sections shot away. Using the automatic pilot, Stephenson turned the Fortress north towards the attackers to bring the upper-turret firepower upon the Ju 88s. He now became the attacker. The Luftwaffe aircraft continued in combat with the injured Fort for 25 minutes. Both top turret and nose gun kept the enemy aircraft away and inflicted damage to one Ju 88 seen trailing white smoke.

During the attack, Sgt. Minery dispatched a Mayday by radio and eventually RAF Spitfires arrived to escort the wounded Fort to St. Eval. Clinton passed out in the ball turret. Both Ward and Minery pulled him out and on to the floor where he regained consciousness. The RAF Sgt. suffered a severe throat wound. After an hour, Stephenson landed at St. Eval by automatic pilot with two engines out and one on fire. Ambulances transported the wounded men to an RAF hospital and included: 1/Lt. Clarence Stephenson, S/Sgts. Eddie F. Walden and Earl H. Westerholme, Sgt. Keith J. Clinton, and RAF Sgt. David F. Mulligan. For this engagement Stephenson received the DFC and all wounded men the Purple Heart.

Combat damaged B-17 at St. Eval. Unknown sitting in Jeep. Front row: J.C. Minery, 1/Lt. J. Ward, T/Sgt. E.N. Westerholm with cap on backwards. Capt Stephenson's head is blocked by windshield bar. 1/Lt. Swiatnicki is between Minery and Ward, talking to Stephenson to his right. Top left of Capt. Stephenson is an unknown airman.
Photo. K. Clinton

Presenting Purple Heart medals to L-R. Capt. Clarence Stephenson, Keith Clinton, Earl H. Westerholm, Eddie E. Walden. *Photo K. Clinton*

Rear fuselage damage inflicted by Ju 88 attack. *Photo C. Ashcroft*

Faced with inclement continental weather, Eighth Air Force turned to H2X radar to discover a means whereby heavy bombing operations against Germany in the spring of 1943 could be maintained throughout the winter months of 1943-44. After experimentation and development, H2X or more correctly AN/APS-15, evolved to detect a target through clouds, smoke and haze.

The 482 BG (P) activated at Alconbury in August 1943 and equipped with B-17 and B-24 aircraft, served as a pathfinder unit using both H2S and H2X radar known as Mickey. From its inception until the spring of 1944, the BG led many successful bombing formations to targets in Germany. They also developed an experimental schooling unit to train navigators who received assignments to pathfinder aircraft within individual BGs throughout East Anglia.

Eighth Air Force required radar approach maps of specific enemy targets to assist blind bombing through heavy overcasts. Pinetree initiated a program to marry H2X to the Mosquito airframe. Project officer 1/Lt. Hillard T. Grusin recalls Eighth Air Force planned flying the Mickey Mosquito along the invasion coast prior to D-Day, from Heligoland to the Bay of Biscay, taking radar photos to detect concentrations of enemy armor. Unexpected radar returns would also reveal camouflaged fortification, which low-level PR flights using color film, inspected to determine what the unexpected radar return entailed.

Original 654 BS Mickey aircrew. Except for Larkin, and Balser, the others are listed POW, KIA or WIA. *Photo J. Morrison*

The 482nd also dispatched single B-17 aircraft on night Mickey sorties obtaining radar screen approach photos of peripheral targets in occupied Europe. The Mosquitos would supplement this task with radar approach maps of targets deep inside enemy territory. The 482nd prototyped Mosquito MM308/G to carry H2X. The suffix "G" indicated a guard required at all times to protect the radar equipment. As modifications progressed, training 802nd radar operators began 20 April at Alconbury.

The MM308/G prototype aircraft nose section was removed, an AN/APS-15 scanner installed and a special bulbous nose constructed around it. The nose radar connected to a cathode ray scope (CRT) in the rear fuselage. The radar operator/navigator sat aft of the bomb bay bulkhead, facing forward towards a radar screen and navigational equipment including Gee. An H2X equipment shortage in the UK required removing sets from 482 BG aircraft for the Mickey Mosquitos. A second aircraft MM311 also underwent Mickey conversion. Preliminary flights demonstrated H2X suitable as a navigational aid.

USSTAF requested six Mickey Mosquitos assigned to Fifteenth Air Force, MTO. They planned transferring third and fourth H2X-modified Mosquitos from de Havilland, and a fifth aircraft in 30 days, then one every six weeks until a total of six were delivered. Unfortunately, Mickey conversions for Fifteenth Air Force would delay the Eighth Air Force Mosquito Mickey program. Spare components proved critical requiring cannibalization of radar sets for needed parts. Also, Air Ministry would not guarantee additional factory H2X Mosquito airframes, thus the Fifteenth Air Force plan never reached fruition.

Mickey operator's perspective of equipment. H2X indicator PPI, with Gee mounted on top. At far left is a small remote scope with camera to photograph screen images. *Photo J. Morrison*

Mickey Equipment Operations

Firstly, photo technicians installed a camera with film over a remote scope located left of the main radar screen equipment. Then the navigator squeezed his bulky frame through the small right-rear hatch, while the crew chief forcefully shoved him inside the aft fuselage. He strapped himself to a small seat facing forward. Following engine start-up and preflight check, the Mosquito roared off the runway and slowly climbed for altitude.

The navigator/radar operator using both Gee and DR headed for the target IP. The operator faced a cathode-ray tube (CRT) screen displaying a 180-degree sweep of the terrain ahead. To his left was located the small remote scope with a special designed K-24 aerial camera attached to photograph the screen. He began adjusting the main set and monitoring instruments for voltage fluctuations; if too low the set would not function properly, and if too high, a fuse may blow. Adjustments continued for screen brilliance, focus, and antenna-tilt to ensure a clear radar image return, while tuning to eliminate as many ground returns as possible. Once the photo run began, the camera took automatic pictures at timed intervals. The developed remote scope camera film provided target approach, radar image prints for bomber navigators to reference when operating H2X during blind bombing.

Photographing the CRT scope proved a major problem. The scan moved slower at the center and faster toward the edge. This made determining an adequate shutter speed difficult. Col. George Goddard, the father of AAF PR, arrived in England and solved the problem with special filters allowing for the first time, successful CRT scope photographs.

Camera mount position for the small remote PPI scope to photograph the screen images. Groundcrew installed the camera prior to mission. *Photo J. Morrison*

Mickey aircrew at briefing. Top to bottom standing, Intelligence officer Capt. Finis D. McClanahan; sitting, Robert Balser, Vance J. Chipman, Douglas L. Harris, William Cannon, and Capt. Hubert R. Farrins Intelligence officer. Capt. George Connors providing briefing. *Photo J. Morrison*

On 1 May, 2/Lt. Royal Collard and Capt. Waller (MM311) flew the first experimental Mosquito Mickey test mission. Entering the enemy coast, scope photos began at one-minute intervals but camera failure resulted in only seven pictures. The H2X achieved a 60-mile range from 27,000 feet. Waller lapsed into a semi-conscious state when an oxygen regulator failed but recovered after switching to his emergency supply. Meanwhile, Collard descended to 19,000 feet. Both men flew a second mission to Germany (MM311) on 8 May and the radar functioned perfectly achieving a 50-mile range from 30,000 feet.

The Mickey prototype MM308(G) was destroyed 12 May. The 482 BG crew of pilot 1/Lts. Henry M. Stewart with navigator/radar operator Hillard Grusin, completed a test flight and returned to Alconbury for a landing just as a B-17 began taking off. The tower ordered Stewart to make another short circuit. He raised the undercarriage when the right engine emitted smoke forcing its shutdown. Flying a normal pattern, the landing gear took longer to lower during single-engine operation. He extended final approach until both undercarriages indicated down and fully locked. Stewart touched halfway down the runway and overshot the end. He crashed through a fence, jumped a ditch and road, before coming to rest in a plowed field with both occupants unscathed.

The 482nd transferred the aircraft to 2 SAD for repair, and both MM308/G and MM311 reassigned to 802 RG on 13 May. The 8 AFSC was delegated responsibility to modify four H2X Mosquitos in May and technical drawings transferred accordingly. The 482nd placed engineering officer Capt. Henry E. Clopper on detached service to 802nd for radar equipment technical support.

On late afternoon of 16 May, Lt. Harold Aronson taxied B-17G 43-39777 in preparation for an Atlantic WR mission when the left main tire blew out. The ship lurched and left the perimeter track causing the tail wheel to snap off. And three days later on the 19th, a 388 BG B-17 42-31242 made a crash landing at Watton after a mid-air collision with another bomber.

The 2135 Engineer Avn Firefighting Platoon established on 2 April began operations 23 May commanded by lst Lt. Jack E. Wilkin.

Light Weather Squadron, manned by pilots on detached service from 50 FS, successfully transitioned from P-38s to Mosquito aircraft. An acute shortage of spare parts and personnel trained in Mosquito maintenance remained prevalent. Operational flying began with two ex-RCAF transferee's, Lts. James F. Luma, and Oliver H. Emmel.

Mickey Mosquito MM311 taken at Alconbury 4 July 1944 with serial number censored. *Photo T. Fields*

Side view of Mosquito MM311 with dark blue spinners normally reserved for Light Weather Squadron. *Photo T. Fields*

All were qualified Mosquito pilots, with Luma having shot-down five German aircraft with 418 Squadron RCAF. When matched with experienced navigators, they began flying experimental WR sorties for Light Weather Squadron.

On 23 May, Lts. Oliver Emmel/Iver Fitschen (NS510) took off at 0421 for France. The mission involved gathering weather observation data, patterned on identical RAF "Pampa" WR flights. At mid-channel, heavy icing coated the cockpit windows, then VHF and Gee equipment failure forced the crew to abort the mission. They landed 20 minutes late because of enemy intruder activity at Watton. Next day Watton dispatched the same mission profile at 0250 with Lt. James Luma/Capt. John Walch (NS510) for central France. Flying at 25,000 feet they completed the first successful experimental weather mission. The flight encountered numerous deficiencies. The navigator recommended a more versatile map table with metal clips firmly holding the navigation utensils in an accessible place, and a red light to assist with reading charts.

Lts. Oliver Emmel/Iver Fitschen (NS510) flew the next experimental WR mission to Holland, Belgium, and parts of Germany on 25 May. These WR flights over the ETO became a 24-hour daily occurrence in all weather conditions, "unarmed and unafraid." By months end, Light Weather flew 17 operational missions without receiving damage from enemy action. On 1 June, Watton labeled Mosquito WR missions Bluestocking.

Any command, group, or organization may request a weather mission report from HQ Eighth Air Force during a 24-hour period, designating target, route for coverage, information desired, and radio report deadline. After determining the time period requirement, Pinetree dispatched the proposed mission to Watton Operations. Here a mission plan was formulated and the weather data requirements discussed with an observer-navigator. In the meantime, Operations consulted the 18 Weather Squadron Detachment and discussed the forecast for routes and bases, then dispatched the Mosquito.

8th Rcn (S) Squadron photo personnel at Alconbury. Standing, L-R. John S. Dobrowlski, Henry R. Gundlach, Stevens, Smith, Carl Wanka. Kneeling, Frank J. Lachuisa, Charleston L. Jordan, Stanley W. Belzak, Locke, William. L. Castner, Edward H. Darilek. *Photo T. Fields*

The planned flight included desired and optional routes to and from a target, or to several if required. Cloud patterns and formations with bases and tops were measured or estimated. Orientation of cloudbanks, frontal position, visibility, wind, precipitation, heights of condensation trails, and any phenomena of interest were all noted. If obscured by multi-layered clouds, the pilot descended to determine exact altitude of each cloud level and weather conditions, then climbed to altitude for mission continuation, after which he returned to base.

Meanwhile, Pinetree scheduled a meeting time with meteorologists at the requesting base/command and others desiring access to the information. A secure telephone-conference line linked everyone together at which time the observer provided weather surveillance at designated points. After completing his report, interested participants interrogate him with specific questions to seek further clarification. Watton summarized the data then dispatched this in code to Dunstable for teletype transmission and eventual distribution to requesting units. In this manner, scores of stations shared in the weather information.

Further flying incidents occurred. Major Hugh Bozarth (MM338) performed a test flight the 26th, when the top escape hatch flew off during takeoff. Unhindered, he made a circuit and returned to land. At touchdown all cockpit indicators showed both wheels down and locked, including a second down selection check. The right wing started to lower and application of throttle failed to prevent the wing from touching the ground, causing a swerve towards a parked B-17. To avoid collision, he retracted the undercarriage. The ensuing track cleared the B-17 and slowed the aircraft's belly slide. Also this day, Lts. Otto Kaellner/ James Matuska (NS553) left Watton to practice landings at Snetterton Heath. During flare over the runway approach, the Mosquito stalled and landed hard, blowing the right tire. After a bounce, the pilot completed a smooth three-point landing, but the flat tire resulted in a ground loop, folding the right undercarriage.

Heavy Weather B-17s continued the long Atlantic Ocean flights and closed the month with an eventful mission. This flight began at Watton on 29 May. A previous scheduled Epicure flight returned with mechanical problems and the sortie was scrubbed. Watton dispatched a replacement aircraft: pilot, 1/Lt. Michael J. Lorenz; copilot, 1/Lt. Lawrence J. Ketterer; navigator, 1/Lt. Nathan Swerdlowe; engineer, S/Sgt. Arnold W. Westerlin; radio operator S/Sgt. Joseph Bubulka; machine-gunners, Sgts. Harold Tarvin and Lee Richardson; and Sgt. William Young, RAF met/obs.

After briefing, the crew took off at 1707 in B-17F 42-3279 from North Pickenham. They flew 200 miles to St. Eval, obtained altimeter readings, temperature, etc. and headed on a magnetic course west out over the Atlantic Ocean. Sgt. Young, the met/obs remembers Swerdlowe, the navigator, in a moment of levity passed him a note, "No Gee, No radio, No compass. No. 4 engine smoking." He read the joke, then put a period after "engine" and entered a "No" before "smoking" in jest.

The crew that ditched in Dingle Bay. L-R. Second row. 1/Lts. Michael Lorenz, Nathan Swerdlowe, Lawrence Ketterer. Front row Sgt. Arnold Westerlin, Sgt. Lee Richardson, S/Sgt. Joseph Bubulka, Sgt. Harold Tarvin. *Photo A. Swerdlowe*

RAF Meteorological observers, L-R. Front row: Peter Smith, Tiny Mentz, Dave Mulligan, William Young. Rear row: Unknown, Mike Shaw, Ginger Sharp. In doorway is F/O Bentell. *Photo K. Clinton*

En route, the Fortress gathered weather observations and data at different scheduled points. Passing position 10, someone peering out a side-window spotted oil streaking backward beneath an engine and notified pilot Lorenz. He immediately went to the radio compartment and viewed oil streaking from number one engine. He decided to continue the mission to position 11 and then turn back.

At position 11, the Fort descended to obtain the necessary readings, and Lorenz applied throttles to climb for altitude. During ascent, number one engine started to cough and belch, so he immediately feathered it and performed a climbing turn to 900 feet. He terminated the mission and informed the crew.

Everything seemed to function smoothly during the return trip home. Lorenz notified copilot Ketterer, that he wanted a cup of coffee. He vacated his seat, allowing the flight engineer to take this position, and proceeded to the rear compartment. Here, coffee brewed on an electric stove for cooking food during long 14-hour flights. While sipping and conversing, a sudden backfire noise from engine three broke the relaxed atmosphere. Lorenz immediately rushed to the cockpit, surveyed the situation and feathered the engine.

By this time daylight still prevailed. The propeller of engine number three remained feathered as the Fort bounced slightly in the rough air currents. With each bounce, the prop rotated a foot or so. Over a period of time, the short turns added up to several rotations. It took nine rotations to pump oil from the prop hub, which permitted the propeller to unfeather – and this is what happened. Now they had an uncontrollable propeller creating drag that slowed the aircraft by 15 mph.

The evening turned to twilight as the Fort slowly lost altitude. Normally, the aircraft carried 3,600-gallons of fuel on board. The main tanks retained a capacity for eight hours. The wing Tokyo tanks held 1,000-gallons, while two huge bomb bay tanks held a further 1,000-gallons. At maximum consumption of 50-gallons an hour, this theoretically provided 18-hours flying time.

Previously, 500 gallons in the Tokyo tanks located on each side of the outer engines, had been transferred to the main tanks. That left two full, 500-gallon auxiliary fueltanks in the bomb bay. To maintain altitude, Lorenz had no option but jettison the huge bay tanks. Now losing height rapidly with insufficient time to open the bomb bay doors electrically – they normally open by long screw bars via electric motors – Lorenz reached for the emergency ball-lever to manually release the tanks and gave it a yank. It wouldn't move. He gave it another hard pull and suddenly felt the aircraft immediately balloon upward. The doors opened and 1,000-gallons of fuel dropped to the ocean below.

The two remaining engines were at full throttle providing a flying speed of 80 to 85 mph. The crew attempted, but failed to reconnect the screw bars to close the bomb bay doors. They remained open, increasing aerodynamic drag and along with the windmilling prop, hindered progress.

Darkness now descended rapidly. Lorenz ordered the crew to throw out everything not nailed down – just save the parachutes. After 15 minutes, the engineer reported everything loose had been jettisoned. The Fort maintained altitude, and at times managed to gain height. The RAF met/obs in the nose section held watch over the darkness with open eyes. Meanwhile, the radio operator transmitted a message of the current predicament, while the navigator determined their position every 15 minutes. He also provided Lorenz with a heading for southern Ireland. It was imperative they gain altitude because of mountains in the center of Ireland.

Total darkness prevailed as the Fort continued flying a further two hours on the remaining two engines. With both engines at full throttle it was only a matter of time before they burned out. The question was how long would they continue operating? Sufficient fuel remained in the main tanks for a few more hours.

Lorenz asked the men to notify him if they spotted anything ahead. Eventually, they reached a point where navigator Swerdlowe informed him they were approaching land. Soon they sighted the faint black outline of a mountain that stretched low in the darkness – it was southern Ireland. Lorenz flew closer and instructed the crew to prepare for a water landing.

Except for the pilot, copilot and flight engineer, all other crewmembers went to the radio compartment. They removed the top hatch and placed it in the rear aircraft then each took positions on the floor. On either side of the doorway, one man sat against the compartment bulkhead, while another pressed against the knees of the first man, and another sat pressed against the second man. Both pilot and copilot unlocked their side windows.

Lorenz made a slow, shallow right turn, maintaining altitude to reach a point heading directly for the mountains. Five miles north of Ballydavid, West Kerry County, Lorenz retarded the throttles and initiated a descent in darkness as if performing a normal landing except with both landing gear retracted. The landing lights illuminated the undulating sea, topped with glittering waves. He safely brought the aircraft down on the ocean surface, tail first. The tail mushed in the water, then the lower fuselage touched. The Fort skidded over the choppy waves as the prop blades thrashed and churned in the sea. The forward aircraft then touched heavy as a wall of water splashed over the cockpit. The plastic nose shattered and water rushed in. Then everything came to a standstill. It was 0315 hours. According to copilot Ketterer:

> "We landed with the tail dragging in the water. The aircraft hit the waves with a heavy impact and the open bomb bay acted like a drogue and caught the full force of the water, slowing the aircraft abruptly. I was knocked out briefly by the sudden jarring we experienced. I sat there stunned and thought I was injured, tasting blood on my lips and face. It dawned on me that it was not blood but salt water. When I became fully conscious, I remember visually it was getting lighter and less dim in the cockpit, especially around the side windows. Then I realized we were submerged and the forward fuselage slowly lifted up and out of the water, allowing in more starlight through the side windows. The open bomb bay caught the full force of the water, and the nose of the aircraft went deep down under water. When I glanced out the side, we were just below the surface of the ocean, but soon the forward fuselage lifted out of the water and we were on top. I opened a side window and started to climb out. I noticed Mike Lorenz sitting there in the darkness with his seat belt on. He had been knocked unconscious. I went back in, unlocked his seat belt and with Westerlin on the outside pulling, helped to twist Mike through his side window."

Lorenz struck his jaw on the control column and briefly stunned. After opening the side window to exit, he attempted to stand. The life vest straps between his legs snagged on the seat. He cut the straps with a knife from his pocket, and with assistance squeezed through the side window.

Lee Richardson in the rear, recalls that when the aircraft landed in the ocean, the open bomb bay scooped up a torrent of water which burst open a small wooden door leading to the radio compartment. It immediately flooded and submerged the occupants. The men surfaced from the deluge and climbed through the open top hatch. The RAF met/obs pulled himself out, crawled forward and stubbed his foot against an aerial, injuring an ankle.

Upon exiting the rear compartment, the men pulled a connection to eject and inflate both rubber life rafts from storage compartments. Lorenz, Bubulka and Tarvin assisted injured Sgt. Young into the left raft, while on the opposite side, Ketterer, Swerdlowe, Westerlin and Richards climbed into the other and cut themselves free. Both rafts bobbed in the water nearby for several minutes as the crew observed the Fort afloat on the ocean. Slowly the black silhouette of the tail rose up high from the water as the B-17 slid beneath the waves. Both crews began paddling towards land in darkness.

Ketterer's raft eventually reached the jagged shoreline:

> "We paddled close to the cliffs. The roaring waves crashing against the rocky shore, rose some 20 feet up the cliff wall and then drop down again. Then, the next incoming wave would take us up again. We had one man at a time step off the life raft and on to the cliff wall with each wave crest. This worked fine for the first three fellows, and I was the last one still in the raft. When the waves crested, I stepped out and missed. The next thing I knew, I was hanging upside-down in darkness. Nathan Swerdlowe had me by the ankles. The water dropped 20 feet as the waves went out. The men pulled me up, and we then climbed the 60 or so feet to the top of the craggy cliff. It was 0400."

The men walked some distance and located a stonewall surrounding a dwelling. They rattled a locked metal gate and the occupants soon opened the front door. Maurice and Hannah O'Connor spoke a Gaelic dialect called Urst. They invited the crew inside and the men hung their socks and flight suits before a fireplace. Meanwhile, the husband left to fetch a daughter nearby who spoke English. After being informed the Lorenz crew was still in a lifeboat, the daughter ran to notify her brother. He dispatched boats to find the missing men. The elderly gentleman later transported Ketterer's crew in a two-wheel, horse-draw cart to the local constabulary in Dingle.

The second raft remained at sea. As dawn approached, the crew heard the roaring sound of surf pounding against the rocky cliff walls. Lorenz detected a gap in the inhospitable coastline and directed his men to paddle towards the area (Brandon Creek) when a long black boat with three men rowing came in view. They threw Lorenz a rope and towed the raft to shore.

The civilians assisted the injured Sgt. Young to the O'Connor farmhouse and provided the crew nourishments. For her kindness, Lorenz gave Mrs. O'Connor his moist British pound notes. A cart arrived from the village with a constable and transported the crew to Dingle to rejoin the other members. The police provided accommodations at local hotel, and a doctor arrived to insert two stitches in Lorenz's chin and bound Sgt. Young's ankle

That afternoon, an Irish Army Major arrived and took charge. Next day the crew boarded two vehicles and an ambulance for Northern Ireland. Traveling along back roads they reached the border next night. The Major spoke with the guard on duty. The convoy drove through and continued another two hours before transferring the crew to a U.S. Army truck for transport to an American base. On 6 June the men flew in a B-17 to London and arrived at a building on Grosvenor Square. Col. Stenglein and Maj. Podwojski signed the necessary papers for their repatriation to duty.

The steel-mat PSP runway laid across the grass field at Watton proved a major disadvantage. The airfield was shared with 3 SAD heavy bombers that made the runways rough and unsuitable for Mosquito operations. The arrival of the 859 Engineering Battalion on 30 May launched building a concrete surface runway. As construction work continued, flight operations transferred temporally to nearby bases with Mickey Mosquitos flying from Alconbury, while B-17s and Light Weather Mosquitos flew from North Pickenham.

A new mission profile began the 30th, when Lt.Col. George Doherty/Lt. Lionel Proulx (NS519) left to scout weather conditions ahead of the bomber formations, thus allowing for possible diversion to secondary targets. Lack of SOP handicapped the mission and a 20-minute takeoff delay resulted in the Mosquito arriving over Bremen as 2 BD formations completed the bomb attack. Next day, a Scout mission fared no better when Capt. Willis Locke/Lt Iver Fitschen (MM388) and 1 BD were recalled due to frontal weather in the Strasbourg area. Watton labeled the mission Scout.

Lt. Elbert Harris details the Scout mission profile. 1. Arrive at target 20 minutes ahead of the bombers while feinting to a non-target. 2. Evaluate weather conditions en route and over the target area. 3. Perform visual search for enemy fighters in the air or taking off from airfields. 4. Report flak and its location. 5. Observe and report effectiveness of smoke screens to obscure the target. 6. Forecast what the bombardiers will see on arrival at the target. 7. Report observations to the Bomber Commander via the bomber radio channel. 8. If primary target is scrubbed, proceed to secondary target. 9. Feint toward a non-target as the bombers head for the IP. 10. Time the return to view the bomb impacts. 11. Report bombing accuracy to mission commander. Scouting sorties became a standard daily activity.

Designed for the purpose of furnishing courier service in the ETO, the 8 Rcn Courier Squadron (P) was officially established 4 March 1944 at Cheddington. Major Wilbur C. Allen continued as CO until replaced by Capt. William G. Brownlow II on 1 April, followed 7 April by Major Joseph A. Stenglein. The unit flew B-17Fs on "Snooper" WR missions covering the British Isles, suffering slight damage from friendly antiaircraft fire over the London area. Major Stenglein was promoted to Lt.Col. on 10 May and appointed deputy group commander. Capt. John Richly then took over as squadron commander. The unit also operated three Stinson L-5s and six UC-64s, and flew daily courier runs to Booker Field near Wing HQ using L-5s.

Airspeed Oxford AB768 for training pilots to British-style flight systems. In background is camouflaged Mosquito trainer with letter T/S on tail and band around rear fuselage. To the right is Stinson L-5. *Photo R. Walker*

B-17 23346 "H"of 8th Rcn 'Courier' Squadron (P). Note 'A' on the top tail painted over. *Photo via L. Proulx*

A cable received 31 May ordered disbanding the unit at midnight and a final WR flight was flown that night. All personnel were redistributed among other squadrons; flying personnel who completed a tour of operations were reassigned within ETO. Special Squadron took over courier duties, including all the light aircraft, and Heavy Weather delegated the short Snooper WR flights covering the British Isles.

Heavy weather flightcrew. L-R. Sgts. Robert Poe, Edward Comer, Donald Yennie, and Sam Odensky. *Photo S. Odensky*

June 1944

June witnessed the 802nd develop into a full-fledged operational unit with great strides as training continued. This included the first mishap on 3 June, when Lts. Harold. L. James/William E. Scharbach (NS518) began a night cross-country training flight from North Pickenham. During takeoff run, the aircraft swung and ground looped, collapsing the undercarriage and tearing the right engine from the mount. Both occupants escaped unscathed. The following day, Capt. Earl Muchway/ Lt. Louis Pessirilo (NS515) flew an experimental Bluestocking to Holland to determine the feasibility of establishing VHF radio contact with Pinetree. It proved successful and secure.

The importance of Heavy Weather WR flights passed unnoticed until early June. SHAEF HQ proceeded to implement Operation Overlord, the invasion of Normandy and the second front. The invasion day depended on recommendations from Allied advisory meteorological staff. A committee comprising two British AM members including senior staff meteorologist G/C J.M Stagg, two from the British Admiralty, and two AAF Weather Information Section representatives headed by Col. Irving P. Krick, were assigned to provide Allied Supreme Commander, General Dwight D. Eisenhower a three-day period (window of opportunity) for D-Day.

This decision depended on two vital factors: a high-tide period to carry landing vessels over man-made obstacles lining the coast, and favorable atmospheric conditions with relatively calm weather for both sea and air support. Tidal conditions occurred only once each lunar month, and assembling the invasion fleet required at least 36-hours notice, further compounding the problem. Daily conferences were held at SHAEF HQ, briefing the commanders on current and predicted weather patterns. These reports were based on data from two critical areas, west and southwest of Ireland, that, "would decide ... whether Overlord would be definitely postponed for at least 24 hours or go ... forward to the assault on the coast of France."[33]

Two luminaries with divergent meteorological views contrasted forecasting D-Day. G/C Stagg relied on current weather data, while AAF Col. Krick developed analogues studying weather patterns of the previous 50 years to prognosticate future trends. Stagg remained cautiously pessimistic, while Krick held optimistic views on the forthcoming invasion period.

Remains of NS518 that ground looped on takeoff on 3 June 1944.
Photo W. Scharbach

33 J.M Stagg, Forecast For Overlord, New York: W.W. Norton & Co. Inc., 1971: 101.

Light Weather Squadron navigators. L-R. 1/Lts. James M. Davis, William F. Scharbach, Donald E. Lee and Louis Pessirilo (KIA). *Photo W. Scharbach*

A low-pressure front near Newfoundland curved southward to the Azores, then upward to another low off the west Scotland coast. Fronts normally move west to east across the northern Atlantic. A high-pressure area forming near the Azores, moving towards the continent would force northward inclement weather conditions associated with low-pressure patterns and bring clear skies to the channel area. Data gathered on 3 June, revealed a weakening of the Azores high-pressure area causing the low-pressure front near Scotland to move southwards with poor weather and winds. This low also facilitated movement of the Newfoundland low towards Ireland and the British Isles, further aggravating the pattern.

Worried about unseasonable storms lashing the north Atlantic, the forecasters waited for additional data. Three aircraft, one each from 517 and 518 Squadron RAF, and another from 8 Rcn Squadron (Heavy) (P) took off 4 June on normal sorties. They had been briefed to meet an Atlantic low that had left Newfoundland close behind the previous depression then off northwest Scotland. The 50-nautical mile positions were steadily marked off. Tufted wisps of frontal cirrus came in to view, but with each excursion to sea level, pressure continued to rise. The evidence became unmistakable: a new ridge of high pressure was developing north of the Azores. Eagerly, UK meteorologist decoded the W/T messages, plotted the vertical soundings and prepared their high-level maps. At 18,000 feet the ridge became an extensive high covering most of the Atlantic. Some hours later, the Allied meteorologic crews returned to their bases, unaware these sorties reaffirmed the date for D-Day.

Armed with this data, G/C Stagg, now in agreement with Col. Krick, anticipated 6 June as having a weather break with favorable conditions for the invasion. Gen. Eisenhower, backed by his Allied commanders, decided to proceed with Operation Overlord. Which of the meteorologists actually risked the precise prediction is controversial.

The Allies captured German outposts in Greenland, Spitsbergen, and Jan Mayan Island, thereby blocked enemy access to vital weather Intelligence information. This forced Germany to rely on two strategically located, north Atlantic-based U-boats, and daily reconnaissance flights from France to the west of Ireland to gather weather data. They also dispatched secret agents to England tasked with sabotage, gathering troop movements, war production information, and above all, instructed to transmit daily weather reports. Thus, the German meteorological service sorrowfully lacked atmospheric data for predicting current and future weather patterns approaching Europe.[34] This disadvantage allowed the Allies to achieve tactical surprise launching the 6 June invasion.

On the afternoon of 5 June, Field Order No. 727, announcing the Normandy invasion and a second dispatch revealing "Zero Hour" arrived at HQ 8 RW. D-Day minus one, found CCU Sgts. Joseph F. Capicotto and Ralph Thiry filming the bombing of V-1 sites at Pas de Calais. That night the unit received notification to film the early morning invasion. Sgts. Lynch and Kurjack were alerted for flights over the beachhead. Lynch would fly with Lt.Col. William Cleveland from Wing HQ, and Kurjack with Lt.Col. George Doherty, both in a specially modified Mosquito aircraft. Meanwhile, ground personnel prepared for the big event by painting black and white stripes around the rear fuselage and wings.

Watton dispatched a scheduled Atlantic WR B-17 that evening. It was the worst weather ever experienced by Lt. Harold Aronson. He could see neither sea nor sky. At the furthest point on track, the Fort climbed to 20,000 feet in dense cloud. As was customary, they tuned the radio to whatever U.S. broadcast station came in. The most powerful station, located on a hotel roof in Chicago, transmitted dance music all night long. Once tuned in, Aronson connected the broadcast via intercom for all crew members to enjoy. It was near midnight in Chicago when an announcer interrupted the dance music with news the Allied invasion of Europe had begun:

> "At that time, dawn approached England while we were far out in the Atlantic covered by darkness. This is how we learned of the invasion. The return flight was again flown on instruments in total cloud. The Fort approached England customarily at low altitude to provide visual identification to shore posts on the Cornwall coast. We broke out of clouds, and in the distance, the entire southwest coast towards the channel was filled with ships gathering from small ports, rendezvousing and proceeding to the invasion area. The channel was filled with ship after ship, as far as the eye could see. I'll never forget that sight. We thought we might provoke suspicion flying over the massive flotilla, so we climbed back in the clouds and continued to Watton."

Both 7 PRG and 802 RG flew missions to the beachhead immediately prior to and during the initial D-Day assault. Lts. Oliver Emmel/Iver Fitschen (NS515) left at 0107 on a weather flight over the Cherbourg Peninsula and Dieppe. Aerial bombardment of the beachhead began at 0314, naval bombardment at 0550, and the first assault waves of infantry and tanks came ashore at 0630.[35] The 802nd filmed ground and air activities over the beachhead with special Mosquitos rigged with several motion picture (cine) cameras: one directed rearwards, one downwards, one each installed in the external wing tanks facing forward, and another in the nose. This special mission was labeled Rooster.

Interrogation of 7 PRG and 802 pilots on return provided the chief source of information regarding activities during the early invasion hours. F-5s of 7 PRG encountered the first enemy air opposition. Two Fw 190s attacked an unarmed F-5, but the pilot managed to escape by diving at full throttle towards the beach area after having his canopy blown off and radio shot out.

Cameramen Lt. Daniel McCarthy, Sgts. George Gamble and Al Mandelaro left early that morning in B-17s to film the invasion landings. At the same time, Sgts. Becky Oliva, Lou Zusser, Ralph Thiry, Bob Arrowood and Walt Lipinski filmed the action and bomb strikes at the beachhead from Liberator formations. The men, flying in the Forts and Libs, flew a second time during the day to film the heavy channel traffic and bombing of enemy installations.

At 0912 a Rooster flight launched with Lt.Col. William Cleveland/ Sgt. Eb Lynch (NS555) escorted by two 352 FG P-51s. Unfortunately, cloud cover and ground haze over the beach restricted the photo results. A second Rooster dispatched at 1518 with Lt.Col. George Doherty/ Sgt. August Kurjack (NS535), again with 352 FG escorts. The crew encountered heavy flak near Caen, but brought film back providing SHAEF HQ with the first visual evidence of the invasion scene.

D-Day plus one. The following morning Capt. Joe Baker/Sgt. August Kurjack (NS535) with 355 FG P-51 escorts, left on a Rooster mission. They encountered heavy enemy flak in the vicinity of Barneville, France, and a P-51 flown by Lt. William Winn Jr. received flak damaged but continued and returned to base safely. CCU sent Lt. Daniel McCarthy and Sgt. Al Mandelaro with the Forts to film the destruction of an airfield at Thury-Harcourt where the Luftwaffe moved fighter aircraft to the invasion front. Sgts. Bob Arrowood and Lou Zusser flew in B-24s to record the attack on Alencon, France. Arrowood reached the target, but the Liberator which Zusser flew in began to tremble while attempting to gain altitude over England. He immediately grabbed his chute pack just as the jump signal sounded: "Before jumping, I looked around to see if the waist gunners were going to bailout, but the poor kids were in a daze. I didn't waste any time getting out the escape hatch." He jumped with two other men before the bomber entered a spin and crashed killing the remaining crew. Zusser descended quickly and landed heavily suffering a broken back, hospitalized, and transferred to the USA.

34 F.H. Hinsley, et all, British Intelligence in the Second World War, New York: Cambridge University Press, 1984: 125.

35 George C. Marshall/Henry H. Arnold/Ernesrt J. King, The War Reports, New York: J.B. Lipincott Co., 1947: 184.

Lt.Col. George Doherty on the right after return from the invasion coast on 6 June 1944.
Photo. G. Doherty

The final Rooster was flown 8 June. Capt. Walter D. Gernand/ Sgt. Ebbet C. Lynch (NS555) with Lt. William Winn Jr. flying a 355 FG escort P-51, left at 1520 for the beachhead. After successfully completing the mission the aircraft landed at RAF Middle Wallop to refuel. Both aircraft left at 1930 for Watton but encountered low ceiling and began ascending. The Mustang lost sight of the Mosquito, went on instruments and climbed up through clouds, alone. Obtaining a homing signal from Steeple Morden the P-51 descended through the clouds and landed at the 355 FG base with no oil pressure.

8 CCU photographer Sgt. Eb Lynch KIA on 8 June 1944. *Photo F. McCarthy*

Capt. Walter Gernan, Beaumont, Texas, sitting in Jeep, KIA 8 June 1944. *Photo. G. Scruggs.*

Three hours later, Watton received a report the Mosquito crashed one mile from High Wycombe killing both men. Witnesses reported they heard an aircraft above the clouds flying in a northerly direction with the engines being continuously throttled down and then opened up. The aircraft suddenly broke through the low overcast in a vertical dive, crashing straight into a railroad embankment in a ball of flame. Watton held a memorial service the following day for both men.

A Scout mission flown 11 June by Lts. Richard Petersen/William Scharbach (MM388) for 2 BD targeted aerodromes in western France. They experienced radio problems and were unable to contact the formations. Later that day, Major Hugh Bozarth/Lt. Louis Pessirilo (NS509) flew another Scout successfully for 1 BD to airfields in western France. Back at Watton, Lt. Robert Shoenhair (MM346) experienced difficulty keeping his Mosquito aligned down the runway on takeoff and pulled off power in an attempt to stop. Heading towards heavy equipment on a runway under construction, he applied left brake to avoid hitting the machinery, but the landing gear collapsed upon contact with a dirt abutment.

Other Bluestocking WR flights from 11 through 15 June, reported poor weather conditions, forcing cancellation of all bombing activity to strategic targets in Germany. Scout sorties from 11 through 17 June continued supporting Eighth Air Force bombing missions across France.

Germany launched the first V-1 attack against England the night 11/12 June. Of 11 dispatched, only four V-1s struck London. By 15/16 June, the Battle of the Flying Bombs began with launch of 300 V-1s against England: 144 reached the coast and 73 struck London.[36]

Prior V-1 Intelligence remained restricted to very high command levels though both Eighth and Ninth Air Force and RAF, previously bombed suspected launching sites for more than a year. Only at top command levels did definite Intelligence on the nature of the targets circulate. The exact state of V-1 weapons development remained unknown but considered a serious threat to warrant special attention to acquire detailed photos of all launch sites. Dissatisfied with both RAF and AAF photo Intelligence activities, General Spaatz demanded reduced time periods between procuring aerial photos and attacks against the secret weapon sites.

The Director of Operations, USSTAF, summoned a special meeting attended by General McDonald, Colonel Roosevelt, and their respective staff. The enemy employed elaborate camouflage measures to conceal launching, storage depots, and transportation conveying V-1 fuel and supplies. The Dilly mission was instituted to locate all V-1 launch and support sites. Initiated 15 June the PR missions would continue for 80 days until Allied ground forces overran the Pas de Calais area.

USSTAF compiled daily target lists dispatched to both 7 PRG and 802 RG. After each sortie, the annotated prints and interpretations were immediately processed and rushed by special C-64 courier to the DI. Low-level F-5 dicing sorties located metal tubular launching ramps and blast marks disclosed the fuels employed. Photo Intelligence identified supply and assembly buildings camouflaged with painted doors and windows to resemble farmhouses, or hidden under natural and artificial foliage. For the first time, AAF employed infrared film in the ETO to detect hidden launch sites and depots.

The first major Luftwaffe action following D-Day occurred 12 June and resulted in heavy enemy losses. Meanwhile, at Watton training flights continued. On 13 June, Lts. Norman Magee/Otto Kaellner (NS535) performed night circuit and landings from North Pickenham. While flying the airplane, Magee experienced trouble obtaining a down green light for both undercarriages. After several attempts, the indicator remained lit and the gear fully extended on both his landings. He transferred control to Kaellner who performed three approach and landings with erratic down lock indicator lights. On the third touchdown, the right undercarriage gradually collapsed over a distance of 300 yards. When the wing made ground contact, the plane swung right causing the left landing gear to snap off.

Two Mosquito accidents involved non-Watton crews. On June 15, ferry pilot, Lt. Hutchins preparing to deliver NS533 from Burtonwood to Alconbury, ground looped on takeoff. Major Charles W. Himes, a 2 BADA pilot, wiped out MM310 when landing at Warton the 17th. Also on this day, Heavy Weather Squadron flew the first Allah WR mission that covered areas in the immediate vicinity of the UK.

Lt.Col. Stenglein conducted six Atlantic missions during June evaluating the performance of technical equipment in the weather aircraft. On two separate occasions, he feathered an engine, and on another, mechanical failure required feathering two engines and initiating an emergency landing at St. Eval. Inspection revealed the squadron received war-weary B-17s withdrawn from combat units. These proved unsuitable for long-range weather flights over the Atlantic. Stenglein initiated efforts to obtain new engines for the aircraft.

F/O Richard F. Clounch/Lt. William Cannon (MM311) flew the Group's first Mickey mission on 20 June. They left Alconbury at 2258 for Nuremberg but aborted halfway there when the H2X faltered. Unknown to Clounch, this was his first and final 802nd mission.

The V-1 onslaught against England increased dramatically, forcing transfer and repositioning all available barrage balloons south and west of London to counter this threat. This led to an unforeseen tragedy. F/O Richard F. Clounch, ex-RCAF, transferred to the 802nd three weeks previously from an RAF unit at Gravesend. He filed a flight plan to

L-R. 1/Lts. Vance Chipman and Richard Clounch (KIFA), at Alconbury. *Photo T. Fields.*

36 Wesley F. Craven and James L. Cate, The Army Air Forces In World War II, Vol. 3, Chicago: University of Chicago Press, 1948: 526.

Gravesend to pickup his personal belongings. He took along navigator 2/Lt. Connor C. O'Connor, a squadron member and boyhood friend from his hometown. The flight was cleared 21 June through Alconbury Operations, who lacked sufficient information to thoroughly brief the men.

They flew NS558 to Gravesend, encountering heavy overcast at 1,000 feet in light drizzle with visibility two to three miles. Clounch circuited the field at 800 feet with wheels down to enter the landing pattern and was not in contact with the control tower. The Duty Flying Control Officer at 1510 witnessed the Mosquito strike a balloon tethering cable that sliced off the port wing. This caused the aircraft to spin-in and crash one mile southwest of the field at Malling Cross killing both occupants.

The balloons floated at 4,500 feet, well above the 1,000-foot undercast, concealing any hint of their exact location or presence. The Alconbury preflight briefing stated balloons were situated two and a half miles southwest of Gravesend. Unfortunately, balloon positions constantly shifted. At the time of the accident, several balloons were moored between the crash site and the field, the closest one estimated within a half to three-quarter miles of the perimeter track.

On the 24th, Lts. Victor Doroski/Jacob Hochman (NS568) flew to the Cherbourg area on the first daylight PRU mission. The Mosquito was configured with two K-19B cameras installed in the forward bomb bay taking photos through portholes and a third K-19B camera in the aft fuselage to accomplish the daylight photo-reconnaissance profile.

First Frantic Mission

In 1943, the AAF ETO bombing campaign experienced heavy losses of men and aircraft. The bomber streams came under constant attack by antiaircraft fire and Luftwaffe fighters to and from the target. German war industry was in the process of relocating further east out of reach of Allied bombers. To shorten exposure time to hostile fire, Eighth Air Force developed the shuttle-bombing concept. On 17 August 1943, a bomber force attacked the ball bearing factories at Schweinfurt in central Germany, while another struck aircraft factories at Regensburg in southeast Germany. The task force attacking Regensburg left the target and flew south over the Mediterranean Sea to land in North Africa. After refueling and with bombs loaded, they returned to England attacking targets en route.

When the Allies occupied southern Italy, Fifteenth Air Force established headquarters in the fall of 1943. This command would thereafter attack the Reich's southern flank, drawing away Luftwaffe units from northern Germany, thus relieving opposition to Eighth Air Force forays.

To attack industrial targets in the east, the AAF established shuttle flights to the USSR. On 2 June, Fifteenth Air Force initiated the first shuttle mission named Frantic Joe to the USSR with 129 bombers attacking Debrecen, Hungary, en route.[37] Plans were then implemented for Eighth Air Force bombers to fly shuttle-bombing flights to the USSR from England.

Watton received notification to provide two Mosquitos for a special Frantic shuttle mission set for 21 June, with post-target landings at either Poltava, Mirgorod, or Piryatin, USSR. Capt. Earl L. Muchway was pilot and navigator Capt. Lionel A. Proulx, while Capt. Willis D. Locke piloted the second Mosquito with navigator Lt. Ralph E. Fisher. The navigators received maps, coordinates, and other pertinent mission information. To their dismay, the maps were published by the Tsarist regime before the Communists took over in 1917. Many geographical features had changed in the interval.

The mission route stretched 1,450 nautical miles: 539 from England to Ruhland, Germany, and a further 812 beyond to Poltava, Ukraine, USSR. On 21 June, 1,311 bombers with 1,190 fighters left in two formations for the target area. The main task force attacked Berlin while a smaller second formation of 114 B-17s from the 13 and 45 Combat Wing with 70 P-51 escorts from 4 FG and 486 FS, 352 FG, attacked synthetic oil refineries at Ruhland, 60 miles south of Berlin. The plant produced two percent of total German oil output.[38,39] Muchway/Proulx (NS519) accompanied the main task force to Berlin while Locke/Fisher (NS515) flew to Ruhland with the Frantic force.

Preparing for first USSR shuttle mission, June 1944. L-R. Capt. Lionel Proulx with Earl Muchway on right. *Photo L. Proulx*

37 Frantic Missions, File 527.451, USAFHRA.

38 Craven & Cate, The Army Air force in World War II, Vol. 3: 571.

39 Archie J. Old Jr., Col.: Report on Shuttle Mission to Russia, 6 July 1944,File. 527 .476, June/July Frantic: 4. USAFHRA

Tight fit in Mosquito cockpit. L-R. Navigator Lionel Proulx, with Earl Muchway, pilot on right. *Photo L. Proulx*

Capt. George Scruggs, 653 Squadron engineering officer, accompanied the mission in a B-17, providing technical support and evaluating the logistical requirements for Mosquito participation in future Frantic missions. As he recalls the events:

"I arrived at 100 Group 2200 hours, June 20 … All Mosquito equipment had been loaded (on the B-17), and I was assigned to fly with Lt. Rice on ship #154, *Our Babe*. As soon as the briefing was over I drew flying clothing … then reported to the ship at 0400 and introduced to the crew. The engines were started at 0445 and takeoff was at 0500 on time … The weather was terrible. There was no ceiling and the morning light still had not penetrated the overcast. Our takeoff was successful and we climbed on instruments through the soup, breaking out at approximately 3,000 feet into a sky full of milling bombers. Below, B-17s were still climbing their way through the heavy overcast.

"By 0700 we had completed our assembly and were on course … flying at 19,000 feet behind the main force, destined for Berlin. Our Combat Wing skirted the heavy flak concentration and proceeded into Germany with the main force in the direction of Berlin. Our fighter escort now consisted of both P-47s and P-51s … By now the attack on Berlin had already begun and cripples were seen struggling back escorted by P-51s. We could see the 'circus' … very clearly, and it was obvious the enemy did not appreciate the visit. Before we reached Berlin, the two Combat Wings destined for USSR headed slightly south, at which time I breathed a sigh of relief.

"We reached our target, an synthetic oil refinery in the city of Ruhland at 1020 as briefed. Our escort left us, and the bomb run seemed to last 'nine years' instead of nine minutes as briefed. We bombed our target by groups from approximately 20,000 feet … Heavy smoke rose to 18,000 feet over the target as the last groups went over. Flak was moderately heavy … *Our Babe* received only one small hole in the left wing. Mustangs from the 4 FG picked us up at 1035. Our force now consisted of two Combat Wings of B-17s with P-51s as escorts."

The Luftwaffe fighter force concentrated their attacks on the larger bomber force to Berlin. The main task force then withdrew toward England as the second wave of 3 BD Fortresses hit Ruhland refineries.[40]

Previously, the Mosquito scout of Muchway/Proulx flew to the Berlin area, overtook the bomber stream and arrived early. The sky then filled with thousands of aircraft milling around, including bomber formations, escorting fighters, and Luftwaffe fighters. Muchway took a look at the scene and said, "We're climbing above this." He ascended to 34,000 feet, circled, and radioed weather and enemy fighter information.

After the Forts completed the bombing run for return to England, Muchway/Proulx continued from Berlin eastwards over Germany to Poland and USSR. Entering Poland, they encountered a turbulent weather front with black cumulonimbus nimbus clouds towering to 45,000 feet. Rain pelted and lashed the Mosquito as rough air currents jostled the crew. Upon encountering sleet they descended in an attempt to locate a better flight level.

After several hours flying in such miserable conditions, the crew was unsure of their exact position. Muchway changed course and flew northeast to reach the Dnieper River located north of Kiev, then turned south. Eventually, the Dnieper River appeared in the distance, identifying it by size and contour. They banked and turned, following it south to Kiev, capital of Ukraine. At this time, Yakovlev-5, single-engine Soviet fighters appeared and made an attack pass. Muchway advanced the throttles and dove low to ground level so the fighters could not attack from below.

The Mosquito continued southeast from Kiev following a railway track. They continued following the rail line until Muchway noticed a spur track running northward that connected with Piryatin. He turned, followed it until some three miles from the airfield, when they heard a weak Soviet radio transmitter. The aircraft landed at Piryatin, a dirt airfield with PSP laid down. The other Mosquito arrived earlier having flown a more southerly course. Besides the P-51s, Soviet Air Force P-39s sat parked on the field.

Mirogorad, USSR. Capt. Scruggs second from left, kneeling with sunglasses on.
Photo G. Scruggs

40 Ibid., 8.

Captain George Scruggs, engineering officer. *Photo G. Scruggs*

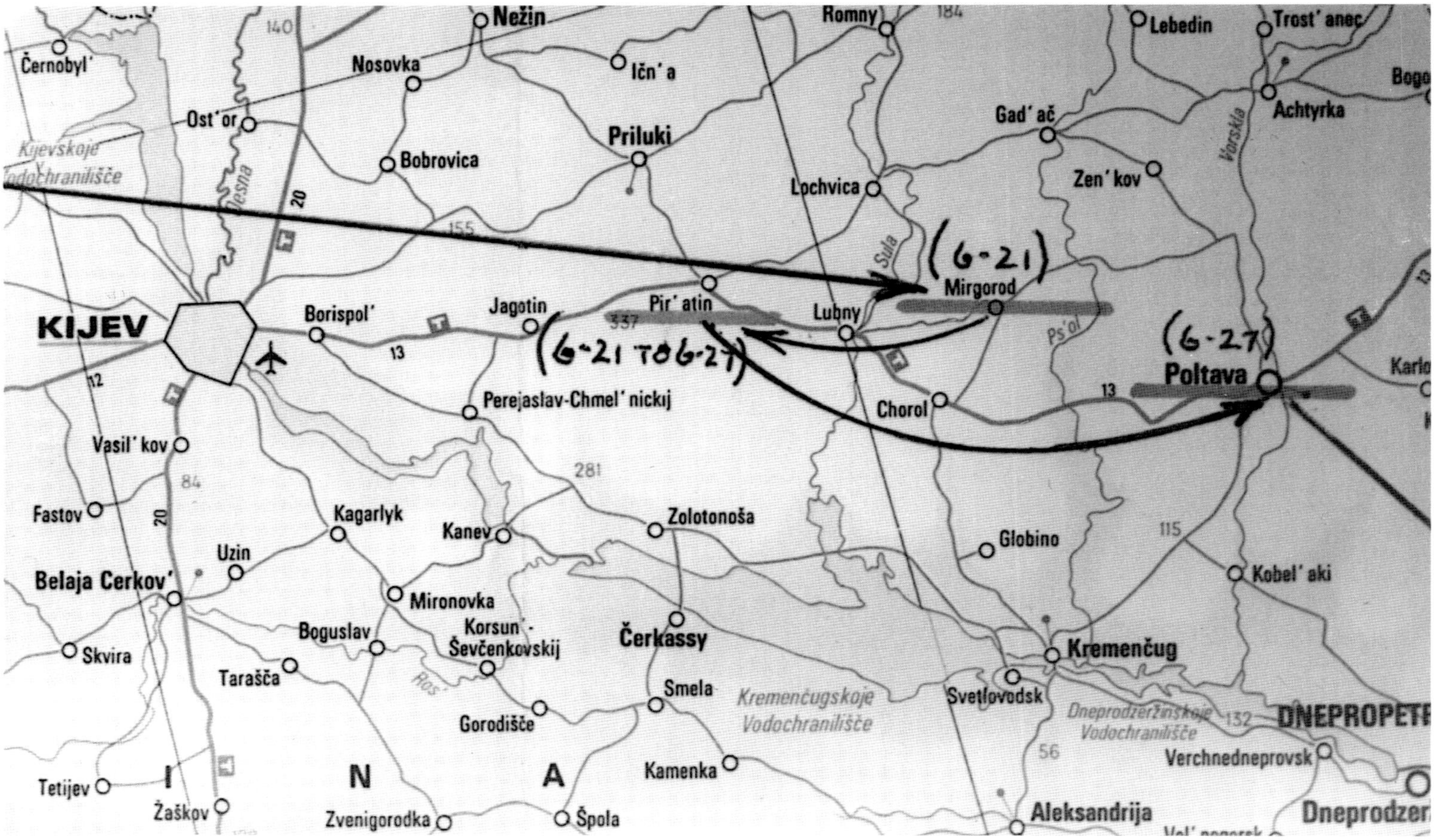

Map showing route traveled by Scruggs in USSR. *Photo Scruggs*

As Muchway taxied to a stop, Proulx pulled up the inner floor door, let down the outer hatchcover and slid out the aircraft. He turned and received greetings from an awaiting interpreter who commented the Soviets had no knowledge of Mosquito aircraft arriving, nor told of the black and white stripes. Both U.S. and Soviet officers attended a briefing. The senior Soviet staff member apologized for the lack of fighter escort and the accidental fighter attack near Kiev. They received notification a German He 177 followed and shadowed the American task force en route.

Meanwhile, Capt. Scruggs lodged in the nose of a B-17 en route to Mirgorod, USSR, recalls leaving Germany:

> "As soon as we left the target (Ruhland), we set course in an easterly direction and headed for Poland ... We passed south of Warsaw, and shortly thereafter, we were attacked by approximately 20 to 25 Me 109s. They approached dead ahead from 25,000 feet, slashing down through our formation. Soon the sky was full of P-51 droptanks ... The attack lasted a few minutes.
>
> "As we neared the Soviet-German lines at 1400 our escort left us. Weather continued to be good with scattered thunderheads ... We landed at Mirgorod at 1645 with ten minutes fuel supply left. The field was already packed with Forts, and everyone was so low on fuel, they were landing three to four planes at a time on the single runway strip ... All crews were very tired, having had no sleep or food. Flying time required for the trip was 11 hours and 45 minutes – eight hours of which was spent on oxygen.
>
> "The 13 CW landed at Mirgorod while the 45 CW proceeded to Poltava. I managed to catch a C-47 shuttle plane in Mirgorod and arrived at Piryatin about 2000. I was elated to learn our Mosquito scouts arrived safely and both plane and crews were in fairly good condition. The Mustang escorts landed at Piryatin and all aircraft for the most part were in good condition. After talking to Captains Muchway and Locke ... I learned that neither Eastern Command nor the Soviets had been notified about the Mosquitos. They knew nothing about Mosquito scouts flown by Americans and asked out pilots upon landing, what they were doing there? Even headquarters at Poltava had no knowledge. Our men were very fortunate they had not been shot down by Soviet gunners."

Meanwhile, events began to unfold at Piryatin. The latitude of Ukraine is considerably north, so near midnight it was twilight and not yet dark when Proulx and Muchway retired to their tent. At approximately 0030, loud explosions and antiaircraft fire awakened Proulx. As he glanced out the tent, it appeared daylight as parachute flares floated down directly over the field. He moved to the opposite cot and vigorously shook Muchway shouting, "Earl. Out of the sack, we're being bombed."

Capts. Earl Muchway on left and Lionel Proulx in lower hatchway with crewchief at far right at Piryatin, USSR. *Photo G. Scruggs*

Now fully awake, Muchway and Proulx raced to a deep pit some thirty yards away, Muchway leaped in not knowing it was an abandoned latrine. He shouted angrily, "Get me out of here." Both men then reached safety of a trench and peered out at the scene, as another wave of Luftwaffe bombers flew over the airfield and neighboring tent village. Muchway noted the Ju 88s couldn't have been more than 6,000 feet up. At this moment, the Soviet defenders hurled an ear-shattering barrage at the enemy aircraft. The bomber formation held firm, miraculously untouched by flak. Muchway burst out, "I can't believe they could miss at that range!"

The attack created friction and animosity between the Soviets and Americans. The Soviets manned weapons from light machine guns to captured German 88s. The Luftwaffe attacked from 5,000 to 12,000 feet, yet the Soviets never 'flicked a feather' off any of them. Col. Donald Blakeslee, in charge of P-51s at Piryatin, almost came to blows with the Soviet commandant. Blakeslee offered to pursue the German bombers with P-51s but the Soviets refused.

Piryatin suffered slight damage and both Mosquitos unharmed. The combined loses at Poltava, Mirgorod, and Piryatin comprised 43 B-17s destroyed and 26 damaged; 15 Mustangs and other Soviet aircraft destroyed; American ammunition dumps set aflame and 450,000 gallons of fuel ignited; one U.S. airman killed and the Soviets suffered 30 fatalities. ATC aircraft eventually transported over 600 remaining aircrew to England.[41]

Scruggs assessed future USSR Mosquito operations:

"I spent the following day working on the Mosquito aircraft. After a thorough inspection, both were found in good condition except for torn fabric caused by hail, a dead battery, a cut in one tire, and a few other minor discrepancies. With aid of the special oxygen fittings brought along, we were able to service them. The only oil available on the base was 1120. Grade 1100 is recommended for Mosquito aircraft. Also, there was no English glycol on hand, and we were forced to use an American brand that, if used in large quantities will cause sludge. In the future, it is recommended that a small quantity of grade 1100 oil and English glycol be included in addition to all the other spare parts carried on this trip.

"It must also be considered, that although the engineering facilities in Ukraine are well set up for B-17s and P-51s, there are no tires or jacks for Mosquitos. From past experiences, we have found that very few landings and takeoffs can be accomplished from a steel-mat runway before a tire change is necessary. This is a very serious problem if future operations are being considered, since neither jacks nor spare tires can be carried. It is possible to carry spare tires in a bomber, but since one jack weighs 800 pounds, the weight would be prohibitive."

The second Mosquito crew at Piryatin, 1/Lt. Ralph Fisher on left and Capt. Willis Locke on the right. Mosquito is NS515. *Photo. G. Scruggs*

41 Ibid., 14.

B-17 wreckage, Poltava, USSR, following Luftwaffe surprise bomb attack.
Photo R. Thiry

"There is no one in USSR familiar with Mosquito aircraft, and if future operations are contemplated, it will be necessary to send at least one crew chief for two planes. He can obtain plenty of assistance, either American or Soviet, at the base. There are some P-51 tools available but a crew chiefs' toolkit should be taken. Although this is a Mustang base, all mechanics are B-17 men …

"By the end of the day (22 June), the Mosquitos had been thoroughly checked over and serviced, ready for flight. A German reconnaissance plane was over the field at approximately 1600, and we were quite certain that tonight was our turn. All flyable P-51s were moved to Kharkov just before dark. All that remained on the field that night were a few Soviet P-39s, three or four P-51s, and our two Mosquitos. Soviet soldiers brought truckloads of tree branches and covered the remaining aircraft, making a good natural camouflage. That night at exactly 0020, the Luftwaffe dropped a flare at the edge of our field. Enemy planes circled overhead for nearly half an hour. Approximately 20 to 25 parachute flares were dropped and much ammunition expended by the Soviets.

"At around 0050, the bombardment began and lasted for approximately one-half hour. Many fragmentation bombs were dropped, a few heavy explosives, and hundreds of antipersonnel (Butterfly) bombs. After the excitement was over, it became obvious the Luftwaffe had mistaken a neighboring wheat field for the airfield. The only damage incurred was to a portion of the gasoline dump and the Soviet motor pool. No German planes were knocked down by antiaircraft, although they flew very low.

"During the next three days – June 23 to 25th – there was very little to do except keep the Mosquitos and crew on alert. Every day at 1600, Photo Joe appeared over our field. One attempt was made by four P-51s to intercept him, but this failed. On 25 June, Jerry obtained a vector from our homing station and buzzed the field at 20 feet. The Mustangs flew to Kharkov each night and returned to Piryatin each morning."

Three 8 CCU cameramen, T/Sgt. George Howell, S/Sgt. George Gamble and Sgt. Ralph Thiry accompanied 3 BD Forts on the Frantic mission to obtain motion picture coverage. Howell and Gamble filmed strikes on Ruhland then flew to Mirgorod. Thiry's group also attacked Ruhland but landed at Poltava Airfield. That night Thiry's Fort received a direct hit and his cameras, film, and photo equipment completely destroyed:

> "They really laid their eggs right on us. The Soviets were taking the engines off the Forts while the bombs were still falling. A truckload of Russian women going out to fight the fires during the raid received a direct hit, yet that didn't stop the rest of them from trying to salvage as many motors as they could. I saw Lipinski on the field and he told me that his ship was also destroyed. I was lucky to catch a ride back to England with the ATC via Iran, Cairo, Casablanca, and a few other African stops."

Howell and Gamble returned on the shuttle flight. They gave Thiry a motorized bomb-spotter camera as he attempted to find a new plane for the return flight. Before leaving, Thiry took 20 color slides of bomb-gutted Fortresses with his small personal camera.

Scruggs recounts the Mosquito's departure:

> "On June 26, approximately one-half of our original force of bombers took off for Italy. I was not allowed to accompany them since only space was available to take pilots who lost their aircraft. Shortly after the bombers departed, both Mosquitos took off to scout the raid, at which time the Soviets were provided an excellent exhibition of the Mosquito's buzzing ability."

The Mosquito with Locke/Fisher scouted the task force attacking Drohobycz, Ukraine. Muchway/Proulx scouted Debrecen, Hungary, and landed in San Severo, Italy. Howell and Gamble filmed the strike at Drohobycz. After hitting the synthetic fuel refineries, they crossed Czechoslovakia, Hungary, and Yugoslavia, before arrival at Foggia, Italy. After a brief respite, CCU men received alert for a mission to Arad, Romania. On 3 July the bombers attacked the marshaling yards and returned to Italy, the third CCU shuttle sortie. And on 5 July, the shuttle fleet flew over Corsica and bombed the marshaling yards at Beziers, France, en route to England. Muchway/Proulx climbed to 25,000 feet to scout ahead when the port propeller ran away. Unable to gain control, they returned to San Severo where an RAF unit performed repairs. Both reached Watton the next day.

And finally, Scruggs departed the USSR:

> "On June 27th, I left Piryatin on a C-87 and flew to Poltava … The evacuation of the combat crews had already begun, and by the time I arrived, nearly half had been flown out. At about 1630 the same day, five C-46s arrived from Tehran. The pilots were ordered to takeoff as soon as possible and were not allowed … to remain overnight. Each aircraft was immediately loaded with 35 men plus all flying equipment and personal clothing, and we took off at approximately 1800."

Over several days the C-46s progressed to Rostov to Armavir, at the Caucus Mountains, to Tehran, and finally Cairo, Egypt. Scruggs developed chills with fever and hospitalized for 13 days. After discharge on 12 July, he experienced considerable difficulty obtaining transportation without proper travel orders. After much convincing, he boarded ATC aircraft and eventually arrived at Watton 60 hours later.

8th Rcn Squadron. (Special) crews at Alconbury. L-R. Capt. Douglas L. Harris, Lts. John J. O'Mara (POW), Walter W. Thompson (KIA), Carl C. Edgar (KIA), Vance J. Chipman (POW), William G. Cannon, (POW), and Russel C. Whitmer (KIFA). *Photo T. Fields*

Publisher magnate William Randolph Hearst Jr. arrived in London to report the war for his newspaper syndicate. HQ Eighth Air Force placed him in contact with Col. Elliott Roosevelt who arranged for Hearst to view the V-1 onslaught. Watton assigned Lt. Richard Geary to pilot a Mosquito for Hearst. At noon 28 June, Geary and Hearst left in MM342 for RAF Newchurch on the channel coast but blew a tire on landing. Lacking proper repair equipment, Geary called Watton for a replacement aircraft. Meanwhile the men received accommodations in a tent city; both 3 and 86 Squadron RAF, flying Hawker Tempests operated from the field. Early morning 29 June MM388 arrived and Geary took Hearst airborne. Hearst reported the following:

> "Our plan of operation was to cruise around at several thousand feet above the level at which the pilotless buzz-bombs generally come in, listening in by radio to the flying control officer as he directed the fighters, then joining the chase. Fighters passed us in both directions. Suddenly behind Dick's shoulder, I saw aircraft diving. We banked steeply and there below us I saw the little thing, its tail aglow, going like a bat-out-of-hell. As we dove on it, a couple of more fighters did likewise. Our air speed indicator showed 360 miles per hour. As I watched, I saw the fighter closest to it hit the buzz-bomb a glancing blow. They call it a strike ... Suddenly came a spurt of orange flame. A couple of seconds later, the bomb proper struck the ground with a terrific flash, and a huge cloud of smoke slowly rose from the field where it had fallen.
>
> "We could hear the men talking over the radio ... We now turned and resumed our patrol and only then realized how far we had come. The chase itself had only lasted a few minutes, but at that speed you cover a lot of ground. The fighters returned to their bases, but our ship had sufficient fuel for several hours, so we went back and waited for the next buzz-bomb ... We wanted to get closer next time, so we decided to stooge around at a higher level in order to pickup even more speed in our dive. No sooner had we decided upon this course, than we heard the radio voice telling us that another buzz-bomb was on its way over, just about where we were.

RAF Newchurch with Mosquito MM342 in foreground has starboard wheel and tire removed. Note airbag supporting wing for free access to undercarriage. *Photo R. Geary.*

Rear view of MM342 at RAF Newchurch with airbag to raise and support wing while tire is changed. *Photo R. Geary.*

At Watton is 1/Lt. Richard Geary on left with William Randolph Hearst Jr. on the right. *Photo R. Geary.*

"Dick saw it first and down we dove. This time we're doing close to 400 miles-per-hour. The little thing was only about a quarter of a mile ahead. We could see it quite distinctly. Two fighters were on its tail, another just off our wing. Dick had the throttles wide open and the plane's motors were literally screaming. We kept off a little to the side, the better to see the show. Someone must have scored a hit, for it slowed up, enabling us to pull along side of it. The pilotless doodlebug was now only about a hundred yards off our right wing on my side. As I looked out of my window, I could see flashes from the fighters' guns pecking away at it. It was probably one of the most exciting moments of my life ... There it was, the bomb itself with its wings attached, and above it, a little cigarette-holder shaped, jet propulsion mechanism. Even as I watched, one of the fighters hit it.

"The thing suddenly banked sharply towards us! There was no time or need for consultation on strategy. Dick pulled the plane up and then banked it vertically over to the left. The bomb passed underneath us on its downward last journey. We kept banking around steeply to keep it in view. Down it went and we saw the great flash it made as it exploded."

At the end of June, Heavy Weather Squadron terminated flying WR flights over the British Isles. The missions involved too many men, requiring too much time to prepare for the brief flight. Light Weather Squadron absorbed this duty. Also, transitioning new Heavy Weather flight crews changed. The new pilot, copilot, navigator, and met/obs, flew three operational flights with the old crews as an introduction to the mission. On the third flight, the new crew flew the entire mission themselves while observed for proficiency. They were then ready to assume and maintain the daily schedule of Atlantic WR flights.

Taken 30 June 1944. L-R. Pilot 1/Lt. Robert P. Walker, Sgt. Burgess Scott, and 1/Lt. Roy Conyers, navigator. Mosquito NS559 landed on a PSP strip near Cherbourg for refueling from Jerry cans. They transported in the nose section to Watton, *Yank* magazine combat artist Sgt. Scott. *Photo R.P. Walker.*

July 1944

The Allied ground and air campaign resulted in the German infantry movement of men and supplies under safety of darkness, making night photography of great value. Group HQ had earlier requested four B-26F aircraft for night photo work. This created a problem because all Eighth Air Force operational B-26 Groups transferred to Ninth Air Force in October 1943.[42] On 1 July, a Martin AT-23 arrived at Watton to familiarize and train crews to fly B-26 aircraft. Also on this day, Lt.Col. William H. Cleveland assumed 802nd command relieving Lt.Col. George O. Doherty who returned to the USA.

Top view of NS590 B-Baker. Note roundels are painted over and U.S. insignia applied. It later carried tail call letter S/B. *Photo. R. Walker*

Serendipity occurred July 4th. Watton scheduled a Dilly Mission and Lts. Robert Tunnell/Harry Anglum (NS568) found the Calais area concealed by 10/10ths clouds. That afternoon, Lt. Richard Geary/Capt. Richard Wright (NS568) flew another Dilly mission, to Boulogne and St. Omer, but again thick clouds obscured the area. Geary descended through the undercast to a thousand feet and skimmed the cloud base en route home over the Calais area. For some unknown reason the intervalometer became actuated and continued taking photos of the flight path. After landing, S-2 found the developed film in focus. Ground crew members had mistakenly adjusted the camera focus for 1,000 feet altitude instead of 18,000 feet as planned. Visiting Wing HQ photo interpreters examined the prints and located two previously unknown V-1 launch sites. Pinetree expressed appreciation for the excellent find and launched air attacks against the new targets.

During late afternoon 5 July, Capt. James M. McNulty made his first flight in a visiting 7 PRG Spitfire BL680. During the takeoff run, he encountered a bump on the grass strip and bounced. The propeller caught the ground and flipped the aircraft nose over onto its back. Now trapped inside the cockpit, he turned the fuel and other switches off, then waited for the crash crew to arrive and extract him.

The following day, Capt. Michael Lorenz in B-17 42-102521 returned to England from an Atlantic WR mission. Flying at 5,000 feet through thick clouds over the Atlantic the Fortress emerged into a clearing at the same time as a Ju 88. Both pilots spotted each other simultaneously and each immediately sought refuge in nearby clouds. That same day, cameramen Sgts. Virgil Mullen and Louis Oliva flew in B-24s photographing target strikes at Fully, France, while Sgts. Al Mandelaro and James Zipperer flew in B-17s filming a secret target in France. Meanwhile, Lts. Peter Dustman/Harry Anglum (NS568) flew the first Seaweed mission obtaining coastal photo strips in the Belle Isle area of France. Seaweed involved a PRU mission objective to photograph the Atlantic coast from Bordeaux, France, to Holland.

Martin AT-23 trainer aircraft at Watton to transition aircrew for B-26 night photo work. Numbers 210 appear on tail and under wing. *Photo T. Tate*

42 Roger Freeman, The Mighty Eighth War Manual, New York: Jane's, 1981: 177.

Tragedy struck on 7 July. Cameraman Sgt. Mark Ortega left with the Forts attacking the Messerschmitt factory and chemical works at Leipzig. To provide weather Scout support for this mission, 1/Lts. John J. Mann/William L. Davis (NS537) departed Watton, and on reaching Southwold at 0830, climbed toward the continent. Over the North Sea the Mosquito suffered right engine failure with a runaway propeller. Mann turned back toward England. At 0825, he transmitted a distress call as the navigator prepared for bailout at 1,000 feet. Five minutes later he reported the navigator safely jumped with the aircraft now at 600 feet. Watton received no further radio contact.

Two patrolling Emergency Rescue Service (ERS) P-47s spotted a body floating in the sea and radioed the position as 25 miles east of Southwold. Two air-sea rescue boats immediately dispatched at 0930 sighted a motionless form in the sea supported by an inflated lifejacket. However, before they could reach the man, he disappeared beneath the waves and never seen again. The search continued until 1700. Watton listed both officers MIA.

The other Scout supporting the 3 BD mission to Leipzig and Merseburg was crewed by Lts. Gene Goodbread/Donald Lee (NS509). Flying at 30,000 feet, they noted 40 single-engine enemy fighters east of Leipzig at 28,000 feet. One aircraft broke away and closed to one-half mile when Goodbread applied power and left the fighter behind.

On 8 July, a Fw 190 attacked a Mosquito scouting a 2 BD mission to Rouen. Lts. Ronald Nichols/Elbert Harris (NS508) flew west of Rouen at 28,000 feet. A Fw 190 came in on their tail from 29,000 feet and closed to within 2,000 yards. The Mosquito then outran the fighter which broke off the attack. Meanwhile, cameramen Sgts. Janowitz, O'Neil and Lacho rode in Liberators attacking Rouen and Parish Auxerre, France. Ron O'Neal's group made three runs on the target before releasing any bombs. This allowed German antiaircraft batteries to obtain accurate range, and a piece of shrapnel tore through O'Neal's left elbow and finished his combat flying. Upon landing at Thurleigh, an ambulance rushed him to hospital where both Sgts. Mark Ortega and Al Mandelaro donated their blood.

Navigator 1/Lt William L. Davis KIA on 6 July 1944 mission.
Photo Davis

By 11 July, Heavy Weather squadron completed 19 Atlantic WR flights. Due to new runway construction the Forts took off from North Pickenham located five miles away, but landed at Watton on return. The construction also force four Mickey Mosquitos and three crews to fly missions from Alconbury.

Early that morning, cameramen Sgts. John Cunney, Joe Beach and James Zipperer left for Germany in Forts to record 1 BD maximum bombing effort of Munchen-Riem Airfield near Munich. Lts. Richard Petersen/R. C. Smith (MM393) flew Scout. Over the target, four Me 109s turned in on their tail at 30,000 feet but did not attack. Two Me 410s flying 2,000 feet below also did not attack. Meanwhile, Lts. Ronald Nichols/Elbert Harris (MM391) scouted Munich for 2 BD. According to Harris:

> "Everything that could fly the Eighth put up for this mission. All Europe was under clouds low to the ground when we left Watton. En route, we passed bomber formations all the way to the target. It was a maximum effort. When we completed the assignment and returned to Watton, bombers were still taking off and forming over England to join the mission. I became convinced on that day that we would win the war."

Next morning, Capt. Willis Locke/Lt. Donald Lee (NS515) flew Scout to Munchen-Riem Airfield for 3 BD. From 30,000 feet, Locke sighted 12 Me 109s flying at 27,000 climbing towards the Mosquito. When they reached his altitude, Locke made a 180-degree turn and left the 109s behind. Munchen-Riem Airfield was the target again on 13 July, but a solid cloud cover prevented CCU Sgts. Larry Barbier and Al Mandelaro from obtaining photo coverage. Also on this day, Heavy Weather Squadron experienced changes. AAF personnel from 18 Weather Squadron replaced all RAF weather observers flying Atlantic WR missions.

Capt. Harold Aronson piloted a B-17 returning from a daylight Atlantic WR flight when the right-outer engine oil gauge began fluctuating. For safety he shutdown the engine and feathered the propeller. An hour later, the left-outer engine ran rough and slowly began to fail. He continued its operation for additional power to preserve altitude and gain mileage to England. Eventually it became necessary to also shut it down and feather the prop. The two inner functioning engines strained to maintain altitude with hours to landfall.

En route, the Fortress gradually lost height but managed to maintain flight at 2,000 feet without the need to jettison equipment. To ease the strain on the two good motors, he decided to unfeather the right-outer engine, since it was shutdown because of fluctuating oil gauge, and restart it. After operating for 15 minutes it began to fail. He terminated its operation and feathered the prop again. The crew then discovered you cannot feather the same propeller twice! The engine contained an oil reservoir tank to permit feathering the propeller. In the act of feathering the first time they consumed the entire reservoir, therefore no oil remained to allow a second feathering.

The propeller began windmilling uncontrollably, creating enormous aerodynamic drag. To counter this effect Aronson applied full power on the two functioning engines. Lacking oil, friction caused the windmilling prop hub to become red hot. Finally the propeller broke off and hurled away from the aircraft into the ocean. No longer impeded by drag, the Fort continued on two engines to affect a safe emergency recovery at St. Eval.

Operation CADILLAC involved a mass airdrop of supplies to Maquis units. On 14 July, Lts. Kenneth Dunaway/James Keiger (NS508) scouted 3 BD drop zones southeast of Chalon-sur-Saone and Vercors plateau areas of France. Cameramen Sgts. George Howell and George Gamble were assigned to cover a 96 BG supply drop to French underground forces. Gamble recalls:

> "We took off at 0340 and climbed to 11,000 feet over Snetterton Heath. After assembling, we headed for our Combat Wing rendezvous. We completed assembling at 0600 and headed for France. We crossed the invasion coast at 18,000 feet. Near the IP, we descended to 500 feet over enemy-occupied territory in our B-17F. Our first run was an attempt to distinguish the target area that was identified by bonfires and people waving flags. We returned to the IP and started on our cargo runs. It came off on schedule and the lead parachute landed in the middle of the markers. We were flying near the formations' tail end, so I photographed the other Forts releasing their cargoes. If a plane wasn't sure of hitting the markers, the pilot then made an individual run. I observed trucks drive out and pickup the supplies. We did not encounter any opposition but received a radio message that some groups were attacked by enemy fighters. However, our fighter escort was perfect, and we didn't experience any trouble."

U.S. forces captured St. Lô with its major road junction between Normandy and Brittany on 18 July. The following day CCU Sgt. Louis Oliva returned to the air for his 24th mission documenting the bombing around the Caen battlefield while Sgts. Robert Arrowood and John Cunney filmed target strikes on Strasbourg, Germany. That evening, Watton held a formal stag party at the Officers' Club for Colonel Roosevelt to celebrate his engagement to actress Fay Emerson.

Shortly after the 802nd moved to Watton, Lt. Vance J. Chipman arrived for duty to assist with Group development. Previously he flew RAF night-fighter Mosquitos before transfer to USAAF. Sporting a trim mustache, Chipman spoke with an English accent and became quite popular among base personnel. Ralph Tipton became a good friend:

> "I was going to London on leave … That day, Chipman was … on his way to London also, so we teamed up for the journey. Getting a hotel room in London was almost impossible. I had made friends with a desk clerk at the Russell Square Hotel, and he was able to arrange rooms for Chipman and me … The next morning I went to his room to awaken him. To my amazement, he began to speak with a typical American mid-western accent. I asked him point blank, 'Just where in hell did you come from?' Laughing, he replied that he was a former racetrack driver from Chicago. He joined the RCAF when war started in Europe.

Light Weather Squadron NS594 with invasion stripes. It was involved in an accident on 28 January 1945. *Photo C. Ashworth*

Ground crew replenish the oxygen supply of MM391 E-Easy. It was involved in an accident on 9 September 1944. *Photograph L. Mann*

"At breakfast that morning, we agreed to meet in the hotel pub at opening time. At 1100 hours I was there, but not Chip. About 30 minutes later, in he strolled with a monkey on his shoulder. Chip explained that as he was walking past a pet shop, this monkey saluted him, or so he thought. The monkey seemed to be searching for a long lost friend. Chip bought him for 20 pounds, and for the remaining two days that monkey accompanied us on a tour of pubs, causing quite a commotion …

The monkey, now named Mickey, was a novelty at first but became a nuisance to all concerned. It was not toilet trained and defecated anywhere. Once it suffered a rather large bowel movement on one of the lounge chairs in the Officers' Mess. Men playing cribbage were interrupted when Mickey grabbed a peg and scampered along the top of the lounge chairs lining the center of the room. He jumped on the black curtains covering the windows and climbed to the top with his prize.

"A few months later in July, 1944, while at dinner, some chap seated next to me asked if I knew anything about some monkey romping through the Officers' BOQ. The deputy Group CO, Lt.Col. Leon Gray was on the warpath. It seems the monkey had gotten into Colonel Gray's quarters and opened tooth paste, shaving cream, talcum powder, etc. and decorated his room in a most non-military manner! He added something about Lt.Col. Gray was looking for a Chip or Tip … and that was enough for me. I left my food, quickly eased out the Mess … As I passed the Manager's Office … (he) asked me to come in his darkened office and watch the show in the bar. Chipman was sitting at one end of the bar drinking Scotch with the monkey perched on his shoulder. Chip's personal pewter mug was filled with Ale. The monkey would jump down, take a large gulp of Ale, and leap back onto Chip's shoulder, much to the amusement of the bar patrons.

"Apparently, the officers threw a party for Col. Elliot Roosevelt in the base BOQ. After a few polite drinks with the senior officers, Elliot retired to the bar where the lieutenants hung out. He preferred their company when serious partying was in order. He soon became involved in verbal bantering with Chipman, whom Roosevelt had met on previous occasions. He was much admired by Elliot because of service … in the RCAF. At one point, the bantering stepped over the line, and language 'unbecoming an officer' was exchanged. Chipman had survived tight situations before, but never facing down a colonel. Instead, his hazy mind concocted a deserved symbolic act. He went over to the bar, found a pair of scissors and returned to cut off the colonel's tie. The two now had a considerable audience who held their collective breath waiting Roosevelt's response.

'I challenge you to a duel,' was his counter.

'Fine,' Chipman replied. 'Your choice of weapons!'

'Fire extinguishers at 12 paces!'

"In that location, fire extinguishers were the hand pump variety, filled with water. Back-to-back they paced off six strides in front of the bar, to the cadence count of the spectators. Four, five, six … and as if by silent agreement, the two duelists turned and began pumping madly with nozzles aimed at the audience, who of course were in full dress uniform. The place became bedlam with Chipman and Roosevelt roaring with laughter, the only two dry officers in the place. The party ended with the colonel returning to High Wycombe.

"Chipman could have faced court-martial for destroying government property. Instead, Lt.Col. Gray banished him to 30 days in a pup tent. 'Contributory cause' in any investigation, the annals of the 25th include a different version, quote, 'a pet monkey owned by Lt. Chipman set off a flare in the bar, and in attempting to put it out, Lt. Chipman used an extinguisher, the contents of which did considerable damage to club furniture.'"[43]

NS619 arrived at Watton 9 July 1944 in RAF markings. It was involved in an accident 22 March 1945. *Photo L. Proulx*

43 Bulletin Board, 25 BG Association newsletter, 18:3 (August 2004): 8-13.

Invasion stripes initially encircled MM393 rear fuselage of F-Fox. Later the top fuselage was painted over PR blue leaving only lower stripes. Note different color fuselage where stripes were painted over. In background is NS569 N-Nan being serviced. It was in an accident, 4 April 1945. *Photo D. McCarthy*

The 20th marked an unsuccessful assassination attempt against Hitler at his Wolf's Lair headquarters in East Prussia. The 376 Engineering section also was unsuccessful, in obtaining much-needed Mosquito components. More aircraft were deadlined from parts shortage than for any other reason. Supplies ordered from British sources received no compliance. Inexperienced mechanics to maintain the Mosquito aircraft proved another major cause of operational delays. Most had never worked on Mosquitos prior to arrival at Watton. Many were removed from mechanics school before finishing their training course, transferred overseas as a basic soldier and then reclassified as airplane mechanics. A large number required attendance at DH company schools to learn Mosquito function and maintenance.

Major Paul G. Davis, Pinetree weather officer, suggested a new WR route over the Atlantic. This required flying a tract further mid ocean to the Azores to increase weather coverage of prevailing westerly winds and fronts. Col. Irvin Krick visited Watton and consulted Lt.Col. Stenglein and Major Podwojski regarding the proposal. This resulted in an experimental flight dispatched in B-17G 42-102970 including Stenglein, Podwojski, the Group navigator, communications, and administrative officers. The Fort made a brief stop at Bovingdon to pickup Major Davis before proceeding to St. Mawgan to refuel.

Leaving just before midnight on 17 July, they flew a 260-degree course for 700-nautical miles. At 48.02 degrees North and 22.5 degrees West, the Fort changed course to 180 degrees for 600 miles. At position 8, the Fort climbed to 18,000 feet and flew to position 14 with the crew on oxygen, making weather observations every 50 miles. The navigator utilized all forms of navigation, i.e., German Console system, astro-navigation, long-range radio direction finder, DR, and compass. Reception of Azores' QDMs and fixes proved effective and a localizer beacon for letdown in overcast, proved dependable for landing at Lagens at 1110 the following morning. The Navy facilities provided good messing, same day U.S. newspapers, bananas, and silk stockings. After determining available logistics, the men returned to Watton on 22 July and obtained en route weather data. The new route proved feasible and eventually labeled the Sharon flight.

2/Lt. Vance J. Chipman shaking hands with Colonel Elliot Roosevelt. *Photo USAFHR*

Sign in front of a pup tent reads, "These quarters occupied by 2nd Lt. Vance Chipman, the No.1 fire extinguishers, fluid dispenser and buzz artist in the group. Date of Occupation 26 July 44." *Photo T. Fields*

The BD Forts flew to Augsburg on 20 July with cameraman Sgt. Joe Beach filming the raid. At the same time, Liberators bombed the marshaling yards near Eienbach where Sgt. Virgil Mullen filmed the smoking target and heavy flak thrown up to his formation. At Watton, Lt. Mahlon Funk ferried B-17G 42-39871 to North Pickenham as a standby WR aircraft. He executed a safe landing but 700 yards down the runway an undercarriage retracting motor failed causing right gear collapse and ground loop.

Lt. Harold Aronson flew an Epicure mission on the 21st. En route, loss of oil pressure and elevated temperature at position 13, force shutdown of number three-engine and aborting the flight. After feathering the propeller, he experienced considerable difficulty preventing it from moving in the wind. Flight engineer S/Sgt. Okey E. McGroskey stripped wire cable from the B-17 interior to lasso the prop and prevent any further serious trouble.

Four new B-26s Marauders arrived at Watton the 21st for hazardous night photography. The B-25 carried only 12 PFB while the B-26 increased this number to 20. Experiments confirmed the B-26 performed favorably in the low-level aerial photography role. Eventually one Marauder was reassigned to 7 PG as a communication aircraft.

Watton's new concrete runway finally reached completion and awaited installation of the lighting system for night flying. Four Mickey Mosquitos at Alconbury and the B-17s at North Pickenham returned to Watton.

During July, fifteen Black Widow P-61s arrived at Burtonwood. General Doolittle allotted four aircraft to Watton as potential Mickey aircraft. Pinetree desperately required radar approach photos to assist the bombing campaign against Germany. Three aircraft remained at BADA while 42-5556 arrived at Langford Lodge, Northern Ireland, where it underwent turret removal and fuel tank installation for exhaustive long-range, high-altitude testing. The final results and analysis are unknown, but no P-61s were assigned to Watton.

Experience gained in North Africa utilizing night photography prevented the Axis powers from concealing troop and supply movements under cover of darkness. Therefore, Colonel Roosevelt proposed on 14 March 1944, modifying Mosquito aircraft for night photo or "Joker" missions. This included the following: 1. Install flak protection armor plates on bomb bay doors and both bay sides of the flash bombs if practicable. 2. Install one K-19B night camera in the bomb bay, forward of the PFBs and clear of the bay doors, and two K-19 cameras in the aft fuselage. 3. Install wiring for camera and bomb release mechanism. 4. Install a fabric enclosure around each camera, connected to a heating duct to warm the cameras. Heat obtained from a Merlin radiator exchanger passed through the ducts to the cameras in both forward bay and aft fuselage.

The initial concept combined night photo and H2X role in the same Mosquito. When flash bombs, cameras, and auxiliary fuel tanks were installed, displaced H2X equipment would be relocated alongside the pilot in the cockpit. The camera operator/navigator remained aft of the bomb bay. The extra weight proved too high a price to pay utilizing H2X as a navigational aid when conducting night photo missions. Roosevelt canceled these plans. Both H2X (Mickey) and night photo (Joker) roles were assigned to different configured aircraft.

Two night photo variants were proposed. The long-range version had one 58-gallon fuel tank removed from the forward bomb bay area and bomb shackles installed to hold six M-46 PFBs. Each PFB produced a maximum of 700-million-candlepower, employing the type M-111 time-delay fuse. In the short-range version, both bomb bay fueltanks were removed and 12 or more PFBs installed. Mosquito NS582, previously configured for daylight PRU missions, flew to Langford Lodge, Northern Ireland, to serve as prototype.

1/Lt. Clayton Hackman in front of Mk.III Mosquito trainer. Note different color spinners. *Photo. C. Hackman*

By 12 July difficulties obtaining proper shackles hindered further development and only the short-range version reached fruition. To compensate the loss of bomb bay tanks, two 100-gallon jettisonable fuel cells replaced the 50-gallon underwing tanks. The first two Mosquitos, NS554 and NS551 direct from the DH factory, arrived at Langford Lodge equipped with 100-gallon underwing tanks. Subsequent 802nd aircraft, after Joker modifications, returned to Watton for 100-gallon tank installation.

Conflict arose with PFB installation in NS582. Two Watton crew chiefs, T/Sgts. Lesley Hassell and James Walsh inspected the prototype Mosquito at Langford Lodge. The 20 PFBs were installed facing rearwards with tail fins towards the nose. Both crew chiefs surmised that upon release, the wind stream would force the PFB backward, striking the rear PFBs or the aft bulkhead and possibly cause detonation. Both objected to the installation as dangerous and vowed they would refuse certifying for flight any aircraft so configured. They requested the PFBs face forward. This adamant position infuriated the engineering staff who ordered both men returned to Watton. Confronted with the dilemma posed by the crew chiefs, the engineering section was forced to comply and the 20 PFBs reinstalled facing forward.

Special Squadron requested volunteers to perform the test phase and Lt. Richard Geary came forward. Unknown to him, great risk and danger involved carrying the highly sensitive PFBs. In early 1943 the RAF developed the technique of night photography using PFBs and 140 Squadron delegated this duty.[44] Geary went by B-24 to RAF Northolt to obtain firsthand information from the crews involved in this work. The RAF lost three night photo Mosquitos during operations from midair explosions that killed all, except one crew member. Geary interviewed the survivor who reported the blast force from the explosion blew him free from the aircraft. The information proved disturbing and Geary had second thoughts involving himself in the development program. Worst yet, rumors of the RAF Mosquito explosions filtered among Watton flight personnel and navigators were reluctant to fly with him. Finally, Lt. William Miskho volunteered, and both men discussed and planned the first test flight.

Mosquito NS582 returned to Watton configured with two cameras in the forward bay and one aft.[45] A daytime drop test performed 21 July, carried 20 inert PFBs tucked neatly in the bomb bay. Geary and engineering officer Capt. Jessie Abbott flew over *The Wash* and began a mock photo run. Unfortunately, the intervalometer malfunctioned and force canceling the test. A cable release in the cockpit attached to the shackles failed to salvo the PFBs, so Geary landed with the load intact. The engineering officers examined the shackles. Pulling the cable release again still failed to jettison the load. The PFBs were too tightly packed preventing proper shackle function. The solution: reduce the load to 12 PFBs. Having accomplished this task, Engineering scheduled another evaluation flight on the 23 July using live PFBs.

Fearing a potential explosion, the planned night test was flown from North Pickenham, who already had men and equipment available for any emergency situation. After preflight briefing, Colonel Stenglein drove Geary and Miskho to the Mosquito. Both men climbed in, and before closing the lower hatch, the crew chief handed Miskho a role of pennies as a souvenir. The aircraft had the name *Bad Penny* painted on the nose.

Geary lifted off slowly, then took to the air and flew over *The Wash* to initiate the photo run at 18,000 feet. All bombs released and ignited at the preset altitude. Though the intervalometer indicated all PFBs discharged, Geary pitched the aircraft forward and back, then rocked sideways several times, hoping anything hung up inside would fall through the open bay. This action may have seemed redundant but he did not wish to take any chances. After return to Watton, he received instructions to write down everything performed during the flight to serve as SOP for future pilots flying Joker missions.

The test revealed deficiencies. The prototype aircraft had two K-19B cameras in the forward bomb bay, and one in the rear. Results using high-speed night film produced large-grain photos, yielding poor details after enlargement.

Therefore, Col. George Goddard, the father of AAF aerial photography, recommended the original installation plan be followed with one camera forward and two in the aft fuselage to allow utilization of daylight-type film to correct the problem. All night photo Mosquitos were duly configured: two modified K-19 cameras with 24-inch focal length K-17 cones, mounted at a six degree split-vertical to provide 30 degree side overlap, located aft behind the bomb bay. Two camera photoelectric cells, each covered by a Plexiglas blister, sat mounted on each side of the fuselage. The third camera, a 12-inch focal length K-19B with photoelectric cell attached to the base mount, sat in the forward bomb bay ahead of the PFBs. It had no camera port, taking photos through the open belly compartment during the photo run. The squadron utilized this configuration for high-altitude employment. For medium-altitude operations, two standard 12-inch focal length K-19B cameras replaced the modified, long focal length cameras in the rear fuselage.

To hasten concealment from each PFB ignition, all night photo Mosquitos had all under surfaces and tail section painted glossy black. The upper forward fuselage and top wing remained PR blue for aiding pilot visualization to reference flight attitude in darkness.

On 24 July, CCU cameramen Sgts. Nowell, Mandelaro and Zipperer filmed the bomber raid on enemy troop concentrations in the St. Lô area. The following day, U.S. Army forces mounted a breakout from the beachhead at St. Lô and Avranches, and pushed the lines to the Meuse River. Preceding the ground attack 1,500 heavy bombers and hundreds of other combat aircraft dropped bombs on enemy positions along a narrow front. Sgts. Janowitz and Lacho filmed the formations, bomb strikes, and smoking targets. Unfortunately, while observing the strike from the ground, a misdirected bomb from his own air force killed Lt. General Lesley J. McNair.[46]

20 July 1944, B-17G-5-VE 42-39871 landed at North Pickenham when right undercarriage failed. *Photo Blue/AAF*

44 Sharp & Bowyer, Mosquito, 130.

45 AAF files list NS554 and NS551 delivered to Langford Lodge. Geary's log book lists NS582 and NS592 received from Langford Lodge and flown during the test program. Therefore his log book entries are quoted only.

46 David A. Anderton, Aggressors: Interceptor vs Heavy Bomber, New York: Zokeisha Publications Ltd. 1991: 28.

In the early morning of 25 July, Lts. Gene Goodbread/Warren Barber (NS510) returned from a weather flight over the continent. During the mission, Goodbread suffered anoxia from oxygen equipment failure. Adverse weather conditions at Watton required diversion to North Pickenham. Upon touchdown at 0250, the Mosquito swerved from the runway and wiped out the landing gear. Later that day, ground crew members attached a cable around the aft fuselage to drag the aircraft by tractor from the field and broke the aircraft in two.[47]

Watton received notification to provide air support for Project Aphrodite. Conventional bombing failed to destroy massive constructed submarine pens at Heligoland and other highly protected V-weapons sites in the Pas-de-Calais area. The Germans had three secret weapons: the V-1, V-2 rocket, and V-3, a gigantic cannon also called the "London Gun" to fire rounds at England. It comprised a 400-foot-long barrel buried underground with only the muzzle protruding, near the French town of Mimoyecques, France.[48]

Total destruction of V-sites could only be assured by precisely delivering an enormous quantity of high explosives directly on these targets. The Aphrodite program would determine both effectiveness and technical aspects utilizing remote-controlled, war-weary B-17 and B-24 aircraft as weapons. Stripped of armament and non-essential equipment, they were packed with 50-pound boxes containing nitro-starch explosives: 18 boxes in the nose, 28 in the radio compartment, 215 in the bomb bay and 144 in the aft fuselage; all tied down together and shimmied with wood to prevent movement.

Capt. Willis Locke (NS516) flew to RAF Fersfield for a briefing on the project. He then took project officer Capt. Lloyd Humphries for a weather test flight over the Calais area at 0250 on 27 July. Also that night, Lts. Ford Porter/James Davis (MM393) performed Bluestocking WR duties 30,000 feet over Rouen, France. The crew suddenly experienced a terrific brilliant flash followed by an enormous force that instantly lifted their aircraft 500 feet upward. Porter suspects an RAF bomber nearby received a direct hit and exploded beneath the Mosquito.

The Bomb Divisions initiated experiments using P-51s as scouts. The 28th saw both Mosquito and P-51 scouting for 3 BD. At the target Lts. Richard Petersen/R.C.E. Smith (NS509) orbited for five minutes and made positive identification, then notified the bombers to make a visual run. Two minutes later, the P-51 reported the target obscured by 10/10ths overcast. The Mosquito could not see the P-51, evidently lost off course. The formations followed the Mosquito's instruction for a visual bombing run on oil refiners in the Leipzig area.

Alsothis day, Lt.Col. William H. Cleveland taxied L-5W 42-98997 in preparation for a flight to Wing HQ. The aircraft collided with a Tow Tug on the perimeter track. Neither pilot, nor tug driver were aware of each other's presence.

Lts. Richard Geary/William Miskho (NS554) flew the first Mosquito night photo Joker mission on 29 July. They left North Pickenham after midnight, crossed the heavily defended Pas de Calais area and encountered flak between Ostend and Dunkirk. Unfortunately, heavy clouds obscured the ground negating photo results. The crew requested weather Scouts cover the target 45 minutes prior, but this recommendation never reached fruition. Bremen was the target for Eighth Air Force bombers that afternoon with CCU Sgt. Virgil Mullen filming the attack.

Mosquito front and rear bomb-bay areas loaded with six Type M-46 PFBs each without the Type M-111 fuse installed. *Photo USAFHRA*

47 Bulletin Board, 11:1 (February 1997): 8.
48 Philip Henshall, Vengeance, UK: Allan Sutton Pub. Ltd., Stroud, 1995: 70.

On the 30th, Lt. Joseph Brannock delivered F-8 43-34962 to Watton when the tail wheel sheared off on landing due to faulty construction of the strut housing.

The F-8 Predicament

Already committed to accepting the F-8, HQ USAAF ordered they be delivered to England. These numbered 25 F-8s with 23 assigned to Eighth Air Force; two crashed in staging areas and ten arrived at Watton with 13 more expected. Two aircraft at Watton required major repairs – a tail wheel collapsed on landing, while the other required an engine change and Prestone cooler replacement due to engine overheating.

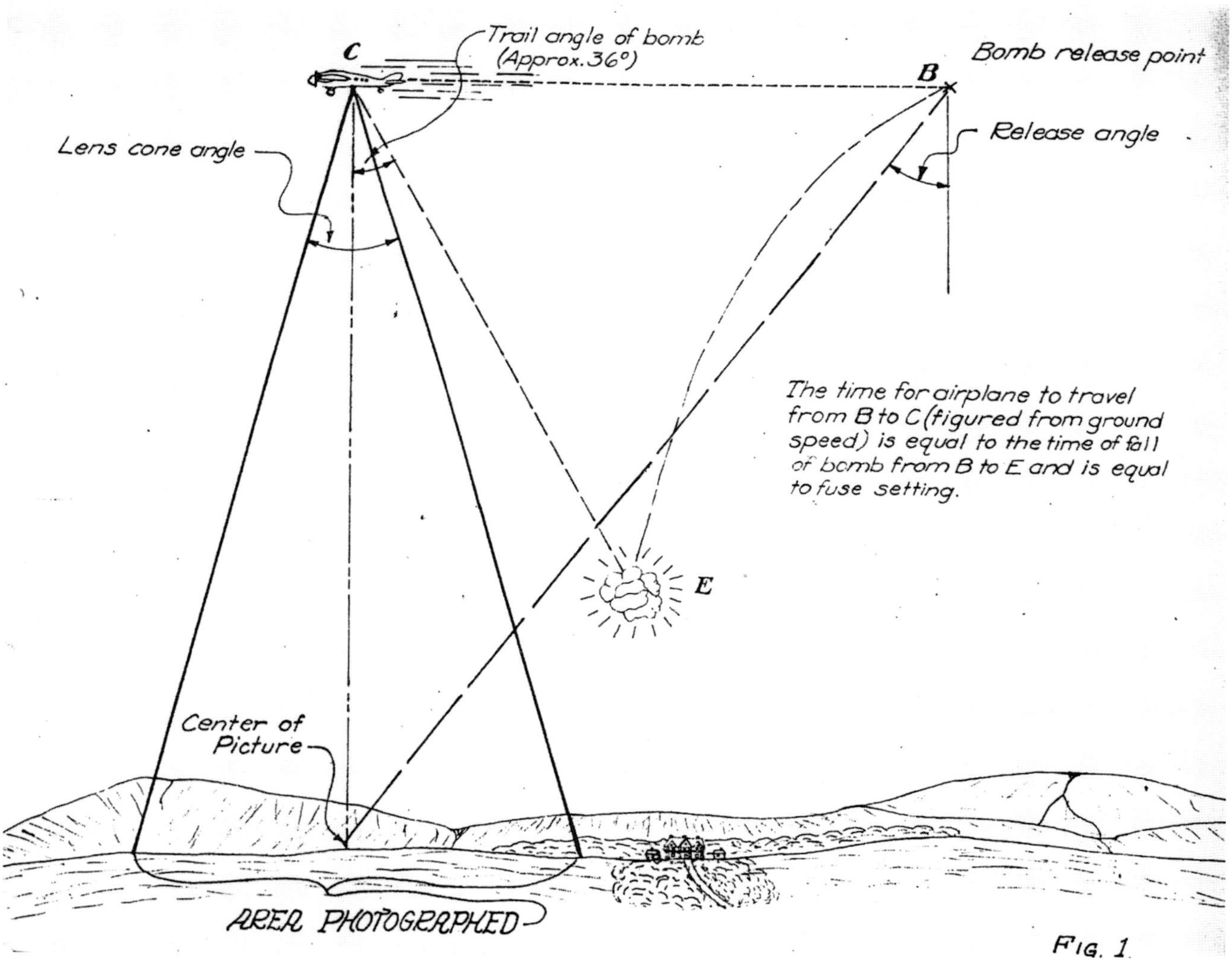

FIG. 1

Diagram illustrates the complex problems involving timing PFB release, ignition, and camera shutter speed at moment of flash, and fuse setting. *Photo USAFHRA*

1/Lts. Gene Goodbread and Warren Barber escape injury 25 July in NS510 at North Pickenham. *Photo W. Barber*

An example test flown at Watton by Capt. Roy Ellis-Brown lacked adequate engine performance for high-altitude mapping work. The F-8 also possessed a distinct cockpit odor. Engine component disparity between Canadian and UK-built Mosquitos hindered the F-8 conversion to much needed trainers. He therefore recommended none be accepted. On 27 July, Colonel Roosevelt cabled Pinetree ordering a halt to further F-8 deliveries and requested reassignment of those already at Watton to another theater.

General Doolittle disagreed and decided to distribute them within the Eighth Air Force:

Each Combat Wing, one F-8	14
Each Bomb Division, one F-8	3
Composite Command HQ	2
Service Command HQ	2
8 AF HQ flight	2
	23

The 802nd would maintain and service the F-8s for these organizations. Colonel Roosevelt protested and refused to accept their assignment. He bypassed Pinetree and dispatched a cable directly to General Spaatz halting further deliveries stating, "No requirement exists in the Eighth Air Force for the F-8." This action upset General Doolittle and created irreconcilable differences between the two officers.

August 1944

August 1 began with Lts. Gene Goodbread/Warren Barber (MM387) surveying weather over the Haute Savoie region of southwestern France, 15 miles west of Mont Blanc. They reported fairly clear skies at 0348 allowing Eighth Air Force to proceed with a scheduled supply drop to the Maquis. That morning, Lts. Oliver Emmel/Iver Fitschen (MM368) flew ahead to Scout the mountainous region and found that a multi-layer overcast had moved in overnight and now covered the area. Confronted with the difficult decision on whether to recall the bombers, Emmel descended through the various layers and by good fortune located a fairly clear route between the broken cloud levels leading to the drop zone. He climbed up through the cloud tops for altitude and led the first wave of bombers down through the pathway to unload their cargo. Then ascended back on top again to lead the next formation down until all supplies were safely parachuted.

Paris became the opening scene for CCU activities in August. Sgts. Oliva and Lacho experienced poor light conditions to photograph bombing of an airfield in the suburbs of Paris. Heavy clouds force 2 BD Liberators to a secondary target, another airfield visible through a light haze. While Lacho filmed the strike, Oliva's plane aborted because of engine trouble. He kept his camera grinding taking scenes of crewmembers throwing all loose equipment from the bomber to gain altitude for the channel crossing.

Eighth Air Force renewed attacks on German oil storage centers early morning of 3 August. Sgts. Cunney and Mullen accompanied B-24s to Douai, France, to film strikes on the depots. Large columns of smoke rose from the battered target. Next day, cameramen Sgts. Mandelaro and Lipinski flying in 1 BD Forts continued to pull the long rough missions with filming the attack on Anklann, Germany.

Numerous changes took place at Watton on the 4th. Major Marshall Wayne/Lt. Iver Fitschen (NS515) flew a Gee Whiz mission evaluating Gee over the continent. That night, Major John Hoover/ Lt. Jacob Hochman flew the first B-25 night photo Joker mission to Mayenne, France. Flying 12,000 feet over Avranches they observed an unidentified aircraft maneuver to a position on the Mitchell. Hoover turned into it and lost the intruder by evasive action. At Granville they were boxed-in by flak as shrapnel penetrated the fuselage, but the crew returned safely and landed at Mount Farm.

1/Lt. Daniel McCarthy in front of F-8 parked in storage at Watton July 1944. *Photo D. McCarthy*

Unwanted F-8 Mosquitos parked in open storage area at Watton. *Photo R. Conyers*

The Aphrodite Project took center stage. The mission profile called for two "Mother" control B-17 aircraft to takeoff from RAF Fersfield and climb to 20,000 feet. The second Mother provided backup should problems arise. The B-17 "Baby" took off under control by a pilot and technician, and escorted to a designated control point by a B-24 navigation ship. Flying a proscribed 50-mile rectangular course, the robot crew set the controls and equipment, allowing the Mother to fly the Baby by radio command. After the Mother ship successfully acquired remote-control, both Baby crewmen bailed out over England. The Mother aircraft then guided the drone by radio command to the objective and dove it into the target.

Three Mosquitos carrying two cameramen flew to RAF Fersfield to film the mission. Since a cameraman occupied the navigator's position in the Mosquito, Lt. Ralph Fisher instructed them on basic navigation skills using both DR and Gee. Other cameramen would film scenes from the Mother ship. Joseph Capicotto relates that:

> "We were secluded in an old wooden barracks and kept separate from everyone else. Mission briefing was provided using target models. These were exact replicas constructed from photographs, displaying the coast and distinguishing features for identification. Our target was easily recognizable by railroad tracks leading to it.
>
> "The mission was very dangerous. We were taken to a Baby, removed our shoes and socks then climbed inside for an inspection tour. They did not want any static electricity to cause an explosion. So, we walked in our bare feet through the drone from the rear to the cockpit. It was packed with crates of explosives piled and wired together. This is an experience I will never forget. We now knew the project's importance and what we had to do."

The mission was set for Friday, 4 August. The two task forces involved were each assigned one Mosquito to film the events. Other CCU men, Sgts Augie Kurjack, Ralph Thiry and Joe Capicotto flew in the B-24 Mother ships. Capts. Willis Locke/Lloyd Humphries (NS516) were the first off at 1121, scouted Manston-Gravelines-Saint Omer and transmitted back target weather data. Then two Mother ships of each task force left 45 minutes before the Baby, and climbed to altitude. The first two Baby B-17s took off at 1300 hours at five minute intervals escorted by navigation ships. Both Mother aircraft established contact and control of their respective Baby and guided it around the checkout course in preparation for targeting Siracourt, France. Lt. Dean Sanner/Sgt. Mark Ortega (NS581) left at 1330 and joined a Baby at Orford. Sanner flew 150 yards slightly behind and starboard as Baby neared Woodbridge. Ortega viewed the events from the nose section:

> "One of the B-17 men reached the bottom escape hatch. We expected him to jump, but he just sat there with legs dangling in the wind stream for the longest time. I kept filming him sitting there and ... eventually he jumped."

At this moment Sanner noticed the Baby became erratic:

> "All of a sudden the nose of the Baby started to rise which worried me. I hauled back immediately on the control column and banked sharply to get out of there."

The overwhelming effects of centrifugal force pinned Ortega to the cabin floor until Sanner recovered from the turning dive. Apparently the parachute equipment lodged technician Sgt. Elmer Most in the Forts' escape hatch. Ortega watched him "sitting there" for the longest time. The drone pilot Lt. John W. Fisher forcefully shoved most through, at which time the B-17 went nose up and stalled. It then fell on the left wing and dove to the ground killing the pilot in a massive explosion, three miles northeast of Woodbridge. Kurjack recorded the crash and explosion from a B-24 Mother while Sanner flew over to allow filming the smoke column and blast scarred forest area below.

Sanner returned to RAF Fersfield to allow Ortega to reload fresh film and took off again at 1530. Too late to rendezvous with the second Baby he headed for Le Touquet, France, and the target at Wizernes. Unable to locate the Baby, he backtracked to Dinginess unsuccessfully, then returned to Le Touquet and flew low over the target, but heavy flak force evading by climbing for altitude. He returned to RAF Fersfield. Unknown to him the second Baby flew to a cloud-covered target. The high-flying Mother released the Baby concealed beneath the clouds and it exploded 500 feet past the objective.

1/Lt. Ralph Fisher (KIA) on right with cameraman Sgt. Walt Lipinski on left looking at map. Fisher instructed 8 CCU men on Mosquito navigation for the Aphrodite missions. *Photo Mark Ortega*

Meanwhile, Lts. Robert Tunnell/Daniel McCarthy (NS559) left at 1350 and followed their Baby from Woodbridge until 40 minutes from the enemy coast, then climbed to 10,000 feet with the robot in sight. Clouds force their descent to 8,000 feet keeping Baby to port. They observed the robot circle 300 feet directly over the Watten, France, target. Apparently, flak damaged the robot's receiver, hindering efforts to crash it into the target. The Baby circled wide around the area for several minutes, then impacted 15 miles east of Watten and exploded 200 yards southeast of Fort Philipi. Tunnell flew on the deck for several minutes through a hail of 20mm fire providing McCarthy improved access to the scene. According to McCarthy:

> "I lost the loop on my first roll but got plenty of action scenes of one Baby exploding in a large field, just missing a canal. We banked and came back at 30 feet and took shots of the town that had been pasted time and time again. There were hardly any roofs on the houses."

Back at RAF Fersfield a malfunctioning camera delayed Lt. Dean Sanner/Sgt. Mark Ortega (NS569). Once off, they attempted to locate the task force. At Watten they encountered flak at 10,000 feet, so Sanner climbed to 13,000 feet, took evasive action and returned to England. After viewing the Aphrodite film, CCU decided the risk involved overshadowed the results of the mission; too many cameramen were committed to the assignment. The Mosquito proved an excellent platform to film future Aphrodite operations.

While Aphrodite was being filmed, cameraman Sgt. Robert Arrowood flew his tenth mission to Germany. He went with the same 446 BG, 707 BS crew three times before and each time forced to abort. Finally on the crews' 30th mission, flak struck B-24H 42-95190 over the target area with number four-engine smoking. The bomber made a sharp turn to port and departed the formation. At about 1340 the aircraft made repeated calls for fighter protection. The crew bailed out and the Liberator crashed at Wittefield near Rieste at 1430. Arrowood landed safely at Welplage near Hunteburg and became a POW.

The weather Scout Mosquito reported deteriorating target conditions so Task Force One returned to England and a second Baby dispatched. The drone pilot was Lt. Cornelius A. Engle, 385 BG, who later transferred to fly 25 BG (Rcn) Mosquitos. After rendezvous, setup, and bailout safely accomplished, the Mother craft directed the Baby to the supergun site at Mimoyecques, France.

Meanwhile, Tunnell returned to Dinginess and while patrolling for the next task force, four V-1s passed beneath him. He located the Baby at Tenterden and followed it to St. Ingleveret, France, where it exploded 1,500 feet short of the target. Tunnell went to 800 feet providing McCarthy a better filming view. On return he flew low over the area, trailing tracers all the way to the coast. Scattered flak holes peppered the port wing and a 20mm round struck the port motor, forcing return on single engine.

Cameramen in the B-24 Mother plane located a Baby and followed it to France. Kurjack recalls:

> "I managed to get some shots of the Baby in flight over France and also shot scenes of the crash. I had one hell of a job trying to photograph the Baby. I had no intercom connection and was crowded in the nose with an angle of view restricted to vertical scenes with only a slight forward angle. The Baby flew ahead of the Mother and was within sight for only a short time."

Thiry complained of similar obstacles:

> "The Baby was out of sight most of the time. I did take footage over England and filmed the scene of the parachute jumper hitting the silk at the coast of England. I had the Baby in my finder when it exploded on a target of opportunity."

That night, Major John Hoover took Colonel Roosevelt as copilot on a B-25 night photo mission over Mayenne, France. They encountered trouble with the bomb release mechanism but completed the mission and returned safely to Mount Farm with no battle damage. Roosevelt transferred the Mitchell to Watton for operations in future Joker missions.

Another Aphrodite was scheduled Sunday, 6 August at RAF Fersfield. After briefing, Capts. Willis Locke/Lloyd Humphries (NS515) left at 0830 to reconnoiter target area weather at Watten, France, V-weapons site. Two Babys took off 20 minutes apart and led to the control point by a navigation ship. Lts. Robert Tunnell/ Daniel McCarthy (NS581) followed at 1047, one minute behind their Baby and watched the crew bailout over Woodbridge. Mother directed Baby out over the North Sea. Tunnell flew at 9,000 feet, and when two miles from the enemy coast, the robot executed a tight left turn and crashed out of control into the sea. The Mosquito returned to England for another Baby. Unable to locate the second task force, Tunnell flew to Watten, France. Two single-engine enemy fighters climbed to the Mosquito's altitude west of Gravelines, but Tunnell kept them trailing behind.

An enemy 20mm round struck NS559 port engine on 4 August 1944 Aphrodite mission, forcing return on single engine. Lt. Robert Tunnell on left while on right Lt. Daniel McCarthy examines the round. *Photo NARA*

On 6 August, Lt. James Matuska/Sgt. Ray A. Berge (NS557) flew to Langford Lodge, Northern Ireland. Upon entering the traffic pattern, the wheels were lowered and landing occurred at 0900 hours. After 40 feet of travel, a severe bouncing and chattering occurred. The left wing dropped causing the Mosquito to swing off the runway towards a group of parked A-20s. To avoid a collision, Matuska retracted the undercarriage. The Mosquito flopped to the ground and slid to a safe stop with damage to both engine nacelles, props, fuselage, port undercarriage, and tail assembly.

8 August 1944 accident shows Mosquito NS557 on a flatbed trailer. Note red serial number on vertical stabilizer and early paint scheme with top PR blue extending to this area. *Photo F.A. Law via RAF Burtonwood Association*

Next day, Lts. Ford D. Porter/James M. Davis (MM368) taxied out for a weather sortie. Porter attempted lowering the flaps but accidentally activated the landing-gear handle, and the undercarriage retracted causing minor damage.

Lts. Walter W. Thompson/Earl C. Edgar (MM311) left Watton 7 August at 2215 on a Mickey mission with targets at Salzbergen and Magdeburg. They never returned. At Watton no one knew the circumstances of the disappearance except the crew were MIA. According to a Luftwaffe report the Mosquito was shot down 8 August at Klausheide, seven kilometers east of Nordhorn, County Bentheim at 0110 hours and crashed in a swamp area. It was 90 percent destroyed and the bodies of two airmen recovered. One was identified as Lt. Earl Edgar and the other according to pieces of clothing as an unknown American. The Germans buried both men in the military cemetery at Lingen/Ems.

CCU Sgt. Al Mandelaro flew his 24th mission on 8 August filming the attack on enemy troop concentration at St. Lô. A piece of flak came up through the Fortress nose and struck him in the ribs. Luckily, his flak suit retarded the fragment but left a large chest bruise:

> "I wasn't on intercom so I asked the navigator if he would kick me when the bombs were to be released. I was to film bombs away and target strikes and I didn't want to miss this. When the shrapnel hit me, I turned around and told the navigator 'You didn't have to kick me so hard.'"

Lts. Daniel McCarthy and B. Jones, with Sgts. Roy Randolph and Joe Beach also filmed the attack but complained, "We could see enemy artillery in action as … tiny pinpoints if anything at all. You can't obtain good pictures from 18,000 feet." During the mission, a formation of bombers misidentified their target and bombed Canadian Army troop positions south of Caen, killing 65 soldiers and wounding over 300 others.[49] And Capt Willis Locke took NS515 to Mount Farm and flew General G.A. Anderson to witness his division formations bomb near Caen, France. Known as Special Scout or Division Scout mission, Watton later labeled it Redtail.

On 9 August, the 325 Photographic Wing (Rcn) formalized and the previous designation 8 Reconnaissance Wing (P) dropped. The 802 Reconnaissance Group (P) became the 25 Bomb Group (Rcn). The change in designation did not affect the function or operations of the Group, except the 50 FS from Iceland ceased to exist on this day.

The 25 Bomb Group (M) designation previously was assigned to a Caribbean area antisubmarine unit flying lightly armed Douglas B-18 aircraft carrying depth charges. They later converted in August 1943 to the B-25H having a 75 mm cannon and depth charges. Comprising 10, 12, 35 and later 417 Bomb Squadrons at Borinquien Field, Puerto Rico, the squadrons were dispatched to various bases along the Antilles Island chain, British Guinea, Curoaco, Venezuela, and later Panama Canal Zone. The Group became defunct in December 1943 and transferred (on paper) to Alamogordo AAF, New Mexico on 24 March 1944 for disbandment and the personnel shipped to various assignments, many in connection with training B-29 crews. The 25 BG (M) designation was reassigned to the newly reconstituted 802 RG (P) becoming 25 BG (Rcn)(S).[50]

To avoid confusion, a summary detailing the 802/25 development is recapitulated in the accompanying summary.

On 7 August 1944 while taxing, the pilot accidentally pulled the landing gear lever when attempting to lower the flaps. *Photo G. Scruggs*

49 C. P. Stacey, The Victory Campaign, Operations in Northwest Europe 1944-45, Vol. 3, Ottawa, 1960: 223.

50 Dan Hagedorn, Alae Supra a Canalem: Wings Over the Canal, Paducah, Kentucky: Turner Publishing Co. 1995: 110.

8 Reconnaissance Wing (P) and 325 Photo Wing Development *by Louis E. Schueler, Historian 325 PW (R)*[51]

It is quite possible that many members of the 325 Photo Wing, Reconnaissance may not remember what outfit they were in if, in fact, they knew at the time. There was no 325 Wing until August 9, 1944, sometime after the invasion of France.

It all started with the 8 Reconnaissance Wing (P). P stands for provisional and will occur frequently in the following. The Wing was established by Eighth Air Force General Order No.69 on February 19, 1944 and headquarters was established as AAF Station 113 Cheddington on February 23 with the following units assigned:

HQ & HQ Squadron, 5 Combat Crew replacement Center Group
5 Replacement & Training Squadron (Bomb)
HQ & HQ Squadron 6 Combat Crew Replacement Center Group
6 Replacement & Training Squadron (Bomb)
HQ 7 Photo Group, Reconnaissance
13 Photo Reconnaissance Squadron
14 Photo Reconnaissance Squadron
22 Photo Reconnaissance Squadron
27 Photo Reconnaissance Squadron
381 Service Squadron (78 Service Group)
1274 Military Police Company (Aviation)
55 Station Complement Squadron

The 7 Photo Group, Reconnaissance with its four squadrons and attached units were all legitimate organizations and had been operational for some time. The assignment of these units to the Wing affected them only in the matter of channels of command, and all organizations remained at AAF Station 234 Mount Farm.

Since a provisional unit has no legal basis for personnel or equipment, other units must be provided to furnish these. In this case there were the 5 and 6 Replacement and Training Squadrons (Bomb) which were ordered (without personnel) to Cheddington on February 20. A little later, on March 12, the 50 Fighter Squadron from Iceland via Nuthampstead was ordered to Cheddington with personnel.

The 5 Combat Crew Replacement Center Group Headquarters and Headquarters Squadron provided personnel for positions in Headquarters and Headquarters Squadron of the 8 Reconnaissance Wing (P). If you were in either the 8 Reconnaissance Wing Headquarters or its Headquarters Squadron you were probably a member of the 5 Combat Crew Replacement Group Headquarters or Headquarters Squadron.

On March 4, three provisional squadrons were formed – the 8 Weather Squadron (Heavy)(P), 8 Weather Squadron (Light)(P) and the 8 Courier Reconnaissance Squadron (P). Somewhat later the 8 Reconnaissance Squadron (Special)(P) was formed and the 8 Courier Reconnaissance Squadron (P) disbanded.

These squadrons, in turn, were made up of personnel assigned to the 5 and 6 Replacement and Training Squadrons and the 50 Fighter Squadron.

Since we now have several squadrons sort of floating around, we need a group headquarters and these people came from the 6 Combat Crew Replacement Center Group to form the 8 Reconnaissance Group (Special)(P) on March 22. Shortly thereafter, the unit designation was changed to 802 Reconnaissance Group (Special)(P) on 30 March.

On April 12, the 802 Reconnaissance Group (Special)(P) moved with its provisional squadrons to AAF Station 376 Watton while the 8 Reconnaissance Wing (P) and its A-2 (Intelligence) and A-3 (Operations) sections, which had been at Mount Farm moved to High Wycombe on April 12. High Wycombe was the seat of Eighth Air Force Headquarters (Pinetree).

The only other significant name change occurred on May 31 when the 1 Photo Technical Squadron, which previously had been assigned to the Wing, became the 8 Photo Technical Squadron. The 1 Photo Technical Squadron was transferred to the Ninth Air Force without personnel or equipment.

Eighth Air Force General Order No. 474 on 9 August 1944, made everyone legitimate as follows:

8 Reconnaissance Wing (P) – 325 Photo Wing, Reconnaissance
802 Reconnaissance Group (Special)(P) – 25 Bomb Group (Reconnaissance)
8 Weather Reconnaissance Squadron (H)(P) – 652 Bomb Squadron (Weather Rcn)(H)
8 Weather Reconnaissance Squadron (L)(P) – 653 Bomb Squadron (Weather Rcn)(L)
8 Reconnaissance Squadron (Special)(P) – 654 Bomb Squadron (Rcn Special)(H)

Anybody confused? Many people were transferred from one organization to another during this period. On August 9 therefore the Wing consisted of the following units:

High Wycombe

HQ 325 Photo Wing, Reconnaissance
3 AAF Combat Camera Unit
161 AAF Motion Picture Unit 16 MM
163 AAF Motion Picture Unit 16 MM
1 AAF Photo Intelligence Detachment
19 AAF Photo Intelligence Detachment
942 Engineer Aviation Topographical Battalion
2 Photo Technical Squadron
8 Photo Technical Squadron

Mount Farm

HQ, 7 Photo Group, Reconnaissance
13 Photo Reconnaissance Squadron
14 Photo Reconnaissance Squadron
22 Photo Reconnaissance Squadron
27 Photo Reconnaissance Squadron
381 Service Squadron (78 Service Group)
55 Station Complement Squadron
1274 Military Police Company, Aviation
2134 Engineer Aviation Fire Fighting Platoon

Watton

Headquarters, 25 Bomb Group (Reconnaissance)(S)
652 Bomb Squadron (H)
653 Bomb Squadron (L)
654 Bomb Squadron (Special Rcn) (H)
8 AAF Combat Camera Unit
375 Service Squadron (Separate)
1070 Quartermaster Company Service Group
2135 Engineer Aviation Fire Fighting Platoon

On this day, Lt.Col. Joseph A. Stenglein took command at Watton relieving Col. William H. Cleveland who resumed his duties as deputy commander, 325 Photo Wing Rcn.

Aphrodite held spotlight on August 12. Lts. Robert Tunnell/Dan McCarthy (NS569) flew to RAF Fersfield for briefing. The U.S. Navy planned testing the Castor FM control equipment on a radio-controlled PB4Y-1 serial 32271. The aircrew was pilot Lt. Joseph P. Kennedy Jr. and copilot Lt. Wilford J. Willy. Cameraman Sgt. Joseph Capicotto met Kennedy briefly when the young pilot landed a PB4Y-1 at Watton en route to RAF Fersfield.

Two PV-1 Mother aircraft, one with Joseph Capicotto on board, took off first then followed by the Baby. The Mother fell in behind the Baby and the task force started to fly the rectangular test course. Before reaching the first turn the Mother had control and remotely turned the robot. Two minutes later and without warning, the PB4Y exploded in a fiery fireball: " … there was a blinding flash of light … with fire and flame going straight up and down. A Mosquito camera ship had flown through the center of fire and gone out the other side trailing smoke and now it was circling and losing altitude."[52]

51 Bulletin Board, 3:5 (October 1989): 2.

52 Jack Olsen, Aphrodite: Desperate Mission, New York: G.P. Putnam's Sons, 1970: 239-240.

The Mosquito landed at the nearest field, RAF Halesworth.[53] According to McCarthy:

> "We were flying over England and decided to close in on the Baby. I was in the nose of the plane to obtain shots of the Baby in flight ahead of us. The Baby just exploded in midair as we approached and I was knocked halfway back to the cockpit. A few pieces of the Baby came through the Plexiglas nose and I got hit in the head and caught a lot of fragments in my right arm. I crawled back to the cockpit and lowered the wheels so Bob could make a quick emergency landing."

Whereas Tunnell records:

> "I didn't get a scratch but was dammed near scared to death. We were approaching to take close-ups of the Baby that was under control and flying level. We came in on the right side high and then the dammed thing exploded in midair. The Mosquito went up a few hundred feet and I didn't get any response from the controls. I was getting ready to reach for my parachute but decided to check the controls again. This time they responded and I decided to try to make a landing. One engine was out and the other smoking. We were near a field so I headed straight for it. We made a good landing and then the second motor quit. I had just enough speed to get Mosquito off the runway but couldn't taxi in to a hardstand. I'm sure glad the pictures of our previous mission were good because I don't think we're going to get that close to the Baby again."

Tragedy struck the returning second Frantic mission. Lts. Ronald M. Nichols and Elbert F. Harris (NS533), along with Capt. Joe A. Baker and Major John W. Walch as supervisory backup in another Mosquito, received notification to Scout the second shuttle mission from England to USSR. The crew chiefs would fly in the bombers, and Lt. Ralph Fisher briefed the men regarding his navigating the first shuttle mission.

On Sunday, 6 August, the bomber fleet departed England and flew along the Baltic coast, then turned southeast to Berlin, except for two shuttle Groups (95th and 390th) and their fighter escort. They proceeded easterly targeting a Focke-Wulf aircraft plant outside Gdynia, Poland. The shuttle task force planned flying southerly, then southeasterly to Kiev, and easterly to Poltava and Mirgorod. P-51 fighters from 357 FG escorted the shuttle force.

The Mosquitos took off on schedule, flew astride the main bomber stream towards Berlin, joined the shuttle task force, and proceeded to the target in Poland. Clear weather predominated the shuttle route. After loitering in the target area, Nichols/Harris left the bombers and fighters on a south-southeast heading. The course passed 75-100 miles west of Warsaw. In the process of suppressing a Polish underground uprising, the German Army ravaged the city; huge columns of smoke rising to 30,000 feet could be seen 200 miles away. The Mosquito descended to 6,000 feet at Kiev and remained at this altitude to Poltava.

The U.S. Navy had developed the Loran (LOng RANge) radio navigation system and installed transmitting stations in Britain and Italy. The Navy wanted data on the reception range and accuracy, and a Mosquito assigned this duty. A Navy team arrived at Watton prior to departure, installed the receiver in NS533 and provided operating instructions. They classified the equipment "Top Secret" and ordered the crew to post an American to guard the plane at all times in USSR, and protect the equipment with their life if necessary.

Debris from 12 August exploding PB4Y-1 penetrated Plexiglas nose and injured Lt. Daniel McCarthy, 8 CCU. *Photos. D. McCarthy*

53 Stewart Evans, Aphrodite 2, Fly Past (November 1984):.23.

1/Lt. Daniel McCarthy with Florence Babkewicz of the Watton Red Cross, whom he married. *Photo F. McCarthy*

Nichols/Harris landed at Poltava and were guided to a parking spot. The other Mosquito arrived slightly earlier. The pilot called the tower with a prearranged signal to dispatch a guard to protect the Loran-equipped aircraft. Eventually an MP Jeep arrived with a sentry. The crew went to Operations for debriefing by Major Walch and Capt. Baker. Several hours before sunset, both Mosquito and bomber crews attended a briefing and were placed on alert for a mission the next day.

The following morning, Nickols/Harris flew scout for a bombing mission to a synthetic oil plant near Katowice in Polish Silesia. The mission proved uneventful and bombing pattern accurate. After return, Major Walch arrived with news the Soviets learned of the oil refinery attack and became furious. Their military had planned to capture the facility and move the entire plant to Russia. They ordered the shuttle task force to leave the country. Next day, the bomber formations flew to Romania attacking primary targets near Ploesti. The Mosquitos landed at San Severo, Italy, in the Foggia area, and the men received three days rest. Harris recalls his return flight to Watton:

> "On Saturday morning, 12 August, we departed Italy on the final leg of the return mission to England. Our target en route was the large French aircraft company complex Avion Sud and adjacent Fracazal Airfield at Toulouse, France. Clear weather afforded us a good view of Naples and Rome. We flew across Corsica, past the bombers and proceeded to Toulouse. Having fulfilled our usual duties at the target area, we positioned ourselves at 29,000 feet to view the attack by the last box of bombers. While performing a routine check on the 'blind spot' behind our tail, I noticed two fighters above heading toward us. They were a long way back, but I tentatively identified them as P-51s. I notified Nick and he turned the plane for better visual inspection and confirmed the identification. Nick immediately started calling on the fighter radio channel to have the Mustangs called off but didn't receive any reply. I kept watching and urged him to pickup speed and get us away. He switched to the bomber channel and sent repeated calls for them to call off the P-51s. He continued looking back at them, then pushed the throttles full forward, tilted the nose down and released the wing tanks. He kept calling on the radio. By this time the fighters were a mile behind us and had apparently gone to full throttle and closing fast. Suddenly I saw their guns winking and yelled this to Nick. Tracers passed above us – I didn't expect they would start shooting that far away! Nick initiated evasive action, but the next burst of tracers struck the right engine and we were on fire. The tracers continued streaming all around, hitting the Mosquito.
>
> "I was wearing a chestpack harness and the chute pack lay on the floor. Normally it was located behind my feet, but in the turning maneuvers to observe the Mustangs, the pack slid forward. The fighters were estimated a half-mile behind the last time I saw them. I quickly released the seat belt and bent down to pickup the chute pack. At this moment the plane banked and I was thrown violently against the floor on my hands and knees. I strained to push myself up but couldn't move. I then realized the plane entered a high-speed stall, followed with a tight spin. It seemed I was thus poised forever in this position, and my strength wouldn't hold much longer – my body would be pressed to the floor. I felt the centrifugal force slackening and, shortly thereafter, was able to grab the chute pack and get myself upright on the seat. I snapped the pack to my harness and quickly looked around. The Mosquito was in a lazy spin to the left. Nick was fighting the controls, but the aircraft wasn't responding. Fuel poured from the punctured right wing forming a sheet of flame extending along the trailing edge, from the engine to beneath the fuselage, with occasional bursts of flame under the fuselage toward the tail. Standing up, I could also see flames beneath the left engine. Smoke leaked into the cabin from the bomb bay that contained a full fueltank. Since the plane wasn't responding to Nick's efforts, I assumed the controls were shot away. I tried calling Nick on the intercom but didn't receive any reply. Was the radio out? I can't remember if I checked the radio cable plug to see if it had accidentally pulled out during the violence.
>
> "The Mosquito entry door, located lower right and forward, was rigged with an emergency release for abandoning the plane in-flight. The canopy roof also had an escape hatch that we were instructed to use only for crash landings. In the air, anyone leaving the aircraft via this exit was likely to hit the horizontal stabilizer of the tail section. I immediately thought of alternatives while reaching for the escape kit in my navigation case and shoving it into my shirtfront: 1. Being killed by the tail; 2. Going out the bottom, catching fire and being burned to death. I decided to pull the bottom hatch to investigate the extent of the fire underneath. I quickly tore off my flying cap, oxygen mask and radio gear, then hastily crawled forward to the hatch and gave a good tug on the release handle, but nothing happened. There wasn't much time left before the plane exploded, so I decided not to waste any further effort and take my chances on the tail.
>
> "I straightened up, tapped Nick on the shoulder, pointed to the bottom hatch and shook my head, pointed to the top hatch then pointed to myself and nodded; he responded with a nod. I reached up, pulled the canopy hatch release and the top hatch cover flew off. I climbed out the opening and placed my legs under me to push myself out and down then thrust with all my might and jumped. I must have involuntarily closed my eyes before striking the horizontal stabilizer, then bounced off and fell away spinning. A sheet of flame issued from under the right wing with streamers of fire and black smoke, while fierce flames under the fuselage trailed the Mosquito. I passed through the flames or the very hot slipstream and for several seconds hesitated to open my eyes, fearing my clothing and chute were on fire. Finally I opened them and found nothing burning.
>
> "I left the plane at 12-14,000 feet. Remembering training, I extended my arms and spread my legs to stop the spin. My outstretched arms helped stabilize myself on my back to allow easy access to release the chest parachute. I looked over my shoulder and determined there was a long way to go before reaching the ground, and per training instructions, decided to take most of it in free-fall. A smoke trail from the burning Mosquito descended to the northeast, towards Toulouse, but I didn't see the aircraft or determine if it had crashed, nor was any other parachute seen.
>
> "I was falling toward a cultivated field north of a hard surfaced, east-west road. South and perpendicular to the road spread a long strip of dense woods. I continued glancing periodically over my shoulder at the ground during the plunge to earth. Suddenly the ground seemed to be coming up very fast. I immediately pulled the D-ring and the pilot chute popped out, followed by the main parachute and then the shock of it opening. I spun around and immediately reached up, caught the risers, and spread them to halt the spinning. Still struggling with this task, I glimpse out the corner of my eye that I was about to hit the ground. I had misjudged my altitude but luckily deployed the chute with not a moment to spare. I hit hard and my breath was knocked out for a few seconds. I then arose to collect the chute. My left hip was still partially numb from bouncing off the stabilizer and my right ankle was sprained. With only a light wind, I was able to collapse the parachute canopy and bundle it under my arms."

Watton received notification a Mosquito was shot down over the target area with both crewmen presumed MIA. Their status later changed to KIA. The 357 FG history states an escorting P-51 claimed a Ju 188 destroyed at 1145 hours north of Toulouse at 29,000 feet. There are two sides to every story. According to the P-51 pilot involved:

> "I was flying No. 4 ... when I noticed a single, two-engine ship at 9 o'clock. It circled in the sun, and next time we picked it up, it was at 6 o'clock still in the sun. The ship was about 29,000 ft. and we were at 23,000 ft., and I turned towards him (covered by the wing man). I climbed to his altitude and he started in a shallow dive straight away. I closed very slowly and pulled above him to check for insignia. There positively were no stripes or roundels on the ship. I checked the engines and they were radial. The other pilots said it was a Ju 88 and I called in a Ju 188. I blew a couple of exhaust stacks at this time and had to throttle back – the ship (Mosquito) was pulling away from me. I gave it a burst out of range and set the left engine on fire and he started burning. I got a deflection shot later and saw no markings. The ship went into a spin and I didn't follow it down. I claim one Ju 188 destroyed. Reports have it that there was a Mosquito destroyed in the area. I'm positive of no British markings on the ship and it offered no recognition gestures that I perceived. It appeared hostile by swinging around on our tail in the sun and then trying to dive away from me."

An inquiry was held to determine what occurred. The board members revealed the fighter pilot genuinely believed he had a Ju 188 in his sights. The Eighth Air Force Field Order for the shuttle mission contained the following statement: "(7) 8 Photo Reconnaissance Wing will dispatch two Mosquito type aircraft to report weather conditions to task force. Operational control will be exercised by 3 Bomb Division."

The above lead to the belief that fighter pilots taking part in the shuttle mission were aware of the fact two Mosquito-type aircraft were a part of the task force. Aircraft recognition was not the forte of every fighter pilot. Numerous attacks by AAF fighters and bombers on 25 BG (Rcn) Mosquito aircraft during the war testify to this fact. The Board of Inquiry recommended all Mosquito aircraft accompanying any task force for weather observation be painted with special markings. This would eliminate any future attack by friendly fighters en route or over strange target areas. Therefore, all 25 BG (Rcn) Mosquitos had the tail section painted crimson red to aid identification for friendly aircraft. All organizations involved in future operations were to receive notification, by an Eighth Air Force Field Order, the presence of specially marked Mosquito aircraft.

Harris sought refuge in the nearby forest and eluded capture for several days then established contact with the Maquis:

> "The French insisted the Mosquito was shot down by German Me 109s. They observed a chute bloom out behind the plane, than catch fire and burn. There were large pieces of the plane falling and a badly burned body found near where the Mosquito crashed and exploded.
>
> "I met one of the men who retrieved Nick's body, built a coffin and helped bury him. He said Nick was badly burned with only fragments of his clothing remaining and his body was horribly mangled. He showed me a small burned piece of his 'pink' trouser from the beltline. The body was found some distance from larger pieces of wreckage. The Mosquito exploded and charred pieces were scattered over several acres. I visited Nick's grave in a country churchyard – a fresh mound of dirt with a wooden cross. The Frenchman offered to carve Nick's name on a piece of wood and mount it on the cross. I gave him Nick's name and offered the Frenchman money, but he refused, saying, 'For a man who gave his life for France, I am glad to give my time.'"

The advancing U.S. Army reached Toulouse and repatriated Harris to England. He was reassigned to the USA. While awaiting transport, the late Albert Harris relates his final thoughts:

> "One of my fellow evadee, a fighter pilot from the 357 FG who was shot down prior to the Shuttle, returned to his home base. He checked the records and provided me with information the log of one squadron had a kill claimed on 12 August over Toulouse of a pale blue Ju 188. The claim had been crossed out and nullified. Our Mosquitos were blue, the only planes so painted in the Eighth Air Force. About the only characteristics the Mosquito and Ju 188 had in common were twin engines and roughly the same size.

On left is 1/Lt. Elbert Harris with 1/Lt. Ron Nichols on right. Nichols was KIA 12 August 1944 when a P-51 shot down the Mosquito. *Photo E. Harris*

"I reported this to the Interrogation Officer who said he would relay the information through channels. He didn't seem too concerned. My fellow evadee wouldn't tell me the name of the pilot or squadron number. As far as I knew the rogue pilot was never punished. I became furious at that moment and gave thought to pursuing identification, then going and killing him. My meditation on the subject was cut short by orders to report to Stone Replacement Depot. A week later, a group of us were ordered to Prestwick and thence by C-54 for the trip home. Thus ended my abbreviated but exciting overseas service. As is often said, war experiences may make interesting bar conversation, but not many of us wish to relive the true ones."

Colliers magazine became a popular item at Watton. The 12 August issue featured a war story written by F/O Charles Gorham. Colliers' critics hailed his article as one of the outstanding short stories of the war.

Watton provided support for the Batty glide-bomb project. Lt. Daniel McCarthy, hospitalized after the 12 August Aphrodite mission, became unavailable for further filming duties. In his place, CCU selected Cameraman S/Sgt Augie Kurjack from Pittsburgh, Pennsylvania, to fly 13 August with Lt. Dean Sanner (MM370) on the first Batty mission. This involved launching from a B-17, the GB-4 radio-controlled glide bomb, consisting of a 2,000-pound AN-M622 bomb attached to a streamlined airframe having a 12-foot wing span and twin tails. It had two-axis control using an elevator and rudders. The GB-4 carried a television camera. An operator in the launch airplane guided the bomb during decent. Released 17 miles from the target at 15,000 feet, it glided at 250 to 300 mph and took four minutes to impact. According to Sanner:

"The mission was to hit the E-boat pens at Le Havre, France, with two 2,000 pound GB-4 radio-controlled glide bombs hung under each wing of a B-17. Our duty was to photograph the GB-4 in flight and also any damage inflicted by the bombs. I was warned not to fly between the robot and the mother ship after release to avoid interfering with the radio guide signals. On the way to the IP, Kurjack the photographer, took pictures of the stubby-winged glide bombs hung under the B-17's wings. A P-38 also accompanied the B17. There were no clouds that day.

1/Lt. Ronald Nichols, KIA 12 August 1944 on final leg of shuttle mission. *Photo R. Walker*

1/Lt Elbert F. Harris after return from France. *Photo E. Harris*

Groundcrew working on Merlin in hangar, possibly from NS533.
Photo G. Scruggs

Groundcrew pose near Q-Queen, possibly NS533. Note invasion stripes encircle the fuselage. Did NS533 have invasion stripes when shot down? *Photo. G. Scruggs*

"After the first release, I found it very difficult to follow the GB-4 down and maintain it in sight. It seemed erratic in both horizontal and vertical flight. I followed the first bomb down some distance but lost sight of it, then broke off pursuit and climbed for the second run.

"After the second GB-4 release, I followed it down, zigzagging back and forth, but experienced great difficulty chasing it due to the steep trajectory. I lost sight of the glide bomb at low altitude as it blended against the marshy background. Now very low, I intended to head for open water between the banks of the Seine River, then climb back to altitude over the English Channel. By chance, I flew over the bomb when it exploded and the ensuing blast literally blew me out of the Mosquito. I abruptly found myself hurtling through the air. My right arm was badly injured so I pulled the parachute ring with my left hand. After deployment, I made one or two swings before hitting the muddy marsh.

"When I left Watton that morning, I couldn't find what we called the tuning fork. Its purpose was to fasten together the shoulder and leg straps to hold a pilot in his seat. Ordinarily it was fastened to a lanyard. I flew without it that day and believe this saved my life.

"My first thought upon landing was to evade. I quickly unfastened the parachute harness. My right shirtsleeve and pant leg were shredded and bloody, and I noticed my right ankle was bent inward at an angle. I slowly stood up on my left leg, and when I attempted to step forward with my right foot, down I went in agony. I was paralyzed by excruciating pain. After an hour of suffering, several German soldiers approached with their guns drawn. They signaled if I was armed and I replied no. They transported me sitting on a rifle between two soldiers until wooden planks were located; these were stretched over several rifles for me to lay on.

"I observed smoke rising in the distance and learned it was my aircraft. The photographer Augie Kurjack was in the nose compartment when the explosion occurred. He experienced the full impact of the blast and perished in the explosion.

"I was eventually taken to Le Havre where all remaining clothes except underwear were taken from me. That night I lay on a table with a guard at the door. During the interval a drunken Army Major stood in the doorway waving a pistol. My impression was he wanted to shoot me. The guard refused the officer entry and he finally left. Next morning I was transported to an underground hospital where a splint was applied to my right leg and ankle. The fibula was fractured, and the bone fragmented in both the right ankle and elbow."

On 15 August, the Germans transferred Sanner to a hospital in Rouen. After several more hospital moves he arrived at Oberursel on 24 September for confinement and interrogation before transfer to Stalag Luft One at Barth, arriving 8 October, and remaining until war's end.

On 14 August, CCU Sgt. Walt Lipinski left Thurleigh in a B-17 to film the bombing of Stuttgart. Returning aircraft reported his Fort MIA, but he arrived at Thurleigh near midnight and greeted with a big reception. His aircraft ran short of fuel and performed an emergency landing at a Spitfire field on the southeast coast of England. Meanwhile, Lts. Oliver Emmel/Iver Fitschen (MM387) delivered a courier pouch to General Eaker at his Mediterranean Air Force HQ in Italy. The crew left Mount Farm and flew by way of Gibraltar: Maison Blanc, Algiers, and Manigliona, Italy. The Mosquito experienced engine, directional gyro, and electrical trouble, and flew to San Severo for repairs. Both men later returned the 19th by the same route.

The night of 17/18 August, F/O Russell Whitmer/Lt. Robert Balser (NS625) flew an unsuccessful Mickey mission. While over the Dutch peninsula, 20 miles from Zwolle, several rockets burst all around the Mosquito. Heavy flak encountered at Wilhelmshaven and Wesermunde required evasive action. Then searchlights coned them near Zwischenahn but cloud cover intervened. The starboard generator warning light lit continually and H2X radar failed near Oldenburg.

A single-engine aircraft attacked Lts. Gene Goodbread/Warren Barber (NS635) the 18th, scouting for 3 BD to St. Dizier and Ravieres. A silver Fw 190 with no markings appeared off the Mosquito's left wing at 28,000 feet and turned to approach from behind. Goodbread applied power and out ran the 190.

Capt. Emmet Bergholtz of RKO Studios, Hollywood, assumed 8 CCU command 19 August with Lt. Daniel McCarthy as unit vice. While at the FMPU in California, Bergholtz viewed film shot by CCU in various theaters. He noted the best footage was achieved at low altitudes. Only bombs away scenes were obtained higher than 18,000 feet where all targets appeared identical. He declared a new policy: no filming at high altitude unless specifically requested by Eighth Air Force or 325 RW; all future bombing attacks be filmed either on the ground or from low-flying aircraft.

1/Lt Dean Sanner in Mosquito flies formation with B-17, waiting for Batty glide bomb release. *Photo USAFHRA*

E-Boat pens near end of channel marked target on the Batty mission.
Photo USAFHRA

Release of GB-4 Batty glide bomb from B-17. *Photo USAF Museum*

1/Lt. Dean Sanner became POW on 13 August 1944 Batty mission. *Photo R. Walker*

On the night of 19 August, pilot Lt. Ray L. Musgrove and navigator Lt. Harold Fordham (NS625) took off at 2319 on a Mickey mission to radar photograph the approaches to Bremen, Germany. The aircraft failed to return. According to Fordham:

> "Our aircraft headed over the North Sea climbing for altitude. We passed over the Frisian Islands, then cut back south and west heading for Bremen where I took a long series of scope photos. Suddenly the aircraft went into very erratic movements. I immediately called the pilot and asked what was happening. He replied, 'I don't know, we're caught in lights.'
>
> "Ensconced in the rear fuselage, I was unable to see anything outside. Shortly after, the aircraft fell off on a wing and began a slow downward spiral. I glanced at the altimeter and the needle started winding down rapidly. I attempted to contact the pilot on the intercom, but there was no reply. I sat on my flak suit wearing a B-3 jacket and black RAF-type, fleece-lined boots for protection from the cold temperatures, along with my backpack parachute. The altimeter needle continued dropping rapidly indicating a crash was imminent. I unfastened my lap belt and fell backwards towards the tail section because of centrifugal force from the spiral. We had a dim light inside the fuselage and the escape hatch was to my lower right hand side. I struggled to regain my position and released the escape hatch. When it flew off, I did not observe any fire or flames externally. I managed to place my legs through the cramped opening. By this time the Mosquito lost considerable altitude. When I attempted to squeeze further through, the parachute snagged on the hatch ledge. Stuck halfway out the hull, my legs kept banging against the outside fuselage in the slipstream. Finally, I was able to rock myself again and again over the edge of the opening and fell clear of the Mosquito. We had been flying at around 25,000 feet and I bailed out at between 5,000 and 10,000 feet. Not bothering to count, I pulled the D-ring immediately and the parachute opened with a jerk. I searched all around the area but could see nothing except blackness. There was no crash, no flames, no fire, nothing! Musgrove initially said we were caught 'in the lights.' But now free and descending, there were no searchlights – nothing but pitch blackness surrounded me.
>
> "Here I am in complete darkness and I'm oscillating in the parachute. This was my major concern. Oscillating about 15 degrees to each side gave the feeling I was going to go over the top of the parachute any time. Below lay a black featureless void that provided no clue as to my distance from the ground. Suddenly and without warning, I landed heavily in a field and tumbled over backwards slightly stunned, then lay there to catch my breath."

At Watton, when the aircraft failed to return, both men were listed as MIA. It was the second Mickey Mosquito to mysteriously disappear.

Meanwhile back in Germany, Fordham shred the silk canopy on a barbed-wire fence, then buried the remains. Now west of Bremen and using the north star as guide he proceeded towards the Dutch border. After cutting off the tops of his escape boots, the oxford-type shoes fit poorly, causing both feet to blister and swell. Traveling at night and sleeping during the day, he avoided villages to elude detection. Tree branches formed a raft to keep his clothing dry as he swam numerous canals while journeying westwards.

After ten days without food, starvation proved difficult to overcome. One night he entered a church on a village outskirts and lay on the floor to rest his swollen agonizing feet. A priest entered and assisted him to a living quarters for nourishments. After a brief period, a Luftwaffe officer entered and asked Fordham to accompany him to a waiting vehicle for transport to an airfield. Two days later he arrived at Oberursel, a suburb of Frankfurt for interrogation, then transfer to Stalag Luft III, Sagan, Silesia.

In late March 1945, advancing Soviet troops from the east forced vacating the POW camp. The prisoners marched for several days then entrained in boxcars for Nuremberg to another camp. On 2 April the POWs force-marched again, to Moosburg. During the trek, Fordham hid in bushes and escaped. Heading westward he encountered artillery fire and observed U.S. armored vehicles enter a village located in a valley. He manged to climb on board a Sherman tank and shouted he was an American. After transfer to Mannheim and Paris, then Camp Lucky Strike, he arrived at Le Harve, France, for shipment to the USA.

On August 20, a 3 SAD C-64 43-35392 failed to "S" while taxing and collided with a six-bis vehicle parked in the center of the perimeter track, waiting for a tow truck.

1/Lt. Harold Fordham, navigator on the ill-fated 19 August 1944 mission. *Photo H. Fordham*

Pilot 1/Lt Ray L. Musgrove KIA on 19 August 1944 mission. *Photo L. Musgrove*

Intense and accurate barrage flak boxed-in Lts. Robert Walker/Roy Conyers flying B-26 #205 during a night photo mission to France, the 23/24th. Assigned to cover St. Just-Bretevil, they encountered flak near Le Treport and red rockets with a long tail streaking behind, exploding around their aircraft. Immediately two flares appeared above 50 yards ahead, and two unidentified aircraft – one with twin engines – gave chase. Walker applied full throttle and raced out over the channel only to encounter intense fire from a Flak ship 20 miles northwest of Le Treport. Despite the skirmish, they obtained fairly good photos.

Night-fighters attacked Lts. Carrol Stubblefield/Louis Pessirilo (NS512) on 25 August during a Bluestocking over northwest Germany. At 0102 several searchlights coned the Mosquito at 30,000 feet just before an unidentified aircraft appeared. Ten minutes later another unidentified aircraft at 26,000 feet fired several rounds from three o'clock high. Stubblefield initiated violent evasive action and lost the intruder.

The outstanding change during August involved replacing 652 Squadron B-17 aircraft with B-24s. New flying crews assigned to the squadron received initial training in Liberators and regular personnel underwent conversion training. The differences between both aircraft immediately became apparent. The B-24 flew scheduled flights on average one hour less time than the B-17, climbed to altitude faster, but somewhat difficult to handle in adverse weather conditions. The Lib proved more suitable for carrying large fuel loads, and installation of specialized weather equipment compared favorably with the B-17. On the 26th, 652 BS began flying the B-24 on Atlantic WR.

The month ended with several accidents. On the 27th, Lt. Otto Kaellner ground looped NS590, followed the 29th when Lt. Clayton Hackman in MM364 swung left off the runway on takeoff, hit a mound of dirt and broke off the tail wheel. The 30th saw Lt.Col. Joseph Stenglein perform a local flight to test the installation of a new tail wheel assembly on F-8 43-34962. A high oil temperature and falling oil pressure force him to feather the starboard engine and perform a single-engine landing. He accomplished a wheels-first touchdown. With loss of speed, the tail gradually settled when a loud thumping, scraping noise indicated the tail wheel failed to extend causing the rear appendages to drag along the runway.

September 1944

Adverse weather conditions curtailed 654 Squadron operations the first part of September. The 653 BS continued flying weather flights over the continent and British Isles, and comprised comparatively few missions flown by the entire Eighth Air Force during this period.

The first day of September began with a Scout mission to the diesel engine and ordnance factory at Mainz. The 3 BD leaders ordered a recall because of 8/10ths cloud coverage en route. When Capts. Willis Locke/Lionel Proulx (NS516) reached the target area they found CAVU conditions and remained nine minutes attempting unsuccessfully to recall the bombers.

A 652nd Squadron B-24 44-0233 at Watton. Note nose canopy for weather observer. *Photo C. Hackman*

On the 2nd, Lts. Vance Chipman/Malcolm MacLeod (MM388) flew to Alconbury to transport 482 BG radar engineer Capt. Henry Clopper in the nose section to the DH plant, Hatfield, Hertfordshire. Due to prevailing wind conditions, Chipman landed on a short grass runway, muddy and slippery from rain. Unable to stop from lack of traction, the aircraft went off the end where the pilot swerved to avoid hitting a gun emplacement, causing the undercarriage to buckle.

Lt. Robert Tunnell/Capt. Emmett Bergholtz (NS553) flew to RAF Fersfield to film the final USN Anvil mission Sunday, 3 September. The Navy changed the robot arming method to guarantee crew safety. Manual arming replaced electrical detonation to prevent recurrence of an explosion that previously killed Lts. Kennedy and Willy.

The first off were Capt. Robert Howle/Lt. Ralph Fisher (NS513) from Watton at 1105 to scout the Heligoland weather. At Fersfield, the PB4Y Baby took off and both route setup and bailout accomplished successfully. Two PV-1 Mother control planes guided the Baby across the North Sea followed by several escort fighters and Mosquito with Tunnell/Bergholtz.

After relaying target weather conditions, the Mosquito with Howle/Fisher returned to England and joined the P-51s escorting the task force. As Baby approached the objective, the Mother ship received a poor TV picture and located the breakwater of nearby Dune Island, mistaking it for Heligoland. Eight seconds prior to impact the TV picture ceased indicating a possible flak hit. The Baby started a right turn and the controller dove it into Dune Island. The PB4Y exploded near an airfield causing extensive damage to barracks and surrounding buildings.

As Howle recounts his flight:

> "We flew to Heligoland and observed the submarine pens with tunnels leading into a high cliff. Lt. Fisher reported 7/10ths cloud target weather back to Fersfield. En route to England we met and accompanied the Navy task force back to Heligoland. We kept both drone and control aircraft in view at all times. During approach intense flak encircled the Liberator descending gradually to a target, not too far from Heligoland. It crashed near some buildings and the resulting explosion created a stupendous blast wave and huge fireball."

According to photographer Bergholtz:

> "I guess the toughest part of the mission was sweating out the three abortions before it finally came off. Bob Tunnell is the best pilot I have flown with on a photographic mission. He knows where the picture is and gets the Mosquito in such a position that the cameraman has an excellent view of the subject to be photographed."

The Allies liberated Brussels on 3 September and Antwerp the following day. The onslaught of Allied air activity force the Germans to deliver supplies to their infantry units at night increasing requirement for night photo Joker missions.

Mapping coastal regions and installations continued with Lts. Robert Shoenhair/Floyd Mann (NS593) on the 6th. Solid cloud cover hampered the Seaweed mission to St. Nazaire but two photo runs were flown over Brest at 7,000 feet. A flak burst damaged the vertical stabilizer, terminating the mission. That afternoon, Lt. Albert L. Stokes delivered F-8 43-34950 to Watton in error. The aircraft veered left off the runway, hit a ditch, and sheared off the tail wheel.

The first fatal 652 Squadron accident occurred the night of 6 September when B-17G 42-31736 crashed on takeoff at 2223 hours, one-quarter mile off the end of the runway. After starting the engines Lt. Clarence Stephenson proceeded out from the dispersal area. He slowly taxied the bomber around the perimeter track, came to a stop and performed an engine run-up, then maneuvered the aircraft to takeoff position. Met/Obs 1/Lt. Marvin R. Rogers, gunners S/Sgt. Keith Clinton and T/Sgt. Joseph C. Minery sat in the radio compartment. Minery with headsets on, heard the pilot call the tower and report being ready for takeoff. Minery immediately notified Clinton and Rogers. Then followed the engine roar as the aircraft quivered in preparation for a carrier-type takeoff. This was followed by a sudden lurch forward as the bomber moved down the runway with the usual little bounces initially, then smoothness as wing lift took effect.

Met. Observers position in nose of B-24 44-40882 for atmospheric measurements and cloud pattern observations. *Photo A. Podwojski*

Nose art of B-24 "Mamma's Kids". *Photo A. Podwojski.*

Major Alvin Podwojski with "Short Arm." *Photo A. Podwojski*

On the ground, Pfc. George S. McNerney on duty at an antiaircraft emplacement heard sputtering of engines followed by a loud roar that quickly died out. In the darkness he discerned an aircraft flying 25 or 30 feet off the ground with right wing lower than the left. There appeared a fire or a terrific exhaust flame emanating from the aircraft's left side. The Fort dropped to the ground with sounds of twisting metal as it slid and bounced the last few yards before coming to rest 50 yards from him.

Meanwhile, in the radio compartment the first indication of something wrong was a sudden flash of light clearly visible through the side window, followed by a rough tremor or vibration. Rogers shouted for the other men to brace themselves. Immediately, Clinton jumped up and climbed through the rear compartment door into the waist fuselage followed by Rogers and Minery. Flames shot out from the bomb bay section past the closed door. The aircraft began to shake violently. Clinton fell to the floor and grabbed hold of the ball turret and felt it dragging on the ground. The other men held on to anything available to prevent being thrown around. When the bomber ground to a halt, Clinton gave a desperate lunge and dove out the waist door. He hit the ground and rolled, as the other two men followed behind. All three ran a short distance then turned to make an appraisal of the situation. The Fortress appeared enveloped in flames from the bomb bay forward to the front including the left wing. There was no possible chance to rescue the other crewmembers. The three men moved further away to another field, dropping behind a swell in the ground, fearful the inferno would ignite ammunition on board.

Both rescue and firefighting crews arrived but the flames were not extinguished until 0230. Killed in the accident were: l/Lt. Clarence R. Stephenson, pilot; l/Lt. Eugene W. Swaitnicki, navigator; 2/Lt. Joseph Ward, copilot; T/Sgt. Earl H. Westerholm, engineer; and S/Sgt Leonard L. Licurse, radio operator.

Richard Geary recalls a pilot wager his Mosquito skills:

> "One day at the Officers' Club, while in conversation with his fellow pilots, Vance Chipman boasted he could glide a Mosquito across the base with both engines off. Everyone accepted his challenge and the wagering began. Chipman signed out a Mosquito, took off and climbed for altitude. He dove down to gain speed then shut off both engines as he roared toward the base. The large gathering of men outside the Operations hut stood astounded as the Mosquito, with both props feathered, silently glided past them to the end of the aerodrome. Chipman restarted both engines, circled the field and landed. After witnessing this event, no one ever questioned his flying abilities."

Major Alvin Podwojski/Capt. Lionel Proulx scouted a 3 BD supply dropping mission south of Besancon, France, on 9 September. While flying 500 feet over the hilly area in an attempt to locate FFI ground markers, enemy gunners tracked the Mosquito. Podwojski circled and descended to 50 feet and beat-up the valley to avoid further gunfire. Later, both men received the Coer de Guerre medal presented by a French general.

War weary B-24 44-40233 "Little Lulu" before transfer to Watton. *Photo A. Podwojski*

Refueling 652nd Squadron B-24 44-0882. *Photo T. Tate*

652nd BS B-24 "Bobbet." *Photo A. Podwojski*

On 8 September another mystery Mosquito accident occurred. Previously on the 1st, F/O Russell C. Whitmer/Lt. Robert Balser (NS538) aborted a Mickey mission to Munich when the H2X equipment malfunctioned. The radar underwent bench testing and repair, then reinstalled in the Mosquito. Whitmer performed a flight with radar technician M/Sgt. Raymond C. Armstrong to test the Mickey equipment. At 2145 Watton received notification a Mosquito crashed in an open field one mile from Kennet. The crash completely demolished the aircraft with wreckage covering an area of thirty acres. The area experienced lighting flashes and severe thunderstorm clouds. Either a suspected lightning strike or radar equipment spark ignited fumes from a leaking fuel tank, causing an explosion.

A newly arrived OSS detachment at Watton obtained two surplus ex-Mickey aircraft for a secret project. RCA engineer DeWitt Goddard on assignment with OSS had air scoops installed behind the rear operator's compartment along with exhaust vents fitted in the bomb bay area to dissipate fumes. Watton had all Mickey Mosquitos similarly modified.

Mickey Mosquito NS538, with chocks in place, revs engines. It crashed 8 September 1944 killing both 1/Lt. Russell C. Whitmer and M/Sgt. Raymond C. Armstrong performing a test flight. *Photo A. Lewis*

In Holland, British forces crossed the Albert Canal on 9 September while U.S. troops crossed the Moselle River. In the air, the BDs concentrated bombing attacks against railway targets and oil refineries. Early during the day, Lts. John P. Noble/Charles O. Gorham (MM391) executed a landing at Royal Naval Air Station, Ronaldsway, Isle of Man. When the starboard undercarriage began to collapse, Noble immediately opened throttles and took off. The navigator climbed in the nose to check the undercarriage visually and reported the starboard oleo leg swinging like a pendulum indicating broken hydraulic lines. Noble flew the Mosquito to Watton and executed a belly landing.

Two Mosquitos scouted a 3 BD Frantic mission to USSR 11 September. The crews consisted of Lt. Gene Goodbread/Capt. Richard Wright (NS591) and Lts. Carrol Stubblefield/Louis Pessirilo (NS514).

Le Havre, France, fell as Luxembourg achieved liberation on the 12th. Next day, 13 September, the Allies entered Germany as Lt. Robert Tunnell/Capt. John Blake filmed the final Batty mission. This was Blake's first ETO sortie, having flown 32 missions with Fifth Air Force, SWPA. The Mosquito camera ship accompanied a B-17 escorted by 16 Mustangs. The GB-4 targeted an oil refinery at Salzburg, Germany. The pilot incorrectly released the weapon when ground haze obscured the area making television identification extremely difficult. However, the controller guided the GB-4 bomb toward a large group of buildings barely visible on the screen and scored a direct hit to the dismay of Blake:

> "We followed our instructions to a 'T' and hovered over the target waiting for the strike. The Mother plane that we followed hit a target of opportunity about ten miles away. By the time we got to the new target, all I could see was smoke ... nothing that would make a good picture."

On the 14th, Lts. Ford Porter/James Davis (NS513) flew a night WR to Hanover and Hamburg areas but aborted mid route when the starboard engine began to cut out and the airspeed indicator failed. They returned to Watton when the starboard engine burst into flames and failed to feather. In an effort to land immediately, Porter came in downwind without an airspeed indicator. The navigator hit the fire extinguisher button as the wheels touched the runway, but the Mosquito now traveling too fast, overshot the end and went through a barrier, washing out the aircraft.

A Scout sortie supported a 3 BD Frantic mission to supply arms and ammunition to Polish fighters in Warsaw. On 15 September Lts. George Hand/Schaefer Horne (NS626) left Watton at 1048 and radioed clear conditions at Warsaw, then proceeded to Poltava, Ukraine. After landing they learned the bombers received a recall to England due to heavy cloud conditions over the continent.[54]

MM391 accident on 9 September 1944 with Lts. Noble and Gorham.
Photo F. Klesch

54 Roger Freeman, The Mighty Eighth, London: Arms and Armour Press, 1986: 176.

Lineup of three Joker B-26s with foreground "W" 334193.
Photo R. Walker

Third Joker B-26 aircraft in lineup 334226. *Photo R. Walker*

Watton received a request to perform low-level Bluestocking missions. In preparation, Capt. Robert Lee/Lt. Wendell Biggers (NS635) performed a low level, cross-country training flight the 16th. They flew a triangle run, north then west towards Wales and east to Watton. While cruising along at 500 feet, Biggers donned his oxygen mask.

"What are you doing?" asked Lee.

Biggers replied, "If we're going to stay up here, I'll need my oxygen mask."

Taking this hint, Lee descended to ground level and began buzzing the countryside, leaping over each hedgerow encountered and down into the next field, continuously across England. Surprises do occasionally occur. They dropped into one field, startling a bevy of feeding birds that immediately took flight en masse. Too late. The Mosquito sliced through the flock. When they landed at Watton and inspected the aircraft, feathers and bones filled the air scoops and five bird strikes holed the plywood airframe. The crew chief repaired the damage but changed the aircraft name to *Patches*.

The crew of 1/Lt. Gene Goodbread/Capt. Richard Wright (NS591) returned to England 17 September from the Third Frantic mission. They previously left Watton the 11th along with Lts. Carrol B. Stubblefield and Louis Pessirilo as his navigator in Mosquito (NS514). They scouted weather and target information for a strike force from 3 BD attacking Chemnitz, Germany. The Mosquitos overtook the bomber stream arriving ahead of the formations. Clear weather covered the target but smudge pots emitting smoke partially obscured the ground area. They observed 100 Fw 190s southwest of Chemnitz flying towards the bomber steam. Stubblefield/Pessirilo encountered four Luftwaffe fighters at 30,000 feet while over Brux. The enemy attacked from three o'clock. Stubblefield took evasive action by turning into the attackers and applied power to outrun the fighters.

Goodbread orbited until the strike force reached the IP, then he proceeded to USSR. At the Dnieper River, the Mosquito turned south towards Kiev and descended to 3,000 feet. On both sides of the Dnieper, from north of Kiev and far south as viewed, shell craters littered the area to the horizon. Following a railroad track in a southeast direction took Goodbread to an airfield at Mirgorod. Now low on fuel, he landed on a grass strip with PSP planking. The Soviets refueled the Mosquito for takeoff to Poltava. The Poltava military airfield underwent repairs with many Soviets working on the field. The control tower lacked radio contact with the workforce and instructed Goodbread to buzz the runway. On first pass at 100 feet, only a few laborers cleared the strip. But on a second pass at about five feet, they all vacated the area immediately.

After landing, Brig. General Kessler met the Mosquito, loaded the crew's baggage into his Jeep, and drove the men to the Officers' Mess. Stubblefield and Pessirilo flew an alternate route and arrived earlier. After half an hour, a Russian delegation arrived. The Soviets were very polite and requested a tour of the Mosquito that Goodbread provided.

On 13 September, the bomber task force resumed the shuttle mission, proceeding to Italy. En route, the bombers attacked steel factories at Diosgyor, Hungary. The Mosquitos scouted the target and proceeded to Lucera, Italy. After a brief reprise, the task force departed 15 September for England. Scouting ahead, thick clouds and deteriorating weather increased on approaching France. Radio failure forced Goodbread to abort and immediately return to Lucera. After radio repair, he went airborne to establish VHF contact with the shuttle bombers and gave the recall message. He repeated the radio recall at frequent intervals with only one faint acknowledgment. Upon landing back at Lucera, he learned the task force received the communication for return to Italy. Lts. Stubblefield/Pessirilo experienced Mosquito mechanical problems, delaying their departure.

NS635 "Patches" with Wendel Biggers on left, mechanic, and Bob Lee.
Photo W. Biggers

Flanked by two crewchiefs is Capt. Richard Wright on left with 1/Lt Gene Goodbread prior to Frantic mission to USSR. *Photo R. Wright*

On Sunday morning, 17 September, Goodbread left for England and scouted ahead of the shuttle fleet. Approaching the English Channel, C-47s and fighters flying an easterly course filled the sky.

Goodbread witnessed the Allied air armada involved in Operation Market-Garden. The brainchild of British Field Marshal Bernard Montgomery, this strategy aimed at taking advantage of the chaotic German retreat by concentrating an Allied ground attack in a narrow northern corridor to capture the Ruhr area, and race east to Berlin. His plan: Drop "a carpet of airborne forces" behind German lines along a 60-mile corridor from the Belgium border at the Meuse Canal to Arnhem, Holland, on the far side of the Rhine River. Capturing Arnhem would allow the Allies to skirt the Siegfried line and gain rapid momentum in the level ground of northern Germany to Berlin.

The airborne operation known as "Market" targeted the vital bridges over the Maas, Waal, lower Rhine Rivers at Grave, Nijmegan and Arnhem. U.S. and British paratroopers would seize and hold these strategic bridges to allow British ground forces to drive towards the Zeider Zee and allow the 2nd British Army to eventually sweep into Germany. The land operation is called "Garden."

Also, Germany now launched V-2 rockets against London, Brussels, Antwerp, and Liege. Receiving the label Big Ben, the 200-mile range rocket was fired from the Holland/German border areas. The Market-Garden drive would block movement of further missiles to this region.

The success of Market-Garden required good weather both at airfields in England having gliders and transport aircraft, and the target areas in Holland. The 25 BG (Rcn) received orders to perform low-level weather sweeps over the battle area of Holland. The sorties proved so numerous, 654 Squadron also participated due to shortage of 653 BS aircraft and personnel. Since 654th had more experienced night pilots they performed both day and night flights. This required flying on the deck to determine the low-cloud base. During daylight hours, Watton dispatched 10 sorties at one-hour intervals. On one mission Lts. Victor Doroski/Harry Anglum (NS593) flew low over the Dutch countryside. Skimming at treetop level over roofs and streets in the southern part of Rotterdam, they became the target of streams of tracer shells.

On 17 September, "The greatest troop carrier fleet in history took off, consisting of 1,544 aircraft and 478 gliders, from England ... Only 335 British transports with 323 gliders, 1,044 USAAF transports with 108 gliders reached their objectives."[55] That morning Lts. Richard E. Petersen/R.C.E. Smith (NS620) flew to Bovingdon where Smith exchanged his place with Maj. Gen. F.L. Anderson. Petersen took off at noon with the general as passenger and two P-51 fighter escorts for the purpose of observing the airborne invasion of Holland. For one and one-half hours the Mosquito ranged above from 4,000 to 10,000 feet while paratroopers and gliders landed below. This Mosquito mission was labeled Pee Dee, a contraction of parachute dropping.

The British lst Airborne Division parachuted west of Arnhem tasked with capturing and holding both the highway and rail bridges in Arnhem, six miles from the drop zone. They encountered heavy resistance from SS Panzer Divisions refitting for reassignment, who engaged and hindered the British advance toward Arnhem. The Germans blew the railway bridge as British troops approached. It then took until dusk for the British battalion to reach the northern approach to the road bridge, but stiff German resistance drove back all attempts to cross.

Early on 18 September, Lts. Clayborne Vinyard/Clayton Hackman (MM384) performed a test flight from Watton to determine the performance of a landing gear indicator warning light reported unserviceable. Upon touchdown the warning light registered the undercarriage down and locked. With slowing speed, the left wheel retracted causing a ground loop that snapped the rear fuselage in half.

Late afternoon of the 18th at 1445 hours, 1/Lt. Robert A Tunnell and cameraman S/Sgt. John G. Cunney (NS593) set out from Watton in support of Operation Market-Garden. This was Cunney's first mission in a Mosquito to film the parachuting of airborne troops over Holland. When the crew became overdue, Group HQ sent a cablegram HQ Air Defense of Great Britain to obtain a plot on the aircraft. Because of the low-altitude mission requirements, no plots were recorded. Watton listed both men MIA. It would take 40 years before their disappearance finally became resolved.

S/Sgt John G. Cunney, 8 CCU was KIA 18 September 1944 with Lt. Robert A. Tunnell.
Photo J. Capicotto

55 Wesley F. Craven and James L. Cate, The Army Air Force In World War II, Vol. 3, Chicago: University of Chicago Press, 1951: 589.

The Search: On 5 May 1945, a private from Eureka, California, on furlough in Canada performed a goodwill tour of veteran hospitals located between Toronto and Ottawa, Ontario. During a visit he claimed to have spoken with disfigured 1/Lt. Robert A. Tunnell hospitalized with both legs missing and arms paralyzed. They talked for a few minutes but the private was unable to learn how or when Tunnell reached the hospital. Tunnell previously enlisted in the RCAF as Robert Ainsley Tunnell, R1451219, his adopted name. During childhood his parents died, and – Harold A. Skinner – took the name of a married sister who raised him. The private reported Tunnell now used the alias Harold Skinner. Apparently Tunnell, listed previously as MIA, wanted everyone to believe him dead. He asked the private not to divulge to anyone that he spoke with him.

The private informed Tunnell's sister, Zelma Somma, in California. She then sent a letter to George A. Cunney, brother of missing photographer, John Cunney, detailing the above disclosure. George Cunney contacted the Personnel Services Division in Washington, DC, with this news. A government official spoke by long-distance telephone with the private who answered all questions verifying Mrs. Somma statements. The official initiated action to have the private interviewed by authorities at nearby McClelland Field and obtain their opinion.

Meanwhile, George Cunney flew to Toronto and contacted the Canadian Veterans Affairs (VA). Any USA citizen who enlisted in the RCAF is considered a Canadian veteran no matter which air force he later transferred to and served. Official policy prevented disclosing information if the veteran in question wished to remain anonymous.

Cunney traveled to Ottawa and visited both VA and RCAF representatives. All government files examined uncovered nothing on Harold A. Skinner. Canadian files listed Robert Tunnell as presumed dead by the U.S. War Department. Records are not infallible and errors do occur, but Canadian officials believed this highly unlikely. They urged the private be thoroughly questioned by American authorities to determine if his story is indeed true.

After hearing from George Cunney, both of Tunnell's sisters, Zelma Somma and Helen Churchman, drove to Toronto, Ontario, and started a thorough search of hospitals in eastern Canada. They traveled across Canada, from the Atlantic to the Pacific coast visiting every military hospital and VA institution in between, again with no success. Believing him still alive, the sisters offered a $1,000 reward for information on their brother. Back in California, after intensive questioning by military authorities, the private confessed his story was a hoax, and he never met Tunnell as claimed. This revelation added further anguish already suffered by both sisters and family members.

Aviation researchers in West Germany, reported a Luftwaffe file recorded on 18 September 1944, two Mosquito crashes at the Luftwaffe air field Plantlunne, Germany, near the Dutch border. The first occurred at 0230 hours when searchlights illuminated an RAF Mosquito. It blinded and disoriented the pilot, causing the aircraft to spin in and crash, killing the crew. Later that afternoon, a Mosquito aircraft approaching the Plantlunne area was shot down by accurate ground machine gun fire, killing both crew members. Luftwaffe personnel recovered and buried both bodies in nearby Lingen as unknown airmen.

Parachute supply section. L-R. Unknown technician, F/O Spears (KIFA), Lt. Peter Dustman, Capt. Victor Doroski (KIA), Lts. Robert Lee, and Charles Veldon. *Photo R. Walker*

After the war, a researcher in Holland correlated the Luftwaffe account with a MACR and determined the dead were indeed Tunnell and Cunney. German aviation researchers notified the U.S. government who exhumed the remains for transfer to the American Cemetery at Neuville en Condroz, Belgium, buried in unmarked numbered graves. In 1986 the 25 BG (Rcn) Association notified the Cunney family of the previous findings. In August 1990 this writer traced the Tunnell family members and provided information, leaving no doubt Robert Tunnell is KIA and now rests in Belgium.

A Frantic mission delivering weapons and supplies to resistance fighters in Warsaw left early morning of 18 September. Lts. Richard E. Petersen/R.C.E. Smith (NS630) scouted ahead of the bomber formation. They sent radio transmissions upon encountering a weather front near Warsaw and again in the target area, with both messages acknowledged by the bombers. They observed three unidentified fighters 17,000 feet over Warsaw. When the aircraft began climbing towards the Mosquito, Petersen applied speed to evade and continue to Poltava, Ukraine. At Warsaw, the bomber formations parachuted packages containing weapons, ammunition, fuel and medical supplies. While 130 reached the Poles, the remainder, nearly a thousand packages, landed in German-controlled areas.[56]

Supporting Market-Garden, the Allies dispatched a second airlift of troops and equipment to Holland. Despite rain and thick clouds that spread northward across Belgium towards the Allied area of operations, the weather cleared over the drop zone and 1,306 transports with 152 gliders reached the target areas.

The British at Arnhem experienced a series of radio communication failures with no means of contacting London, nor other nearby units. On 19 September, a third wave of transports left from England and experienced difficulty reaching the target area. Hampered by poor weather, many turned back. The Polish Brigade went in by gliders towed by RAF aircraft and met fierce German resistance.

By 20 September, enemy troops surrounded the 1st British Airborne troops defending the north end of the Arnhem road bridge against a succession of determined German tank and infantry attacks. The sustained onslaught forced British troops to abandon the bridge. Resupply attempts by air to other British positions failed when the material fell into enemy-controlled areas.

Inclement weather deferred further attempts to parachute in Polish troops. Again Watton dispatched low-level weather missions. Lts. Otto Kaellner/Edwin Cerruti (NS552) encountered heavy machine gun fire from an airfield four-miles east of Utrecht. Lts. Robert Walker/Roy Conyers (NS594) entered through France, flew north and crossed Holland and exited by the Frisian Islands, receiving small arms fire and light flak from nearly every village flown over.

Unrelenting weather the 21st prevented further supply aircraft leaving UK bases for Holland. Rain and thick fog hung over the Arnhem area. Two low-level weather missions flew to the Belgium/France areas adjacent to Holland.

While efforts continued in support of Market-Garden, 652 Squadron initiated the first Sharon flight to Azores on 21 September. Lt. John R. Satterfield and crew flew to the Azores in one aircraft, remained 24 hours to rest, then flew another aircraft back. The ship flown to the Azores remained for 48 hours to allow preparation for the return trip. This required a full complement of groundcrew stationed at Lagens to assist with this task. The mission to and from Azores involved flying 4,086 miles in over 30 hours duration. Both physically and mentally tiring, returning after the brief rest period increased hazards and danger to both crew and aircraft. Once sufficient crews became available, plans required they fly the same aircraft down and back, increasing sleep and rest time from 24 to 48 hours.

Groundcrew washing down Mosquito. *Photo H. Steinsson*

56 Norman Polmar and Thomas B. Allen, World War II—America at War 1941-45. New York: Random House, 1991: 101.

On left is 1/Lt. Richard Petersen, then navigator 1/Lt. R.C.E. Smith flew a Shuttle mission. *Photo W. Borges*

Lts. Peter Dustman/Charles Veldon (NS552) left at 1624 for Holland. On becoming airborne the right landing gear failed to retract. After numerous unsuccessful attempts to raise it, Dustman decided to land. After a 300-yard roll, the right undercarriage collapsed, snapping loose the wheel. The Mosquito skidded off the runway directly towards an English civil work party. The bouncing wheel knocked down a worker and the Mosquito then slid over the poor chap. A bit shaken from the ordeal he suffered only a bruised right arm.

Support for Market-Garden continued the 22nd, with Lts. Robert Walker/Roy Conyers (NS554) flying to Arnhem. SHAEF HQ did not know who controlled the road bridge and asked the men to visually inspect and determine its status. Walker descended towards the bridge from 500 feet then flew on the deck to obtain a closer view. He immediately observed German soldiers running to their guns. As the Mosquito flew over the bridge, they encountered tracer fire from both ends of the structure and groundfire continued to follow as they ducked down to tree level. Walker recalls:

> "The brief encounter lasted three to five seconds. The first hit was a direct strike by 20 mm or .50 cal. rounds that disintegrated the left droptank. I immediately punched off both droptanks to prevent any possible fire. The right engine was soon completely in flames. The prop was immediately feathered and fuel to that engine switched off … and the flames slowly flickered out. We then made for the coast flying on the deck but soon received antiaircraft fire from the coastal batteries. Things happened so fast, jettisoning the droptanks, choking the fire off in the right engine, feathering the prop, then quickly estimating how badly we were damaged. Now we were trying to get out of range of enemy guns while staying right at water level. I remember well the water spouts from rounds hitting the sea around us.
>
> When safely out of range from the shore batteries, I climbed into the weather, still flying on one engine. Altitude was required to establish radio contact for a Mayday and directions from the RAF to the nearest functioning field. All England was socked in. There was no way to reach Manston, the emergency RAF base on the Dover coast. There wasn't time for them to fire up FIDO, their fog dispersal illumination system. The nearest base having weather minimums was RAF Bournemouth on the south coast. In any event, I was vectored to Tangmere and landed safely."

Two crews from two different Frantic missions returned to Watton on 22 September. Lts. Richard. Petersen/R.C.E. Smith of the canceled Frantic mission departed Poltava the 19th and landed at Lucera, Italy. Maintenance required the Mosquito be flown to San Severo from where the crew departed the 22nd. The other crew of Lts. Carroll Stubblefield/Louis Pessirilo from the 11 September Frantic mission became delayed in Italy with mechanical troubles and returned with Petersen/Smith.

NS552 G-Golf comes to rest after sliding over an English workman when undercarriage failed 21 September 1944. *Photo G. Scruggs*

Continuous rain and low visibility over England contributed to the second fatal accident of the month. Late evening of 22 September, a Heavy Weather B-24J, 44-40879 (letter H) named *Rockaway Babe* crashed at Watton on return from an Atlantic WR mission. The Liberator approached the field covered by low-overcast clouds with rain. "The visibility was very poor," recalls engineer Arthur Reynolds. "I could hardly see the wing lights." The aircraft began to flare out 100-feet left and parallel to the main runway. The pilot Lt. Robert Menken, pulled the control column back but to no avail. Unable to regain sufficient speed, the aircraft settled to the field. Tower control officer Capt. Oliver Kimball vividly recalls:

> "The left wingtip touched ground and caused the Liberator to cartwheel twice, then slide towards the Griston side of the field where it collided and set on fire two parked aircraft, a B-24 and B-17. I immediately alerted the fire department."

Emergency crews sped to the scene and proceeded to extinguish a blazing B-24 parked on a hardstand, assuming it was the pranged aircraft. Meanwhile, SAD men rushed to save a nearby hangar from flames. By this time, the actual crash B-24 came to rest around the corner from an SAD hangar. Later realizing their mistake, fire department crews attacked the blaze exactly nine minutes after the initial alarm sounded. By then the inferno spread beyond control and impossible to rescue personnel trapped in the mangled wreckage. Casualties were: Sgts. Harry Barnett, Richard A. Knowles, and Cpl. Matthew Chirico; Met/Obs 2/Lt. John H. Macklin and M/Sgt. Richard W. Stoodley. This was a training flight for Stoodley, one of six enlisted weather forecasters who arrived at Watton on 5 September.

Plagued with H2X problems Lt.Col. Cleveland, Deputy Wing CO, summoned a meeting on 24 September with Watton officers. He recommended canceling the Mickey program and four Mosquitos currently undergoing conversion be returned to the RAF after equipment removal. Previously, two Mickey Mosquitos mysteriously disappeared as MIA and many flights aborted because of electrical problems. The successful missions produced barely acceptable scope photos. Wing HQ ordered longnose P-38s equipped with H2X as replacement. Until its arrival, all further Mickey missions were halted. When General Doolittle learned of this action, he issued countermanding orders continuing the radar mapping missions and aircraft modification schedules. The approaching winter meant inclement flying conditions urgently requiring radar maps. There was no choice – the Mickey program must continue.

On 25 September, 653 Squadron dispatched a night weather mission to Holland. Lts. Clayborne O. Vinyard and navigator John J. O'Mara (NS570) left at 0126 and became POWs. Vinyard recalls:

> "The takeoff occurred in fog so thick, we could only see 100 yards in front of us. The climb to 30,000 feet and throughout the flight was in constant cloud and darkness. We depended entirely on Loran for navigation. Apparently we flew too deep into Germany reaching the Frankfurt area before turning back out. On return, we descended to 18,000 feet. Unknown to us, a Luftwaffe night-fighter apparently was on our tail tracking with his radar. We continued flying along, when at approximately 0430, cannonfire raked the aircraft. The Mosquito shuddered as tracers ripped into the right engine setting it ablaze, and rounds penetrated the fuselage striking the armor plating behind our seats. I immediately took evasive action, banked the aircraft left and dove in a spiral dive to 12,000 feet hoping to extinguish the engine blaze by this maneuver. The German fighter was not with me in the dive, though I was afraid he might be.
>
> "The engine fire would not extinguish; flames and smoke still poured from the nacelle. Having a wooden wing, it would not take long before this structure either weakened or caught on fire. I decided the moment was upon us to bailout. The navigator pulled up the inner door of the floor hatch, but the bottom one would not open. I got out of my seat and stomped my foot down on the exit door before it would release. I then moved back to my position and waited for O'Mara to squeeze through first. I followed immediately. When I stuck my feet out the hatch, the slipstream tore off my English flying boots. I fell in the blackness and slowly counted to fifteen. Free-fall is the quietest sensation you will ever experience. There is no sound – absolutely nothing! When I pulled the D-ring, the parachute deployed. Only then could I discern wind in the shroud lines. I passed through clouds and the mist dampened my clothing. It was so dark I could see virtually nothing. I descended from the cloud layer over a farming community. There were no lights visible anywhere with the blackout in effect. I was wearing a seat pack, had my leg and chest straps undone with my arms folded across, holding onto the harness. When I hit the ground and if the parachute was to drag me, I could easily release myself from it.
>
> "Instead, I came down in the tallest pine tree you can imagine. I grabbed and held on to the branches, released the parachute, grasped a hold on the trunk and started going down. The limbs covered only the upper tree and I slid down the trunk a long distance. I eventually touched ground in my stocking feet. The pine trees covered a hill surrounded by farmland. There was no underbrush to hide in with only pine needles littering the ground. It was dark of night and I had no idea where I was, no place to go, and without any bushes to hide in. I was free for four hours. Then several civilians arrived, having one individual armed with a rifle. They found me about the same time I saw them. They took me to a little town having about 20 houses. I stayed there until a military guard arrived, then transferred to another town where I joined O'Mara. We were placed in a cell until the next morning, when we boarded a train for Frankfurt."

The Dalag Luft report on the capture of Vinyard and O'Mara states that, "One unknown dead was buried at the crash site." Apparently, the Mosquito crash killed an unknown individual on the ground at Gerardshefan, Neustadt/Aisch, near Erlangen, Germany.

On the night of 24 September, after a long, hard-fought battle the British 1st Airborne received orders to evacuate and retreat from Arnhem. And on the 26th, the survivors reach Allied lines.[57] By the 27th, the struggle for Arnhem ended. Of some 10,000 Allied soldiers, 1,200 died and 6,600 taken prisoner during the nine-day battle. Arnhem would not fall until April 1945, a month before the war ended.

The night of 29 September, B-26G 43-34205 returned from a Joker mission. Pilot Major Clarence E. Short noted the landing gear failed to lock down. The crew operated the manual emergency system to lower and lock the undercarriage and flaps. The Marauder touched down but no braking action occurred, and the emergency air brakes failed to respond. The Marauder went off the runway end and struck a ditch, causing a nose wheel collapse. Crew: Capt. Douglas B. Harris, 1/Lt. Donald C. Winters and T/Sgt. Wayne J. Hall.

---

57 David Eisenhower, Eisenhower At War 1943-1945, New York: Random House, 1986: 477.

Unusual NS569 653 Squadron has "N" in white circle on a crimson tail.
It received damage on 12 August Aphrodite Mission. *Photo. L. Mann*

B-26 43-34205 with nose gear collapse, 29 September 1944. *Photo R. Walker*

It soon became evident employing B-25 or B-26 aircraft for night photography was impractical. The long distance to the battlefront and limited radius of action from Watton restricted their operation. HQ implemented plans to secure Douglas A-26 aircraft to continue this work. Roosevelt returned the B-25 to Mount Farm for 7 PRG use as a utility transport.

On the 29th, film star Marlene Dietrich and her troupe visited Watton. All personnel not required for operations received permission to attend the show. Even other nearby bases received invitations. Thousands watched Dietrich and her entertainers perform during the afternoon in a huge hangar. Everyone enjoyed the one hour and 30 minute performance. And finally on this day, the engineering section was relieved of responsibility for 10 F-8s.

As Canadian troops liberated Calais, the end of September saw the final chapter of Scout missions. Eighth Air Force experienced the breakup of Fighter Command with transfer of Fighter Wings to each Bomb Division. The Bomb Division (BD) now became known as Air Division (AD). After experimentation the AD ceased employing Mosquito scouting aircraft. Each division now dispatched escorted P-51s to perform weather scouting for their bomber forces.

A change of commands took place during the month. On September 25, Lt.Col. Joseph A. Stenglein became deputy commander, 325 Rcn Wing. Lt.Col. Leon W. Gray assumed position as commanding officer, 25 BG (Rcn).

September marked the end of the F-8 program. The following recapitulates these events.

The Bell Aircraft Company, Niagara Falls, New York, received a contract in January 1944 to perform F-8 fuselage modifications for camera installations. By 29 March, 38 F-8s had arrived with two promised by 31 March. Completed aircraft were flown to Hunter Field, Georgia, for camera installation, and equipment tests either performed locally or at Robbins Field, Georgia. The F-8s then went to Romulus AAB, Michigan, to await UK delivery.

The northern Atlantic ferry route involved staging through Rome AB, New York; Grenier Field, New Hampshire; Bangor or Presque Isle, Maine; Goose Bay, Labrador, Canada; thence to Greenland, Iceland, and the UK. Between October and May, twin-engine aircraft were not allowed to fly this route because of icing and foul weather conditions. Instead they flew the south Atlantic ferry route to Morrison Field, West Palm Beach or Miami, Florida; thence to Puerto Rico; Waller Field, Trinidad; Georgetown, British Guiana; Belem and Natal, Brazil; Ascension Island to Dakar, French West Africa.[58]

In 1943, four F-8s were dispatched to North Africa. One Mosquito left Florida 17 October on the south Atlantic ferry route and arrived in Dakar, Africa. Another returned with engine trouble while the third crashed at Ascension Island.[59] One F-8 flown via the North Atlantic ferry route in September 1943, safely reached North Africa. During 1944, 3 Ferry Group flew one F-8 over the southern ferry route via Brazil on 2 July. Nine others delivered over the North Atlantic left for England on July 4, 5, 13, 14, 15, 24, 25, 30 and August 4, 1944.

The Canadian government previously stated the DHC F-8s allocated to AAF were lend-lease in "reverse" with no contract drawn between Canada and USA. They now requested reimbursement for the 40 F-8s and parts. The Roosevelt administration pondered this question and finally advised AAF to process a contract in the regular manner for F-8 aircraft payment.

On 8 March 1944, HQ USAAF cabled the AM requesting 50 Mosquito Mk XVI for weather reconnaissance in the Far East. The British refused to divert this model at the expense of operational RAF squadrons and suggested AAF modify the 40 DHC F-8s for this role. AAF replied the F-8s were unsuitable because of limited engine capability.

In June, General Spaatz indicated the P-61 proved unsatisfactory as a night-fighter against most German bombers due to speed limitations. Spaatz wanted to substitute F-8 Mosquitos in night-fighter units but extensive modification ruled out conversion to this role. Instead, AAF negotiated with the British to secure 36 night-fighter Mosquitos.

Meanwhile, AAF continued the F-8 experimental test programs. A DHC Mosquito arrived at Aberdeen Proving Ground, Maryland, on 7 June for two days, then to Wright Field for installation of Norden M-9 bombsight, standard radio altimeter, and engine exhaust smoke-ejecting equipment. Ferried to Muroc AAB, California, the F-8 participated in a high-altitude bombing program.

Flying accidents further complicated F-8 operations:

On 16 February 1944, F-8 43-34933 participated in a flight evaluation program at Eglin Field, Florida. While performing his first transition flight, Lt. Lovick P. Thomas with passenger Lt. Clarence E. Hertlein, ground looped on landing, collapsing the left undercarriage.

On 11 April 1944, Lts. Charles J. Puzzo and John J. Garrett performed a local test hop in F-8 43-34953 at Hunter Field, Georgia, when the aircraft suffered a ground loop on landing, washing out both undercarriages.

F-8 43-34930 ground loop at AAB Fort Dix, New Jersey, by Major Frederic A. Borsodi on 14 March 1944. *Photo NARA*

58 GP-FE-3-HI. History 3 Ferrying Group, USAFHRA

59 File 307.01. History of Caribbean Wing, ATC, July 1943-June 1944., USAFHRA.

The remains of F-8 43-34946 on 21 April 1944 following a ground loop at Robbins Field, Georgia. *Photo E.V. Cruise*

Side-rear view of F-8 43-34961 circa 3 May 1944. *Photo W. Bodie*

On 21 April 1944, Lt. Howard D. Thoenig with passenger Lt. Ellis V. Cruse flew F-8 43-34946 from Hunter Field to Robins Field, Georgia. Shortly after touchdown the aircraft ground looped, wiping out both landing gears.

On 14 March 1944, Fort Dix AAB, New Jersey, pilot Major Frederic A. Borsodi made ready for Wright Field in F-8 43-34930. He started both engines at 1640 and taxied to the runway. Both Merlin's were runup and found satisfactory prior to brake release for takeoff. At 90 mph the right engine backfired twice, then quit completely. He retarded both throttles and applied hard left brake; however the airplane started to swing and veered 45 degrees to right of the runway. The left wheel collapsed and left wing dug in as the Mosquito ground looped.[60]

On 3 May 1944, F-8 43-34961 arrived at Detachment 37, Miami Army Airfield, home of a small but special HQ ATC Flight Test Unit performing cruise control flights to create charts for long-distance ferry flights. After completing the test program, both F-8 landing gears malfunctioned. Jack pads were constructed and bolted to the aircraft and then mated with jack stands. During the initial retraction/extension gear cycle, the jolt of the undercarriage hitting the uplocks caused the airframe to slip from the jack pads. Both jack stands impaled the wings with large holes. On 15 June, during an attempt to hoist and transfer the F-8 to a flatbed truck using a truck-mounted crane, the cable snapped. The aircraft crashed to the concrete floor, extensively cracking the plywood fuselage and wings.

On 15 May 1944, F-8 43-34943 piloted by 2/Lt. Harry A. Peattie left Romulus AAB, Michigan. The aircraft ground looped on takeoff damaging both landing gear, left wingtip and propeller. An inspection revealed the right brake completely worn through causing the mechanical failure.

On 21 May 1944, F-8 43-34955 was at Romulus AAB. After 16 hours and 55 minutes on the aircraft it became necessary to change the left engine because of coolant loss. Mechanics properly installed the coolant drain plug, but the adapter to which the drain plug fit became loose and fell out. There was no provision to safety the adapter to prevent this.

On 16 June 1944, F-8 43-34932 piloted by Capt. Joseph E. Taylor with Lt. John C. Piebes left Hunter Field, Georgia, for a routine engineering flight test. During a feathering check, the left engine surged to a high rpm forcing its shut down. During single-engine approach the landing gear failed to lock down. The pilot applied throttle to circle around when the aircraft moved towards a left-hand spin. He throttled back, raised the unlocked undercarriage and performed a belly landing. Flames from the burning engine consumed the aircraft.

On 14 July 1944, F-8 43-34943 piloted by l/Lt. Frederick Schrom left Romulus AAB, Michigan. One engine quit on takeoff and the other coughed intermittently. He safely landed on the Willow Run-Detroit Express Way with no further damage to the aircraft. Inspection revealed the accessory section of one engine blown off completely and the intake air ducts on both engines broken off.

On 15 July 1944, F-8 43-34939 flown by l/Lt. Donald H. Kline departed Rome AAB, New York. The left engine quit momentarily on takeoff and the pilot lost control. A crash resulted, destroying the aircraft and causing serious injuries to both crewmembers.

Meanwhile, F-8 42-34928 arrived at Wright Field on 17 July 1944 for tests involving installation of AN/APS-13 equipment.

On 21 July 1944, F-8 43-34957 piloted by 1/Lts. William H. Munn and navigator Charles Drimal took off from Goose Bay, Labrador, when the right engine quit, requiring return to the field. The left undercarriage failed to extend completely and the resulting landing damaged the propeller, engine and destroyed the right undercarriage.

On 21 July 1944, F-8 43-34948 flown by 1/Lt. John A. Joice was halfway out from BW1, Greenland, when he encountered excessive coolant temperature on one engine. He feathered the motor and returned to BW1. After landing mechanics discovered a hole punched in the coolant expansion tank released all pressure, causing the excessive temperature. A mechanic at a previous base lacked proper British tools to unscrew the coolant filler cap. He used a hammer and screwdriver that slipped and punctured the tank.

On 7 August 1944, F-8 43-34933 1/Lt. William F. Cass with navigator 1/Lt. Orlando F. DalVecchio were to ferry the aircraft from Rome, New York, to Bangor, Maine, the second of two F-8s that left the ramp for takeoff. After 12 minutes on the ground, Lt. Jack Williams piloting F-8 43-43936 took off first and circled the field. On takeoff, Lt. Cass' aircraft trailed a stream of white vapor from the right nacelle. He climbed to 500 feet when the starboard Merlin burst into a ball of fire and became engulfed in flames. The F-8 lost altitude and crashed in a field, fatally injuring both men.

Circa 12 August 1944, AAF abandoned F-8 43-34955 after a crash landing in North Africa. An RAF unit retrieved and repaired the aircraft, then transferred it to RAF strength.

On 15 August 1944, F-8 43-34937 piloted by 1/Lt. Perry Collins received extensive damage at Grenier Field, New Hampshire, when it veered on takeoff and the right landing gear collapsed resulting in a ground loop. The base commander claimed Collins had knowledge of the fatal Rome, New York, accident, and fearful of flying the ferry mission, willfully crashed the aircraft. An accident board convened with charges pending against him. Ground crew sent to remove the aircraft noticed pieces of the left brake had broken loose and later found on the runway. Further investigation of the damaged undercarriage revealed rust marks at the point of break indicating it had been cracked for some time. Humbly, the base commander exonerated Collins of all charges.

On 15 August 1944, 1/Lt. Charles B. Conglon taxied F-8 43-34936 at Prestwick, Scotland, to position for an engine check when total brake failure occurred. He attempted using differential throttle to turn the aircraft momentum away from an iron baggage cart and portable loading ramp a short distance away. The decreased turning radius forced cutting the switches and moving the mixture control to idle cutoff position, but the slow turning propeller blades completed a final revolution before just touching the obstacle.

On 17 August 1944, 3 Ferrying Group, ATC refused to ferry any additional F-8s because of safety hazards: 1. Unsatisfactory engine operation and power plant failure on takeoff. 2. Coolant failures. 3. Unsafe loading conditions, claiming, "The aircraft suffers from extreme tail heaviness in a fully loaded condition and does not contain sufficient nose heavy trim characteristics."

A survey of F-8 accidents revealed general problems: pilot reports' indicate hydraulic, fuel, and brake difficulties; servicing and maintenance personnel were unfamiliar with the airplane and unable to perform a proper inspection. Because of these unsatisfactory conditions, the aircraft was considered unsafe to fly. RAF reports note the DHC Mosquito had higher accident rates from mechanical problems than UK-built examples. This suggests DHC experienced quality control problems with aircraft production.

At this time, the F-8 complement at 3 Ferry Group consisted of 25 Mosquitos**:** 14 aircraft delivered; two aircraft at BW-1 (one for maintenance, one awaiting parts); one aircraft damaged at Grenier Field, New Hampshire; four aircraft at Rome, New York; one aircraft at Romulus, Michigan, and three damaged beyond repair.

By 17 August, HQ Eighth Air Force experienced difficulty giving away the F-8s. In view of new projects, they considered converting them to dispensing chaff ahead of the lead bomber formations, but lack of high altitude engines negated this suggestion.

On 10 September, Spaatz cabled Arnold that no further requirement exists in the ETO for F-8 aircraft and requested their transfer to the AM. Increased RAF operations placed several light-bomber squadrons under strength and F-8s, after modifications, would serve as replacements. Due to this urgency, all F-8s in the UK were loaned to the RAF until receiving authorization for transfer under the Lend-Lease Act. The remaining F-8s in the USA were included, subject to RAF taking delivery in the USA.

60 Major Frederick A. Borsodi was killed 31 January 1945 demonstrating a YP-80 in England.

The program received a final blow 21 September when all remaining projects on the aircraft were canceled, bringing to a close the F-8 debacle. But in early October, two F-8s arrived at NACA, Langley Field, Virginia, without cameras. Mosquito 43-334960 flew directional stability and control experiments, while 43-34928 stood by as a parts source. The program confirmed the tail-heavy characteristic. All F-8s were reassigned to the RAF on 27 October 1944 in an "as is, where is" condition.

October 1944

October ushered in a new phase of PR. This involved low-level aerial photographic survey of all marshaling yards in Belgium, France, and Holland, to estimate work required to repair the devastated rail transportation systems. Beginning 2 October, the number of daytime PRU flights increased dramatically with Brussels, Louvain, and Charleroi the main targets.

During an evening Bluestocking WR on 3 October, Lts. James Spear/Richard Taylor (NS519) encountered an unidentified aircraft near the Holland coast. It followed the Mosquito for ten minutes out over the North Sea but did not attack. Later that night, Lts. George M. Brooks/ John W. Rhoades (NS554) flew a Joker mission to photograph railroads and road junctions at Utrecht. Three PRUs were flown on the 5th, and four on the 6th, covering the railroad system previously mentioned. The flights to Luxembourg, Thionville, and Chambly Lille, when flown at 4,000 feet, produced excellent photo coverage.

F-8 43-34960 at Langley Field, Virginia, for NACA flight tests in October and November 1944. *Photo NASA*

Most frequently published photo of USAAF Mosquito, NS591 with flaps down. *Photo L. Proulx*

NS519 participated in the first Frantic shuttle mission to USSR. It was destroyed in a takeoff accident 27 December 1944. *Photo L. Proulx*

Rear view of "F" 654 Squadron PRU Mosquito with single camera port in aft area. *Photo Ken Kenyon*

Eighth Air Force initiated experiments with Loran navigation equipment carried by the Mosquito. The tests determined range and accuracy with potential application to pinpoint aircraft for night photography. The mission profile received the label Skywave. Watton previously evaluated Loran in August during the second Frantic shuttle mission. The first scheduled Skywave left on 6 October, when Major John Larkin/Lt. William G. Miskho (NS582) flew to San Severo, Italy. They performed further Loran testing over the Balkans east to the Black Sea, before returning to Watton on the 8th.

That morning, Lt. Robert Grimes (MM345) performed a test flight for the engineering department that required lowering and retracting the undercarriage numerous times. After retracting the gear, excessive pressure developed in the hydraulic system causing the engine-driven hydraulic pump shaft to break. He used the emergency hand pump system to lower the landing gear and approach without flaps. Upon touchdown, the brakes failed and the speeding aircraft left the runway end causing the tail wheel to shear off.

In early September 1944, P-38-J-10L0 42-67903 arrived at Watton, the first of several for transition flying in preparation for P-38 "droop-snoot" H2X-equipped missions. Newly arrived 2/Lt. Stanley V. Hazlett flew the P-38 at every opportunity to increase his flying proficiency.

Early morning of 11 October, Lt. Richard Geary visited the flightline and was introduced to Lt. Morton Hunt, a newly assigned pilot. Hunt originally trained as air Intelligence officer, interpreting aerial photos, then transferred to pilot training and flew the P-38. He asked Geary for a check flight ride in a Mosquito. New pilots required a specific number of hours dual flying before performing a solo flight. Geary agreed and took off with Hunt in NS553. Previously, Geary flew a dual-control Mosquito Mk III, with Capt. Rodney Peeke as observer in dogfights against several visiting P-51 Mustangs and embarrassed their pilots. The Mk III lacked armor plating and guns, stripped down to bare essentials, making it light and extremely maneuverable.

Geary spotted the P-38 and asked Hunt if he thought Hazlett would engage in a dogfight. He wanted to demonstrate the Mosquito handling qualities in mock combat. Hunt replied, "I'm sure he will." Geary waggled his wings, the P-38 responded in similar fashion, and the dogfight began. Geary flew in front of the P-38 providing him an advantage to jump on his tail. Then he performed a tight turn around Hazellet like the 'turtle and the hare.' They started out around 4,000 feet then went to treetop level performing all kinds of high-speed maneuvers. The aircraft then chased each other up to 4,000 feet in tightening turns. This model P-38 lacked boosted aileron controls, and required considerable strength by the pilot to effectively use ailerons in a dogfight. Geary recalls observing the P-38 come out of a turn and level off at 1,800 feet and make several slow gentle nose dips, then suddenly plunge straight towards the ground. He said to Hunt: "What the hell is he trying to do?"

2/Lt. Stanley V. Hazlett who was killed in the crash of a P-38 on 11 October 1944. *Photo P. Dustman*

11 October 1944 P-38 42-67903 crash scene with smoldering remains. *Photo W. Borges*

A shocked Hunt recalls:

"My recollection is that while we were tailing him in a tight turn, suddenly and without any warning, the P-38 dove towards the ground. I don't recall my words or Geary's, but the import of it was, 'What the hell is he doing?'

"We watched in astonishment. The whole event took very little time; diving at 400 or more mph, he would cover the slightly more than half-mile in a matter of seconds. It all seemed to happen so quickly, that before I could imagine what Hazlett was doing, there suddenly appeared this blooming ball of flame. Overwhelmed by the moment, I experienced a sickening feeling in my stomach from knowledge he became utterly obliterated, destroyed in an instant without a trace of him left."

The P-38 crashed five miles east of Watton near several homes. A board of inquiry speculated two possible causes. First, that Hazlett blacked out from the extreme G-force of the tight circle and lost control. Second, the aircraft received a proper overhaul, but the glycol-coolant fairing on the twin-booms had an elevator cable mechanism that did not receive inspection. The cable could have snapped causing the P-38 to enter the dive and crash. The board recorded the accident as due to unknown causes.

During the month, harsh weather over England and the continent curtailed flying operations to a minimum. On days when Eighth Air Force grounded its fleet, Watton always dispatched aircraft on weather missions over the Atlantic and western Europe. In many cases Eighth Air Force operation reports were completely negative except for eight or ten planes dispatched from Watton. Bad weather greatly hindered PRU flights and Pinetree expressed deep concern with the unavailability

Joseph Capicotto, 8 CCU cameraman standing next to NS570. *Photo J. Capicotto*

of photos. Headquarters expended all efforts to secure coverage of priority targets, but the unrelenting conditions made this practically impossible.

Lt. James Matuska/Sgt. Joseph Capicotto (NS559) and Lt. Otto Kaellner/Sgt. Walter Lipinski (NS619) were alerted 15 October for an Aphrodite mission with Heligoland the target. According to Capicotto:

> "We took off at 0635 in fog and went to Z field for briefing. I was flying with Lt. Matuska. We were briefed at 0900 with the fighter escort provided to protect us. The colonel in charge ordered us to photograph Baby going into the target, the strike and explosion. Well, we took off at 1029 and caught the Baby and Mother plane just as they were leaving the coast of England at Great Yarmouth. The slow pace of the heavily laden bombers required we zigzag back-and-forth between one Baby task force and the other because of our speed. The ceiling was 1,500 feet and we flew from 800 to 1,200 feet over the North Sea. We were constantly on the alert for enemy fighters. The Baby went out of control a few times and we photographed its crazy maneuvers."

Lipinski flew with Lt. Kaellner and took off two minutes after Capicotto:

> "We had trouble finding our Baby plane but located it after flying all over the sky for 20 minutes. We lost it in a rainstorm but 15 minutes later found it again. Flak hit the Baby which exploded midair at 500 feet, one and one-half miles from the target.[61] Then we tried to get pictures of the plane that Joe had been following."

Capicotto continues:

> "I photographed the Baby going in and it exploded left of the target. We followed behind, and from 800 feet Matuska went down low over the submarine pens for me to film. He then gave the Mosquito full speed, hauled the yoke back and made a right bank. He continued climbing straight up towards the clouds with German antiaircraft fire exploding a split-second behind us where the tail of our Mosquito had been. The steep climb rammed me back against the bulkhead near the instrument panel. We got the hell out of the area and emerged from the overcast into brilliant sun. Lipinski's Mosquito followed right behind.
>
> "On return, Matuska flew at 9,000 feet along the Holland coast. Near the Frisian Islands we glanced towards our left and observed three Me 109s a half-mile away flying in the same direction. Matuska looked at me and said, 'We'll look at that.' The 109s continued cruising alongside – they never changed course, nor did we. Meanwhile, I obtained a Gee heading for England. After several minutes Matuska decide to stop 'fooling around' and advanced the throttles. The Mosquito pulled away, then banked right and turned for England. The 109s made no effort to pursue."

Capt. Robert Lee/Lt. Wendell Biggers (NS632) flew a Skywave mission to San Severo, Italy, on the 20th. The men spent several days testing Loran over Yugoslavia, Bulgaria, and Greece, with excellent results.

U.S. ground forces captured Aachen, Germany, on 21 October. The following day, Lts. Malcolm MacLeod/Edward Fitzgerald (NS514), who previously diverted to Tangmere, began a return flight to Watton. The aircraft bore down the runway when the right engine sputtered and lost power. The tower observed blue smoke emit from the engine with backfiring. Both throttles closed and brakes applied failed to slow the aircraft. MacLeod retracted the landing gear as the Mosquito overshot the end and plowed into a four-foot earth embankment and snapped off the rear fuselage.

A night-fighter attacked a Joker Mosquito on the 24th. Lts. Richard Geary/William Miskho (NS554) were at 25,000 feet over Cologne when Miskho spotted the intruder. It kept pace to their left but did not attack. Some 20 minutes later, an aircraft trailing a long flame rapidly approached. Geary turned sharply towards it and the intruder passed overhead and disappeared. Two other aircraft were seen below at 22,000 feet but did not follow.

Tagmere, 22 October 1944 is the accident remains of NS514.
Photo A. Paquet

61 8 CCU history, Maxwell AFB: Sgt. Lipinski states at debriefing the Baby he followed crashed into the sea, whereas the official Aphrodite project summary reports it was shot down by flak.

Tripod was not just any old dog. For most of the 25th's stay at Watton he was known as the freeloader of East Anglia, but more especially of that area between the flight line and the Consolidated mess hall. As the morning shift of flight line personnel moved from the barracks toward the hangars and shops, Tripod would join in. Hobbling pathetically and with purpose he would pause every few yards and sit looking up at the faces going his way. Without fail some pitying soul would reach down and carry him along. When chow time came the dog would repeat his performance but with a change in direction of 180 degrees. This time his trip ended at the mess hall where it assumed he received good sustenance. You see Tripod had his right foreleg amputated near the shoulder that resulted in a canine handicap of great proportions. He was a personality with three good legs and a winning way. *Photo L. McKinely*

On 25 October, Lts. George M. Brooks and navigator Richard C. Taylor (NS582) left Watton at 0215 on a Joker Mission to Duisburg, Germany. The plane swerved left on takeoff, but Brooks applied sufficient braking to straighten the aircraft and took off climbing into the night. Taylor provided Brooks the first heading as they crossed the English coast outbound above Claxton at 15,000 feet.

During the channel crossing Brooks reported something wrong with the superchargers. The altimeter indicated 18,000 feet and Taylor informed him they cut in at 21,000 feet. Brooks continued climbing and at 21,000 feet the superchargers began operating. As they crossed Ostend, Belgium, at 22,000 feet Brooks experienced difficulty leveling the aircraft properly and the horizontal indicator registered the right wing low. Brooks thought a landing gear was extended and intended swinging around over the channel to raise the wheel. Taylor gave Brooks a 309-degree reciprocal heading.

Then Taylor felt the aircraft mushing (stalling) followed immediately by Brooks shouting, "Get set to bailout." Taylor knocked his table down, reached up and pulled the release handle on the canopy hatch and it instantly flew off. The aircraft then seemed to lunge forward in a steep spin. Taylor reached down for his chestpack but centrifugal force pinned him to the floor. He managed to connect one side of the pack to his harness, turned from his position on the floor, and attempted to reach the canopy hatch. He turned and observed Brooks exit through the upper hatch in a sitting position facing the nose. Taylor squeezed through the hatch and found himself momentarily facing the tail. The next thing he recalls is falling through the air. During free-fall to earth, he successfully connected the other hitch to the parachute pack. After pulling the D-ring, he blacked out momentarily then regained consciousness. During descent, he spotted the burning Mosquito wreckage in the darkness below

After touchdown in a farmer's field, he sat on the parachute for several minutes, then walked to a nearby road to a farmhouse two miles away. The occupants invited him in and immediately contacted the police at Dixmude. Sometime later the police arrived by car and informed him they found Brooks' body on the pavement approximately a mile away. His parachute was unopened. After bailing out, Brooks struck either the vertical or horizontal stabilizer and either killed instantly or knocked unconscious.

The police transported Taylor to Ypres where he reported the accident to a Canadian security unit and requested Watton be notified. The Canadians provided a Jeep and returned him to the crash scene. The aircraft was totally destroyed and all photo bombs had exploded. An RAF ambulance and fire unit in Dixmude recovered Brooks' body. Taylor retrieved the personal items and dog tags; one remained on the body and the other with the Canadian security detachment at Ypres. The Canadians drove him to Brussels where he contacted an American unit at Melsbroek and told his story. They provided transportation to Honington by B-17, and then by truck to Watton.

1/Lt. Richard C. Taylor navigator who survived the 25 October 1944 mission that killed George M. Brooks. *Photo R.C. Taylor*

1/Lt. George M. Brooks, KIA on 25 October 1944 accident.
Photo L. Bodnar

654 BS Photo Section, 4 October 1944. Top: Clark, Graver, Dushman, Reynolds, Daley, Roberts. Center: Tom Fields, Brown, Al Lewis, Muench, Gundlach. Bottom: Briscoe, Jay, Harold, Wilcox.
Photo T. Fields

Watton launched five Joker missions that night: Lts. Victor Doroski/Harry Anglum (NS554), Peter Dustman/Charles Veldon (NS559), Major Clarence Short/Capt. Jacob Hochman (NS583) and Lts. Robert Walker/Roy Conyers (NS581) to photograph Cologne and its roadways; and Lts. Norm Magee/John Rhoades (NS569) to Bonn. En route, Walker/Conyers experienced Gee failure and during the bombing run, the third PFB hung up requiring salvoing the remaining load, after which an electrical fire filled the cockpit with smoke. Prepared to bailout, the crew returned and affected an emergency landing at Manston.

In the latter part of October, civilian experts from Telecommunication Research Establishment at Great Malvern brought news that a Ground Control Approach (GCA) AN/MPN-1 would arrive at Watton in December. Three units were earmarked for ETO, two for the continent, and the third assigned to Watton would enable proper approach and landing in zero-zero weather.

After two scrubs, Aphrodite was scheduled 30 October. Sgts. Ralph Thiry and Larry Barbier were flown to RAF Fersfield. After briefing, Capt. James Matuska/Sgt. Ralph Thiry (NS560) left at 1114. After route setup, control and bailout were completed safely, the Mother guided the lead Baby towards Heligoland. Ten minutes prior to arrival, two squadrons from 486 BG hit the target with fragmentation bombs. Heavy clouds with rain showers limited TV visibility and obscured the submarine pens. The Mother lost dive control, preventing efforts to plunge the robot into the target. The Baby flew past Heligoland and a decision then made to abandon it. A 135-degree course was established, and the robot was last seen heading for central Germany with sufficient fuel to reach an area between Berlin and Leipzig.

1/Lt Ben Shaver in front of Joker Mosquito "Hey-Penney." *Photo B. Shaver*

8 CCU cameraman Larry Barbier in front of Joker night photo Mosquito. *Photo L. Barbier*

Poor target weather force canceling Task Force Two. Meanwhile, Task Force One returned to collect the second Baby. En route to Heligoland, flight control was temporary lost, but Baby continued on course for 15 minutes before the Mother reestablished contact. Now off course, the Mother maneuvered Baby in a turn, heading 347 degrees back towards Heligoland. Approaching the target, it failed to respond to dive control. After repeated unsuccessful attempts, it was decided to abandon Baby skimming across the North Sea, ten feet above the waves. The task force returned to England, but the Mosquito failed to receive the recall order. Matuska/Thiry continued trailing Baby another hour on a northerly 348-degree heading when low fuel forced their return to base.

November 1944

Although the U.S. Navy employed the Loran navigation system in the Pacific theater, it was virtually unknown in the ETO. The 25 BG previously conducted calibration trials evaluating its feasibility on test flights over Europe during Frantic missions. A Loran Training Unit arrived at Watton in November to familiarize navigators with this navigational aid. Unfortunately, monotonous manual operation failed to provide any practical application of how Loran functioned in flight. Until a better instructional apparatus could be devised, training continued with the assigned equipment.

Nose art of "Hey-Penney." *Photo B. Shaver*

Camera repairmen of the 654th BS exchange and remount cameras for the next mission. L to R. Charles Austin, Herbert Caudle, Paul Sachs, and Milton Greenberg. *Photo. T. Fields*

B-26 Joker crew Front: Sgt. Adam J. Wunderlick, top turret; Lt. Charles A, Veldon, navigator; Rear: Temple B. Lowe, tail gunner; Lt. Peter D. Dustman, pilot; Sgt. Angelo Spirito, Engineer. *Photo R. Walker*

Night photo B-26 43-34226 "T" with both engines running. *Photo T. Tate*

On 1 November, Lts. Vance J. Chipman and William G. Cannon (NS677) left Watton at 1930 on a Mickey mission to Koblenz, Schweinfurt, and Darmstadt. The crew failed to return and listed MIA. The following are the crews recollections. According to Cannon:

> "We flew a heading southeast to Schweinfurt simulating a bombing run. I started the cameras 50 miles away to obtain scope photos of the image generated by radar. After completing the run directly over the city, I gave Chipman a northwest heading for home. We were at 20,000 feet with a speed of 200 mph. He banked and turned, then leveled out, when the aircraft received a direct hit (from flak) over Schweinfurt. The Mosquito shuddered violently. Shrapnel penetrated the fuselage piercing my lower right leg and left large toe; fragments also struck the back of my head. All communication with Chipman was gone. At the time, I thought we were hit by a night-fighter.
>
> "Immediately, I released the escape hatchway door located on my lower right, to visually inspect for any structural damage and determine if the engines were impaired. I glanced out the opening, and the right engine was streaming a brilliant yellow sheath of flame and smoke. I quickly moved to the hatch and placed my wounded right leg out first, then followed with the left. I made every effort to push myself through. The chestpack parachute hung up in the small opening leaving me hanging partially out in the air stream, banging against the fuselage. It required a considerable period of time struggling at this effort to release myself. I grabbed the side of the fuselage and slowly pulled myself out until finally breaking loose. During this time the Mosquito kept flying fairly level.
>
> "Once free, I never saw the Mosquito again. I counted ten seconds ... one-one thousand, two-one thousand, etc. ... before pulling the D-ring. The parachute harness was secured very tight, and when the canopy deployed, it jarred me quite hard. Then everything was very still and silent. I didn't know if I was coming down or not. There is no sensation in the darkness; it's as if you are standing still in the open air. Billowy clouds lay far below with clear sky and twinkling stars above. It took forever to descend. I never lost consciousness from lack of oxygen and remained alert during the entire descent. I entered clouds at around 4,000 feet and emerged over a large river. Schweinfurt was entirely blacked out. Below in the darkness, the faint outline of the city gradually loomed larger and larger.
>
> "It was so dark, I couldn't see the ground. My intentions were to land first on the left foot and avoid touching the bleeding right leg. But suddenly I hit the ground awfully hard, and the crushing blow broke my injured leg in three places. I couldn't stand up or remove my harness; the excruciating pain paralyzed me. I closed my eyes and lay there in agony."

Cannon landed near railroad tracks close to an antiaircraft battery. Soon the gun crews found him, removed the blood soaked pant leg and applied a tourniquet. They carried him on a mattress to the battery station. A doctor arrived, bandaged the leg, and next day they placed him on board a train for Stalag Luft 13C in Ebelbach. Here Allied doctors rendered medical attention. The Germans furnished medical supplies and equipment while a New Zealand army doctor set the injured leg in a plaster cast to heal. Much later, rapidly advancing Soviet troops forced camp evacuation in April 1945 with wounded POWs loaded into boxcars for Nuremberg. They remained here until Patton's Third Army tanks tore down the barbed-wire fences of the POW compound.

Chipman's Account: After completing the photo run, Cannon said, "OK, got all pictures, will give course for home." Almost immediately a sudden explosion rocked the aircraft, and he heard Cannon scream on the interphone.

Both men POW on 1 November. L-R. navigator 1/Lt. William Cannon, pilot 1/Lt. Vance Chipman, with crew chief T/Sgt Jordan in front of Mosquito "Chip's Chariot" at Alconbury. *Photo T. Fields*

A long brilliant yellow flame trailed the right engine with the wing tank ablaze. Chipman shouted over the intercom for Cannon to bailout but received no reply. Shrapnel penetrated the oxygen mask, lacerated his tongue and knocked out a tooth. But he continued flying straight and level to allow Cannon sufficient time to bailout.

Smoke quickly filled the cockpit forcing Chipman to seek his own safety. Previously, he had the inner floor hatch cover permanently removed for missions. Reaching down, he unlatched the outer floor cover, then squeezed himself through the opening feet first. He fell clear as the tail section broke off causing the Mosquito to tumble.

During free-fall the harness leg straps felt loose. As a precaution he maneuvered himself in a head-down position before pulling the D-ring. The parachute deployed safely and during descent he watched as the Mosquito exploded on the ground below. The dark countryside approached rapidly as he touched down heavily in an open field. He proceeded to shred the silk canopy and bury it along with the Mae West. A quick physical survey revealed a torn left leg ligament and scraped flesh from under both arms. Slowly he began limping away from the scene.

At sunrise he sought refuge in a drainage ditch and slept until late evening. Under cover of darkness he stole a bicycle from a nearby village and peddled away. Two German soldiers on duty near a near a small town ordered him to halt and took him prisoner. Escorted to an aerodrome, Luftwaffe personnel locked him in a secured room. Next morning the guards allowed him freedom of a hallway with a large window revealing several blue-nose Me 109s parked in a blast bay. Temptation proved too great and he began formulating plans to escape in a 109.

At 0600, sentries escorted him to the guardhouse for breakfast and left him alone. Finding the door unlocked he moved slowly towards the Me 109s. Ground personnel discovered and dragged him back to the guardhouse. They beat him with a rubber hose, then provided transport to the Schweinfurt railway station, and placed on board a train to Frankfurt-On-Main and the Dulag Luft camp. On arrival Luftwaffe staff removed all personal effects and placed him in solitary confinement for six days before interrogation began. A Luftwaffe *Hauptman* greeted him with an outstretched hand, but Chipman stood at attention, saluted, then gave name, rank and serial number. The officer requested information for identification purposes, but received only name, rank and serial number. They returned him to his cell.

Eight days later they brought him for further questioning. The interrogation officer removed a folder titled "25 Bomb Group (Rcn)." Observing for any reaction of surprise, he proceeded to tell Chipman his occupation, aircraft number, past operations, mentioning flying accidents and operational losses. The officer offered him a cigarette containing a drug to induce delirium, and engaged him in conversation. Unaware Chipman enlisted in the RCAF, the officer asked where he trained in the USA. Chipman responded he was a pilot prior to enlistment, then provided a highly fictitious account of secret plans to reach Germany by taxiing modified aircraft under the English Channel. Infuriated by his insubordination, guards hauled him back to solitary confinement, but he continued providing nonsensical answers when interrogated. Disgusted, the Luftwaffe transported him and other POWs to Frankfurt-on-Main to entrain for eventual arrival at Stalag Luft III, POW camp near Sagan.

This was the third Mickey Mosquito loss and no one knew why? As a result, Watton suspended all Mickey missions until further notice.

Lt.Col. Leon Gray and Lt. John F. Mink (NS632) flew a Route Survey mission to San Severo, Italy, on the 2nd and returned next day via Dijon, France, to Watton. Gray wanted to visit former colleagues in Italy and claimed the flight as a route test.

A Skywave mission became airborne Friday, 3 November to San Severo, Italy, with pilot Major Marshall Wayne and navigator Captain Richard Wright (NS620). Unfortunately, the Loran receiver became inoperative, but the pilot decided to proceed and have the equipment repaired in Italy. Then the intercom plug became unserviceable, reducing crew communication to written notes in available spaces in the navigation log.

The aircraft flew south over France, when at 25,000 feet between Paris and Geneva at 0545, they observed, "2 large Blue traces – balls of fire at 0130 highest. 30M." Some 20 minutes later they noted a second blue ball descend eastward, Wright's only sightings of Foo Fighters. Shortly thereafter they encountered turbulence upon entering a frontal zone that cleared over the Mediterranean. Arriving over Rome at 0751, Wright calculated a heading to San Severo with an ETA of 26 minutes. Unfortunately, a solid cloud layer completely obscured the Apennines west side because of orographic uplift ahead of the front.

Throughout the flight, the pilot questioned their current position by writing notes on Wright's log. He did not trust the navigator's calculations and went on radio to locate an open field. After interminable vectoring, including one field currently under water, they found themselves momentarily clear of the clouds west of Florence, but with no sign of any airfields. Fifteen minutes later they flew over solid clouds south of Florence between Leghorn and Arezzo.

Now desperately low on fuel, the navigator jumped at 0932. Bailout occurred 1:15 hours after they should have been over San Severo had the pilot maintained the original flight plan. Prior to bailout Wright stuffed both navigation log and uniform hat inside his flight jacket and zipped it up. The log served him well, because the mere knowledge he still retained it, saved him from potential charges for the loss of NS620.

Bailout occurred in heavy clouds with falling snow that muffled the receding roar of the two Merlins. When able to discern shapes on the ground below, he was above the treetops and discovered himself swinging rather violently. As he swung toward a slope, the impact knocked his breath out. He skidded down a hillside, till the canopy snagged in underbrush, and he came to rest in a mud puddle.

After recovering his breath, Wright stood up and released the harness. Two teenagers approached along a path on the opposite side of the ravine, that circled round to below where he landed. They led him to a dwelling where their parents provided food and clothing while his uniform hung to dry near a roaring fireplace. Using sign language, they indicated observing his descent and dispatched the youths to lead him to the home. The parents drew a map in the logbook indicating they lived south of Arezzo. When the rain ceased and with change to a dry uniform, the youth guided him down a mountain trail to a British 8th Army camp. The British provided escort to the crash site in a flooded field, five miles south of Aires. The aircraft dug a deep hole on impact, destroying the Loran equipment. Next day he visited the pilot suffering a broken leg in an Arezzo hospital. The British notified Watton by cable that the crash totally destroyed the aircraft and of the crew's disposition. They transported Wright by Jeep to Rome where he boarded an ATC flight to England.

Mortal remains of NS568 B-Baker on 6 November. Pilot 1/Lt. Otto Kaellner KIA, and navigator, 1/Lt. Edward Cerruti, injured. *Photo E. Cerruti*

A double tragedy occurred 6 November. 1/Lts. Otto E. Kaellner/ Edwin R. Cerruti (NS568) went airborne at 1737 on a night photo mission to Germany. Upon return from Cologne, the Mosquito diverted along with other aircraft to North Pickenham due to strong crosswinds at Watton. North Pickenham experienced light continuous rain with visibility of four miles and winds gusting to 26 mph. On receiving instructions at 2019, Kaellner brought the Mosquito in to land. The aircraft on touchdown bounced twice and made a follow through takeoff. The aircraft began its climb, but failed to gain altitude immediately, hit several trees and crashed one mile from the runway. Kaellner died impaled on the control column and Cerruti suffered serious back injuries.

That same day 1/Lts. Robert C. Grimes/Clarence W. Jodar (MM386) flew a weather flight to Germany. The crew also encountered adverse conditions on return and diverted to North Pickenham, again because of strong crosswinds at Watton. Flying on starboard engine only because of fuel shortage, the pilot prepared for an emergency landing when contact was lost. On descent, the Mosquito flew into a row of trees shearing off six feet of the left wing. This caused the aircraft to roll over and crash inverted in a field and explode upon impact. Both crewmembers died instantly. The tragedy occurred near the Wending schoolhouse at approximately 2110. Kaellner and Jodar flew nearly all previous missions together as a crew, but on this day both were killed flying with different airmen.

On 17 November, 652 Squadron B-24J 44-40484 left Lagens Airfield, Terceira, Azores, for a return Sharon mission to Watton. The crew: 1/Lt Benjamin Marshall, pilot; 2/Lt James M. Gisriel, copilot; 2/Lt. James H. Gulick, navigator; Sgts. Americo J. Mancuso and Jack Dooley Jr.; Sgts. John B. Smith and R. Maxwell Taylor; 1/Lt. Adolph F. Johnson, met/obs.; T/Sgt. Stanley Zagol, met/obs.

They took off at 1252 and set course on the Sharon route to England carrying 3,500 gallons of fuel. Two hours later they encountered deteriorating frontal weather with extreme turbulence and sent the first meteorologic report to Lagens by radio. Three hours after takeoff at position eight, a box climb was made to 24,000 feet for weather observation followed by a descent to 18,500 feet. The flight plan required flying this altitude for 250 miles northeast to position 13 for a box letdown to 10,000 feet. Concerned with the severe frontal weather, they dispatched a radio message to Watton, asking, "Shall I continue to UK." Watton replied, "Return to Sharon base. Bases unfit in UK." They acknowledged receipt and radioed intentions to complete the mission to the halfway point at position 13, then return to Lagens.

1/Lt Robert Grimes killed 1 November 1944 with navigator 1/Lt. Clarence W. Jodar. *Photo C. Hackman*

653 Squadron navigators, L-R. 1/Lts. Edward G. Fitzgerald (KIA), Wendell Biggers, Warren Barber, Clarence W. Jodar (KIA). *Photo W. Biggers*

At 1725, when over position 11, a Lagens message declared Sharon base unfit for return, and ordered the B-24 to land at Lyautey, French Morocco. The pilot altered course heading to 160 degrees. They encountered enemy jamming on all radio frequencies. Electrical storms in the frontal system distorted radio reception and hindered securing information. With only a heading and airspeed and without any external references, they continued instrument flying. After three hours they climbed to 18,300 but high clouds prevented obtaining a celestial fix.

At 2330, the pilot ordered IFF turned on and an SOS signal dispatched requesting assistance. At midnight, they descended through the undercast to 3,000 feet and viewed fleeting lights and coastline below. After flying DR for over eight hours, their position was questionable: were they over Spain, Portugal, or North Africa? While circling they received no response to SOS signals or red distress flares.

They continued a southerly course parallel to the shore at 3,000 feet observing the coastline. Soon a red glow appeared through clouds below. Turning towards this direction a red flashing beacon appeared, and at 0130 an airfield came into view. The Liberator descended to 1,500 feet, and after firing several red flares, the runway lights came on. The B-24 landed 0145 in light rain at Portela Airport, Lisbon, Portugal.

Prior to touchdown the crew destroyed the IFF. Later, the U.S. Air Attaché took possession of codes, classified information, charts, and Gee equipment. After debriefing, the Portuguese interned the crew in Lisbon. The Air Attaché urged their immediate release on the premise the B-24 took off from and landed on Portuguese territory while performing a meteorological mission supporting Lagens Airfield, rather than a combat mission. Repatriation occurred 13 January 1945 and the crew returned to operations.

Again two tragedies struck the same day. 1/Lts. Malcolm S. MacLeod and Edward G. Fitzgerald (NS630) departed 22 November at 0106 for a weather flight. On return they diverted to another aerodrome, Thompson Field, because of low ceiling and rain at Watton. The pilot informed flying control he could see the runway and would attempt to land. He began to circle the field but hit several trees and crashed at 0455, killing both men.

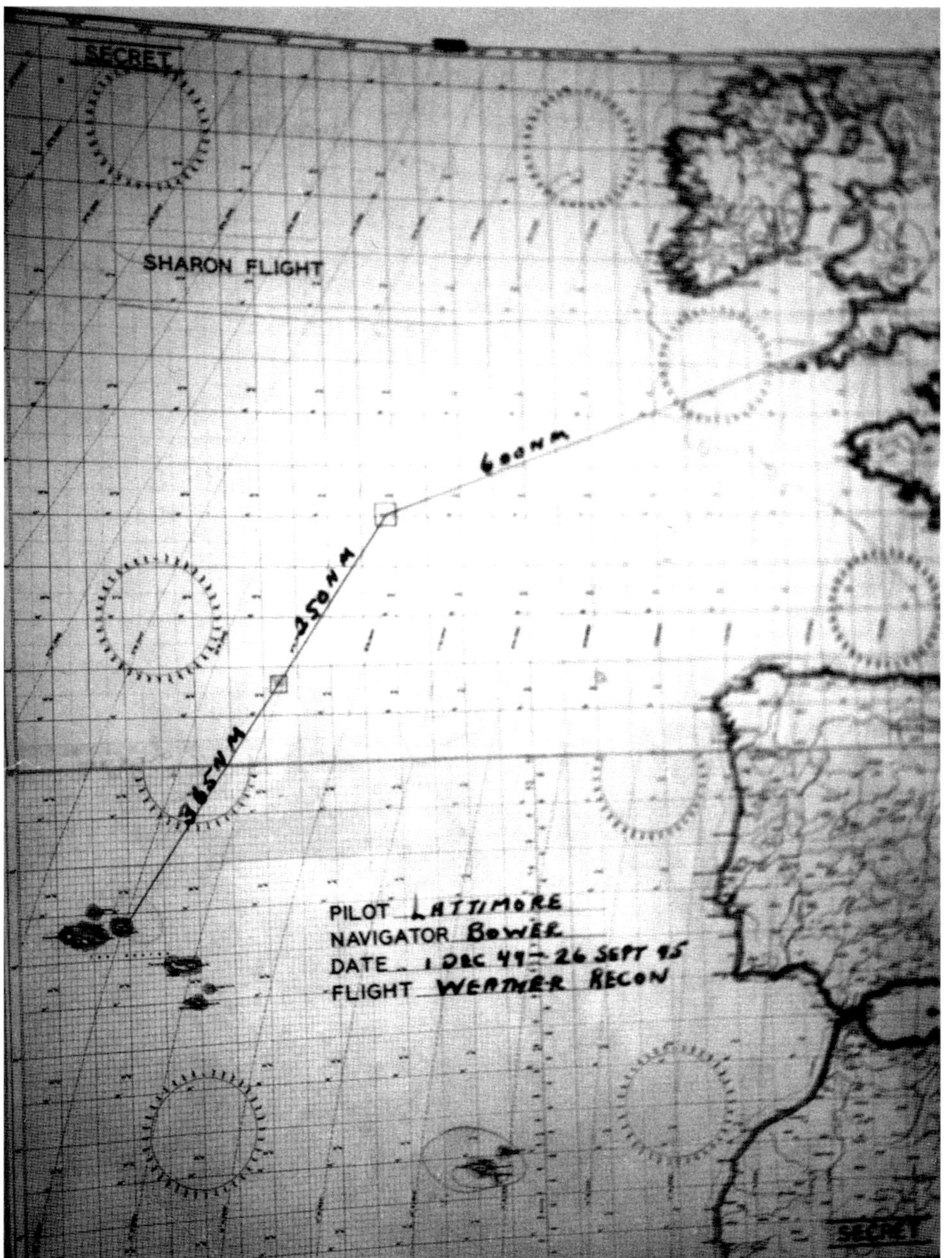

Map of Sharon route to and from Azores. *Photo. O. Bower*

1/Lt. Russell E. Harry and 2/Lt. Milford B. Hopkins (NS515) left at 0322 on a weather flight. Shortly after takeoff the aircraft crashed three miles from Watton killing both crew members. The Mosquito struck several trees and ploughed into a barn, killing a cow and scattering wreckage over a large area. Accident investigator determined the aircraft suffered an engine fire and lost power.

Watton sent Eighth Air Force a request for three additional Mk III Mosquitos for transition training of 45 new pilots expected in December. Instead, Watton received three Oxfords because the Mk IIIs were unavailable from RAF sources. New pilots required a minimum of 28 hours day and nighttime training before flying operations. Experiments with Mosquito tail warning radar devices were found impractical and disliked by aircrews. None were permanently installed.

On 25 November, 1/Lts. Robert Walker/Roy Conyers (NS653) left at 0153 hours on a WR to central and eastern Germany. While flying at 27,000 feet, the oxygen system malfunctioned forcing descent to 10,000 feet. Continuing on course, they checked the equipment and decided to climb once again, but on reaching altitude, Conyers passed out completely and Walker became groggy from oxygen deprivation. Walker descended again to 10,000 feet where Conyers regained consciousness. Instead of abandoning the flight, they switched over to the emergency oxygen system, ascended to 28,000 feet and obtained the necessary weather observations to complete the mission. On the return flight still deep in enemy territory, oxygen depletion forced them to descend once again to 10,000 feet. The crew flew 1,150 miles in extremely adverse weather conditions. Poor weather at Watton force diverting to another base.

1/Lts Milford Hopkins on left and Russel E. Harry on right, KIA 22 November in NS515. *Photo L. Proulx*

RAF camouflage Mk.III trainer LR553 with U.S. markings. *Photo G. Scruggs*

The first Redtail mission was flown on the 26th. This provided AD commanders an opportunity flying in Mosquito aircraft to survey bomber formations en route to the target. Arriving at the respective BG base to pickup a commander, the Mosquito waited until the entire bomber formation formed and flew over the English Channel before taking off. It then joined the formations en route to the target. The AD commander flew back and forth along the bomber stream examining the integrity of box formations. Lt. Gene Goodbread (NS591) flew Lt.Col. John M. Bennett, 100 BG, accompanied by a fighter escort. They observed three V-2 launches from the Rotterdam-Utrecht-Amsterdam area. One V-2 came up through the clouds a half-mile directly ahead. Next day, Goodbread flew another Redtail with Col. William K. Martin (NS591) from 487 BG and noted four rising V-2 contrails in the Saurbrucken area.

1/Lts. Wallace B. Rouse/Allen Morrow (NS596) flew a high altitude, cross-country training flight on 27 November. Extremely high clouds required the Mosquito climb for altitude. They flew according to plan at 30,000 feet, identified Land's End then set course on magnetic heading for Oxford. A doubtful Gee fix indicated they reached Oxford and Rouse began descent to break through the undercast over Watton.

At 1205 they let down through broken clouds over a wooded area believed southwest of Watton and altered course accordingly. The Gee screen displayed distortions obscuring station signals. This indicated they had flown out of effective Gee range where the Germans jammed the signals. Soon the forest area became hilly and a river ran in a northeasterly direction. They followed it seeking a landmark to pinpoint their exact location. Morrow had a gut feeling it wasn't England.

Side view of NS559 at A-83 Denian-Prouvey Aerodrome, France. Red serial number is barely visible due to film type used. *Photo R. Hiatt*

The clouds turned solid overcast again. Morrow attempted contacting ground control unit Backtit for homing directions. Backtit provided a bearing 334 degrees then faded out. They set course 154 degrees and climbed for altitude to achieve better radio communication. Then an unidentified station issued various conflicting vectors. Ripsaw ground control finally interceded issuing instructions to steer 240 degrees. With solid undercast at 7,000 feet the men feared bailing out over the North Sea. A cloud break appeared and Rouse descended beneath flying a 240-degree heading. Morrow recalls:

> "I don't remember how many hours we had been up, but fuel became critically low. Rouse spotted a farm field to make an emergency landing. He performed several passes to practice and then said, 'This is it.' Flaps were down and wheels up. He cut the engine switches, we touched down and started sliding along the ground. The aircraft ground looped and started breaking up; the tail section broke off and the engines tore from the wings.

Martin B-26 Marauder flown on night Joker photography, transferred to France in late-November 1944 for possible operations. *Photo AAF R. Walker.*

> "We finally came to a stop, unfastened our seat belts and stepped out through a large hole in the side fuselage. Unknown at the time, we landed five miles south of Pontoise, France, some 25 miles northeast of Paris. Several farmwomen came running across the field. One had a bottle of 'something' and another had a camera which we used to take pictures. Four French gendarmes arrived to take charge and notified local American MPs to take us to a nearby US base.
>
> "For years I was puzzled as to how we ended up over France when we never flew an easterly heading. It wasn't until 1951, when recalled to active duty as a B-36 navigator on high-altitude SAC bombers, that I discovered what had happened. During World War II, little if anything was known about the jet stream. Now we know the jet stream is a river of air that flows from west to east around the globe in the northern hemisphere. The airflow may reach speeds of 150 knots and a width of 150-200 miles. Crossing it at right angles can place an aircraft off course by 100 miles in a matter of minutes. Heading easterly, it can increase your ground speed substantially and vice versa. In 1944, when we flew over southern England at 30,000 feet, we were literally moved across the English Channel by the jet stream to northeast France near the German border."

L-R. 1/Lts. Wallace Rouse and Alan Morrow with French gendarmes 27 November 1944 in France. *Photo A. Morrow*

A 1944 photo of NS596 that crash landed in France 29 November 1944. *Photo USAF Museum*

Back at Watton, Lt. Simon G. Pilson ground looped MM368 during takeoff collapsing the right landing gear.

Near months end, several CCU officers attended conferences at Pinetree. The Air Inspector investigated both 3 and 8 CCU activities. He requested a CCU officer attend all future briefings at Pinetree to obtain firsthand information on operational activities to assist securing newsreel coverage. The War Room at Pinetree was the setting for the next meeting. Discussing future CCU operations. Cameramen will film only low-altitude missions supporting ground troops, or heavy bomber missions below 15,000 feet. Past filming above this height provided poor results. They planned to film future missions from Droop Snoot P-38s having a spacious nose compartment. Five aircraft were made available for Mount Farm for maintenance and closer fighter escort location. Pinetree would control all future CCU operations eliminating miscellaneous assignments originating in various UK headquarters.

On the 27th, CCU officers visited Fersfield and Bovingdon to inspect modified Droop Snoot P-38s and found the aircraft suitable and practical for obtaining good newsreel coverage. The following day, four cameramen received alert for an Aphrodite mission and reported to Knettishall, but the mission was scrubbed later that night. The month ended with two cameramen alerted for another Aphrodite mission that was again canceled.

As the Allied ground forces penetrated farther eastward on the continent, the night photo Joker B-26s were curtailed due to their limited range. Plans to transfer part of the squadron to an advance base in France from where missions could more efficiently continue were partially affected on 29 November. Six officers and 29 enlisted men found themselves in France with 27 Squadron, 7 PRG under 325 PW operational control providing photo support to Ninth AF. Personnel from engineering, armament, photography, and operations formed the advance contingent whose ranks were to be augmented in December. With establishment of this continental echelon, Colonel Roosevelt planned to supply Army Group Headquarters in France and Germany with immediate interpretations of night photo sorties.

December 1944

The month began with 652 Squadron transitioning from B-24s to new "stripped down" B-17s. The converted Forts cruised 30 mph faster, utilized less runway for takeoff or landing and handled more easily with less trouble to fly than the older B-24, or the non-converted B-17. They even had an improved drift meter installed for taking low-level readings at specified positions during flight.

The 652nd continued flying Atlantic WR flights under adverse conditions; the more atrocious the weather, the greater need for WR flights. The aircrews flew in conditions in which, "Even the birds were grounded." For three consecutive days, dense fog and prevailing bad weather pervaded the UK grounding the Eighth Air Force, except the Heavy Weather B-17s that continued flying even from diversionary bases. The crews of Lts. Steven Luse and Leon Meggenson alone flew six missions in nine days, completing 127 flight hours to maintain the flow of meteorological data.

Enemy Flak defenses inflicted major bomber losses and battle damage. As Allied armies advanced across France and Belgium, retreating German forces transferred all antiaircraft batteries to defend the Reich. This increased the concentration of AAA defenses at major targets and caused grave concern to future heavy-bomber operations. AAF studies concluded that radio countermeasures using Chaff thwarted antiaircraft defenses. The lead bomber elements dispensed Chaff, but remained unprotected from this electronic screen and vulnerable to radar-directed flak. Chaff dispensing aircraft preceding the bomber formation could further reduce losses.

Chaff consisted of large quantities of metalized paper strips cut to the radar wavelength, discharged at a predetermined rate. These created spurious images on enemy radarscopes to interfere with radar controlled AA gun accuracy and also obscured the true size of bomber formations. Loosely packed in cylindrical paper bundles, each contained 2,000 strips. When discharged through a chute installed in an aircraft, each bundle bursts upon impact with the slipstream and dispersed the strips. When properly dispensed, each bundle produces an image on enemy radarscopes equivalent to a single-heavy bomber.

27 November 1944 MM368 ground looped on takeoff by 1/Lt. Simon G. Pilson. *Photo L. Mann*

On 13 August 1944, 3 BD requested eight fighter-type Mark VI Mosquito aircraft for a radar countermeasure program. The Mosquito possessed both speed and performance to accomplish this role. Also, it was the only ETO aircraft capable of defending itself en route to and from a target without special fighter escort. Plans called for each BD to maintain the aircraft and utilize their own aircrews. Pinetree insisted unarmed Mosquitos were satisfactory and attempted to rid themselves of unwanted F-8s by endorsing their application to this role. When consulted, Col. Roosevelt advised the F-8 engines proved inadequate for high altitude and instead advocated the Mk.XVI.

Pinetree deliberated the issue and found the fighter model more desirable over the Mk.XVI. They discussed exchanging F-8s or Mk.XVIs for a limited number of Mosquito fighter models for prototyping. The DH company declared the fighter version could not serve this purpose and recommended installing high-altitude engines on the F-8s. The time factor, added cost, and unavailable Merlin 61 series engines from the P-51 and Mosquito Mk.XVI production commitments, ruled out this suggestion.

To resolve the issue, the Director of Operations at Pinetree decided to recruit Mark XVIs from Watton for the role. In October, Col. Roosevelt urged all aircraft be assigned to one unit for maintenance and operations, and not spread among ADs. Pinetree thus delegated 25 BG (Rcn) responsibility for providing aircraft on call and upon demand of the AD commander. Roosevelt assigned two of three flights comprising 653 Squadron, Chaff dispensing duties. The program required three Mosquitos screen each attacking force for a total of 21 aircraft – seven for each AD – with an attrition rate of 20 percent or four aircraft per month, or as needed. Hence, 21 additional Mosquitos were procured for the project. The 19 aircraft remaining from 102 allocated for 1944 would arrive by December. Pinetree predicated an attrition rate of 86 Mosquitos for the first half of 1945.

Watton sent Mosquito MM371 to Bovingdon for prototyping. This involved removal of two bomb bay fueltanks with replacement by two 100-gallon underwing tanks. Then two large plywood Chaff containers were installed in the bay area, side by side. Two automatic Chaff dispenser stripper heads mounted at the rear – fed from respective magazines – discharged Chaff through sheet metal chutes mounted in the bomb-bay doors. This required installing two control cables for on/off and dispensing rate control, next to the navigator. Between 1,400 and 1,800 bundles of CHA-28 were carried and discharged at a maximum rate of 200-units-per-minute operating both dispensers simultaneously. Total weight of containers and Chaff was 600 pounds.

After completing the prototype, 12 Mosquitos were dispatched to 2 SAD, Abbotts Ripton, for modification. General Doolittle ordered further work halted until the prototype had been successfully tested. A number of trials conducted in November demonstrated aircraft speed and maneuverability suitable for this particular operation. Initial mechanical problems were encountered with the dispensing equipment, but onsite modifications improved their function.

Under supervision of Col. Budd J. Peaslee, CO lst AD conducted experiments dispensing Chaff against enemy antiaircraft defenses. The mission name "Graypea" derived from fusing the names of the 25 BG (Rcn) CO Lt. Colonel *Gray* and Colonel *Pea*slee.

December 4th proved a bitterly cold day when Lt.Col. Leon Gray/Lt. Richard Davies (NS512) flew the first operational Graypea mission. Col. Bud Peaslee/Lt. James M. Davis (NS519) accompanied the flight along with a fighter escort. They rendezvoused with lst AD bombers over the continent and proceeded ahead, dropping Chaff above the Kassel railroad marshaling yards. Discharge began between 500 and 2,000 feet below the mean elevation of the lead group. Because Chaff falls at a rate of 400 feet per minute, effort was made to avoid discharge at the same level as the formation being protected. The Chaff confused enemy radar-directed antiaircraft defenses to fire at the falling metal strips. The success resulted in future bombing forays escorted by Graypea Mosquitos.

This new mission profile increased Mosquito requirements for the first six months of 1945: 654 Squadron – 21 aircraft with 12 for night photo and eight for H2X work, with attrition rate of 10 percent; 653 Squadron – 41 aircraft: 21 for Chaff operations and 20 for Light Weather recon with attrition rate set at 30 percent. Seven Chaff aircraft became available to each AD with all assigned to 653 Squadron for supply and maintenance, since the Chaff dispensing force was considered a provisional unit.

On the 4th, Capt. Gene Goodbread (NS591) flew to Bury St. Edmunds to pick up Col. Frank W. Castle, 3 AD for a Redtail mission. Next day, four CCU cameramen were alerted at 1800 to accompany a heavy bombardment mission to Berlin. The men received prior instruction not to fly unless the weather proved conducive to good photography. A check with operations at Knettishall revealed no conflict with anticipated Aphrodite operations. Sgts. Cecil Brown and Ralph Thiry went with the 96 BG, while Sgts. Joseph Capicotto and Walter Lipinski flew from Bassingbourn. At 0200, CCU received an alert for an Aphrodite mission. Sgts. Brown and Thiry were briefed for 5/10 to 7/10ths cloud on the Berlin mission, so both returned to Watton for the Aphrodite mission only to learn no Mosquitos were available. The briefing at Bassingbourn indicated 3/10 to 5/10ths clouds so both Capicotto and Lipinski flew the mission. Lipinski recalls:

> "Capicotto and I flew as right waist gunner on different bombers. We took off in darkness at 0707 and formed in heavy clouds. All the way to the target our formation flew through 10/10ths cloud extending to 22,000 feet with an occasional break. We ran into our first flak barrage while halfway over Germany, and when we reached the IP, all hell broke loose. The flak was like a carpet and accurate. Three bombers in front of us were knocked out. We were over Berlin when a direct burst hit the wing spar and our bomber fell 3,000 feet before the pilot regained control. He ordered the crew to standby for bailout, but unfortunately the navigator, bombardier and engineer had already jumped. The formation dropped their bombs from 23,000 feet, but heavy clouds prevented any filming. On return, we flew beneath the formation with two escort fighters providing protection all the way to England. When we reached friendly territory, I navigated the bomber back to base."

According to Capicotto:

> "I didn't have it as rough as Lipinski but it was rough enough for me. We received shrapnel hits numerous times but never lost altitude. Prior to, and right after we left the target, I observed our escort fighters battling it out with hordes of Fw 190s and Me 109s. Our fighters knocked down 80 Jerries and provided us with good protection. They remained with the stragglers all the way back to England. We ran into moderate but light flak from a barge about one mile from the coast over the Zeider Zee."

On 7 December, Lts. George C. Hand/Shaeffer E. Horne (MM345) flew a night WR. Throughout the flight, Hand experienced difficulty breathing sufficient oxygen and exhibited mild symptoms of anoxia. Successfully completing the mission he returned to Watton and made an approach at 0519. He set the Mosquito down when it swung off the runway, collapsing the right landing gear. Later that morning, Lt. John Carter ground looped NS590 after a routine training flight.

7 December 1944 landing accident of 654 BS NS590. *Photo W. Borges*

The 8 CCU finally completed *Bail Out and Live* at Knettishall on 10 December, after which the film was sent to Humphries Laboratory, London, for editing and distribution. Captain Emmet Bergholtz submitted a new script titled, *Mozzy Story* which Lt.Col. Gray approved. This production would record efforts of weather Scout Mosquitos in coordinating and effecting better results of bombing missions by Eighth Air Force.

Previously, three Mickey Mosquitos and aircrew were lost during operations, and no one knew why? The decision on whether to continue the flights created friction between administrative personnel. Lt.Col. John Larkin commanding 654th Squadron originally elected flying the H2X missions to Germany at night under the protective cover of darkness. Maj. Roy Ellis-Brown, Deputy Base Commander charged with carrying out Group operations, halted all further flights when the third Mickey Mosquito went MIA on 8 November. Larkin opposed canceling the missions resulting in a heated dispute with Ellis-Brown. The officers brought the disagreement before CO, Lt.Col. Gray, who supported Ellis-Brown's position until a solution could be found.

Luftwaffe aircraft possessed the capability to home in on H2X radar impulses. Thus, Ellis-Brown suspected the Mickey losses were from enemy night-fighters. He consulted 100 Group, RAF at Cheddington who conducted ECM warfare against German air defenses. Aircraft from this unit carried special electronic jamming equipment and screened RAF bomber streams, interfering with enemy radar defenses known as the Kammhuber Line. This defensive position created by *Generalmajor* Josef Kammhuber comprised a belt of searchlights, acoustic locators and radar sites across Germany that controlled night-fighters with or without radar themselves. After discussing the problem, the RAF officials agreed the losses were most probably from enemy night-fighters:

Other squadron members involved themselves with the Mickey dilemma. Claude Moore recalls his involvement:

> "Of the total number of crews employed in such work (Mickey missions), six or seven crews – two remained by the time we had completed a total of perhaps 40 sorties. There was some official concern over such a record and expressed itself in a discontinuance of Mickey reconnaissance flights.
>
> "Meanwhile, three officers who were assigned to fly such sorties, interested themselves to study this subject. Their aim: to hit upon means of making the work more effective and to give a greater margin of safety to the crews. Those three officers were: lst/Lt. Malcolm MacLeod, a pilot; Flight Officer James D. Spear, a pilot; and myself, a navigator. Lt. MacLeod was placed in charge of the squadron's Mickey flights, empowered to revamp the set-up and make recommendations. Shortly after assuming direction, Lt. MacLeod was killed. F/O Spear and I immediately took it upon ourselves to continue such studies as had been already begun.
>
> "F/O Spear had, up to that time, been equally active as Lt. MacLeod in studying the subject and originated much of the work already done. Herewith is a tabulation of some of F/O Spear's activities prior to and following the demise of Lt. MacLeod. 1. He contacted RAF officials to learn all he could about their employment of the Mosquito. In this endeavor he was aided by previous service with the RCAF, which acquainted him with British operating procedure and also gave him entry to certain RAF bases and officials. 2. He learned all he could regarding RAF night flying tactics. 3. He helped open a continuing flow to our base of RAF teletypes regarding their nightly operations. 4. He obtained from the RAF, night detection equipment for possible employment in our operations, equipment not then available to us through American sources. 5. He spent endless time in our Intelligence library, reading on the enemy's air force organization, its employment and tactics; enemy flak and its employment; obtaining an exchange of view with our Group Intelligence section. 6. He remained on the line for many hours with the crew chief and mechanics to learn from them troubles perhaps peculiar to an aircraft so equipped. 7. He spent days with the radar section to perfect his knowledge of radar equipment. 8. He consulted with the squadron engineering officer, to learn his ideas on improvement or modifications that might be effected, and to gain his cooperation in making other changes. 9. He personally testflew the plane, at all altitudes to determine how its performance characteristics might be affected by the H2X equipment peculiar to it. 10. He strapped sandbags in the place originally occupied by the navigator and then gave the plane more, equally exhaustive test flights to determine its fighter affiliation performance.

654 BS members. L-R. 1/Lts. Claude C. Moore (WIA), James D. Spear (KIFA), Warren Barber and Richard C. Taylor. *Photo R. Taylor*

Major Roy Ellis-Brown became involved in Mickey mission debate. *Photo L. Proulx*

"Above all, F/O Spear studied possibilities of increasing the value of such sorties and of improving the chance for the plane and its crew to return from them. After spending some time and considerable effort in such endeavors, and mindful of the need to get Mickey missions into operation, again, F/O Spear personally typed into memorandums the conclusion drawn from his efforts. Included, was a recommendation that consideration be given to daytime performance of such sorties. It was suggested that a few such daytime flights be flown and evaluated. F/O Spear and myself volunteered to fly all such sorties, until such time as their performance permitted conclusive evaluation.

"We made our suggestion to the following parties who were aware of the effort we were engaged in: our Squadron commanding officer, Lt. Col John Larkin; the Squadron operations officer, Major Clarence Short; the Group Intelligence officer, Major Finis D. McLanahan. The first two named officers had become unsympathetic to our endeavors, and decided the Mickey operations should be resumed without any appreciable change in method. Accordingly, they were indifferent to our suggestion and proposal. We then pleaded with the Group Operations officer, Major Ellis-Brown, and in this were supported by the Group Intelligence officer, Major McLanahan. Major Ellis-Brown agreed there might be merit in our proposal, and Group operations authorized and scheduled a sortie.

"On the morning of 12 December, we took off in NS686 on a mission to Augsburg and Lechfeld escorted by P-51s. Going in, we encountered moderate to heavy flak, extremely accurate for height exploding 50 feet off our wingtips. Coming out, we were tailed two miles astern by an enemy plane which the pilot, F/O Spear, thought looked like a jet. Again, we encountered accurate, moderate to heavy flak. The Loran equipment received enemy jamming on return. Halfway across the channel at 6,500 feet, our radio went out.

"Darkness had already begun to settle and weather turned bad. We encountered terrific turbulence, had no vertical visibility with horizontal visibility restricted to approximately 300 feet. We remained on the deck, flying from the south coast of England to East Anglia at 50 to 200 feet. At such low altitude, the H2X equipment did not function well enough to permit bringing the plane in, nor did the weather or our fuel supply permit us to gain the altitude needed for its operation. Using DR, I brought the plane within a mile or two of Watton, but radio trouble and general lack of navigational aids prevented knowing our exact position while in flight. And, of course, weather prevented our seeing the aerodrome near us.

"We had been airborne approximately six hours, maximum for a Mosquito such as ours, when just as our gas supply gave out, F/O Spear found an air base beneath us, and set the plane down at RAF Langham at 1730. Weather prohibited our taking off again. We called Watton and were ordered to remain where we landed. The magazine containing the film was removed and entrusted for safekeeping to the base Intelligence officer, who locked it in his office safe."

Next morning, Watton dispatched a special courier car and returned Moore with the film magazine. The negatives revealed 23 excellent photos of Munich but none on the 20-mile scan. The target was not approached directly from the west, making Wing acceptance doubtful. The mission did prove the feasibility of daytime Mickey missions escorted by fighters.

General Spaatz always expressed dissatisfaction with H2X. When not concealed by clouds, smoke screens effectively obscured the vital targets at Leuna and Politz for visual bombing. H2X bombing techniques had a poor record in terms of damage achieved due partly to difficulties distinguishing the target signal from the radar returns in that area. Spaatz pleaded for immediate delivery of high resolution APC-7 (Eagle) radar equipment to rectify this problem. Until then, H2X photomaps proved vital to support blind bombing procedures.[62]

While the Allies mounted a massive bombing attack on 12 December against German railway installations, at Watton the fighter escort situation improved considerably. Previously, PRU pilots consistently came under attack from German jet-propelled aircraft. Now all PRU missions received fighter escort on deep penetrations. Meanwhile, Capt. Emmett Bergholtz, 8 CCU received temporary assignment to Pinetree to help coordinate activities of both the 3 and 8 CCU with air operations.

Early on the 11th, Lts. Robert Lee/Wendell Biggers (NS632) flew a night WR to Belgium and western Germany. This was Lee's last mission, so he contrived plans to beat up Watton on this final flight. On return from Germany, he executed an emergency landing 0500 at Woodbridge, giving brake problems as an excuse. It was still quite dark, so the men enjoyed breakfast and lingered until dawn, then returned to buzz Watton. Biggers fondly recalls:

"Upon arrival, Lee made several low passes at the Officers' Club, where many air crew lived, then Group and Squadron HQ, and the dispersal area. Overcome with excitement, he decided to make a few low-level runs at the control tower. This was a little two-story building having a 2x4 lumber sticking up with antenna attached. On the first two passes, he aimed one wing at the antenna, then 'just' flicked the wing up and flew on by. But on the last pass he felt a bit devilish. He aimed the fuselage towards the antenna but misjudged his speed and failed to pull up quick enough. We hit the 2x4 which snapped off and bounced along the bottom fuselage.

"Startled over this turn of events, Lee immediately landed and taxied to the end of the runway where two MPs waited in a Jeep. He had been chastised several times previously for low buzzing and now was quite fearful of the consequences, saying, 'They've really got me now!' We taxied to the dispersal area, shutdown the engines, and sat there hoping the MPs would go away, but they didn't. Finally I said, 'Well, we might as well get out and see what they want.' I dropped the floor hatch down, pushed the ladder out and exited the aircraft. As soon as I climbed out, the MPs left. We then examined the aircraft for damage. The Mosquito hit the 2x4 board at the center of the fuselage, snapped off a large stub that bounced along making three or four holes on the bottom hull.

"A weapons-carrier arrived and drove us to debriefing, where we received notification of having completed the 2,000th mission. Surprise! The MPs waited for an aircraft to land as VIP escort to the parking apron, all part of the official ceremony. No mention was ever made of the buzzing or control tower collision. Lee was greatly relieved. Thus began the initial festivities for the 2,000th mission party."

General Doolittle ordered more attention given to combat flying photos and material of human interest for public consumption at home. Therefore, CCU was to cease producing training films and instead issued a list of topics pertaining to air force activities to document on film.

On 16 December, under General von Rundstedt, Germany launched the Ardennes Offensive in Belgium and initiated the Battle of the Bulge. At noon that day, the 2000th mission party festivities began at Watton. An excellent meal was prepared with enough for all guests invited. The officers and enlisted men gathered money for purchasing food and drink, decorations, and entertainment. Previously, 652 Squadron aircraft flew in fresh vegetables, fruit, liquor, and exotic victuals from the Azores. Other KPs and MPs brought in from surrounding bases released practically all Watton personnel from duties. The 7 PRG carried out PR operations where possible, allowing the maximum number of Watton personnel to participate in the festivities. Four dance bands provided music at the Aero Club, Officers' Club and Enlisted Mens' Mess, late into the morning hours.

For several days this enormous "drunk" progressed. One eyewitness described the enlisted mens' quarters appearing like "Custer's last stand" with bodies strewn in stairways, on corridor floors, in lounge chairs of inebriates unable to safely reach their beds. The celebration was heartily enjoyed by all.

One memorable recollection of the event involved 1/Lt. Gene Goodbread:

"The entire 25 BG (Rcn) was removed from operational status for three days to celebrate the completion of 2000 missions. Believe me, I participated. After two days of boozing and partying, I finally crashed into bed about 3 AM. Two hours later at 5 AM that morning, I was awakened and informed that I had been requested 'by name' to fly a Redtail mission with Colonel Castle. (On 4 December, Goodbread flew Colonel Castle on a similar mission.) After a cup of black coffee and two or three Benzedrine tablets, I still could not stand or walk. Members of the squadron operations section bodily carried me to the aircraft, placed me in, and fasten the seat belt.

62 General C. Spaatz to Chief of Air Staff, letter dated 15 December 1944. File 519.161-3. USAFHRA.

"My takeoff was fine but after reaching 1,000 feet altitude, I became very dizzy; the entire outside world was circling around the aircraft. I immediately began flying on instruments and using 100 percent oxygen. The weather was clear and I maintained flying at low altitude. Somehow, I located the air base and attempted a landing. Upon touchdown, the Mosquito bounced about fifty feet into the air, so I pulled up and went around for another try. The second attempt proved more successful. I taxied to where the colonel stood waiting and cut the engines.

"Colonel Castle realized that I was in bad shape and took me to his quarters. He made me drink another pot of black coffee and then asked how I felt. I replied 'OK,' and he said 'Let's go.' The Redtail mission allowed command colonels and generals to survey their bomber formations, and the bombs dropping over the target. It was very effective and probably led to changes that improved bombing efficiency. This mission was very successful in spite of the bad start, and when we returned to the colonel's base, he informed me that he was very impressed. If I would agree, he would request my transfer to his command and make me a squadron commander with promotion to major. I would not have to fly any more combat missions unless I chose.

"Psychologically, I was embarrassed. Physically, I felt terrible. Mentally, I was thinking more rationally, perhaps, than I had in a long time. I had completed two combat tours and wanted to return to the United States. I expressed this to Colonel Castle, and with that, I departed for return to Watton. At this point my oxygen supply became depleted. Though the weather was perfect, on final approach to Watton I saw two runways. Since there normally was only one runway, I decided to land in the middle, between the two. As I descended closer to the field, the two runways fortunately merged into one.

"Although this incident happened exactly as previously recounted, had I received prior knowledge of a scheduled mission that morning, I would have been in better condition and ready to go. Unfortunately, Colonel Castle was killed in action ten days later. He was piloting a B-17 on a mission over Belgium when Luftwaffe fighters attacked and crippled the bomber. The colonel refused to jettison the bomb load because of U.S. troops on the ground below. With two engines afire, he leveled out and remained at the controls while his crew bailed out. He was still inside when another enemy fighter attacked, causing the Fort to crash. Posthumously, he was awarded the Medal of Honor and rank of brigadier general. He was a fine officer and it was certainly a privilege to have known and flown with him as his pilot."

Col. Frederick Castle led the 3 AD mission with the 487 Bomb Group when his plane crashed near Hods, Belgium, after a fighter attack.[63]

A 652 BS Fortress flew a Sharon WR mission to Lagens, Azores, on the 14th. The crew comprised 2/Lts. Virgil S. Junk, pilot, Milton R. Madsen, copilot, Lewis J. Schumacher, navigator; Cpls. Victor J. Acampa, Fineis A. Hodges, Albert J. Grosel, Stanley T.E. Foletta, with met/obs. M/Sgt. Gordon B. Schmidt and Sgt. Conrad W. Schintz.

On 16 December, Lt. "Dusty" Junk taxied out in B-17G, 43-38555 for the return flight to Watton. That day Lagens experienced strong gale-like crosswinds perpendicular to the taxiway, gusting to over fifty knots; empty 50-gallon drums went airborne across the field. The Fort proceeded along the steel mat taxi strip to the active runway using differential throttles to counteract and prevent the aircraft to weather vane in the powerful winds. By the time they taxied halfway, the right brake became very hot from constant operation to maintain directional control, resulting in loss of braking action. Junk stopped the Fort for several minutes to allow them to cool. He then advanced the throttles slowly to taxing speed with less braking, and the Fort advanced gradually forward towards the active runway.

During taxing the tail wheel remained locked to provide added directional control. As they slowly crept along, copilot 1/Lt. Milton "Red" Madsen unlocked the tail wheel in preparation for turning 90 degrees on to the runway. The aircraft weather vaned in the strong wind producing a sharp left turn. Both pilot and copilot were literally standing on the brakes, but to no avail. Full throttle to both left engines had no effect as the Fort rolled over the embankment, and down the sloping side. It ground to a halt standing nose down on its crushed forward section, and the crew immediately evacuated the aircraft.

Transportation arrived and took the crew to the base dispensary. Normally the medics provided a shot of whiskey to calm frazzled nerves, but not this time. Instead, the men climbed into a stretched limo-like vehicle with several rows of seats behind the driver and transported to quarters.

It was raining. En route, a fuel bowser entered the roadway and the limo smashed into it. Someone yelled the tanker was placarded "100 octane" and the crew immediately bailed out the windows and doors, and took flight from the scene. Again, they were transported to the dispensary, but this time the medics gave them an entire bottle of whiskey to settle their nerves.

The crew remained at Lagens until an accident board of inquiry completed their investigation. Meanwhile, heavy clouds, fog, and rain covered all England, preventing return of Sharon aircraft at Lagens. Instead, the Atlantic WR missions continued from the Azores with return to Lagens and all data collected transmitted to England. The Junk crew eventually obtained transport on different returning Sharon flights to England.

Unable to overcome gusty wind conditions at Lagens, Azores, the Fort vaned off the runway, damaging the nose section. *Photo M. Madsen*

652nd BS crew, L-R. 1/Lt Lewis J. Schumacker, navigator; T/Sgt. Victor Acampa, flight engineer; 1/Lt. Milton Madsen, copilot; 1/Lt. Virgil Junk, pilot; and S/Sgt. Albert Grosel, tail gunner. *Photo. L. Schumacker*

63 Richard Le Strange, Century Bombers, Aylsham, England: M. F. Barnwell, 1990: 123.

1/Lt. Jerry M. Roberts and his 492 BG crew, one of several assigned temporarily to Watton, flew Atlantic WR missions. Roberts transferred to 653 Squadron to fly Mosquitos. On the night of 18 December while practicing landings and circuits in MM371, his aircraft touchdown and rolled a short distance before the right undercarriage folded. This caused a right swerve off the runway where the left landing gear collapsed.

A Graypea supporting 1AD was dispatched this day. Just as Chaff discharge began in the target area, Lts. Wallace Rouse/Joseph Levy (NS508) went into an uncontrolled vertical spin. The navigator released the top escape hatch to prepare for bailout. The pilot suffered oxygen loss but regained consciousness at 6,000 feet and aircraft control. The spin caused malfunction of all instruments but the crew managed a safe return to Watton at low altitude.

By 19 December the German Army had advanced 45 miles into Belgium. At Watton, CCU completed a non-operational Aphrodite mission. Four cameramen participated recording simulated target approaches and secret installations. Alerts for Aphrodite during the month were many but at no time did any materialize into missions.

On the 20th, the weather deteriorated so atrociously that a total of only 29 sorties were flown by all Allied Air Forces. Late evening of the 21st, Lts. Wendell Biggers and Kenneth Dunaway returned from liberty at Kingsend. At midnight, they visited operations and learned all alert crews had been dispatched for the night. Operations required a crew to establish a cloud base over the Ardennes region. Rather than awaken other members both men volunteered for the mission.

They flew to Saarbrucken and Metz area with the Battle of the Bulge in progress. The cloud base at 1,000 feet extended to the horizon with no change. Some 15 miles from the target they turned around and flew to the coast, established radio contact and called in their observation. As Biggers recalls:

> "I received a terse reply stating an exact cloud base was required directly over the particular point, not a guess from 15 miles away. We banked and flew back to the target area still covered by low-hanging, overcast sky, and carried out the order. This second weather observation was identical to the first one issued. Satisfied, we returned to base."

On 23 December, Lts. James D. Spear/Carroll B. Bryan (NS638) performed a test flight at 1104. After departure there was no further radio contact with the aircraft. RAF Flying Control at Babdown Farm notified Watton at 1605 that a Mosquito aircraft crashed two miles west of Dursley at 1230, killing both men. There were no witnesses, and with both men dead, no evidence on which to determine the cause.

That afternoon Lt.Col. Leon Gray/Lt. Louis Toth (NS688) escorted by P-51s flew a daylight Mickey mission to the synthetic oil plant at Politz. After successfully obtaining radar photos, Gray dismissed the fighter escort and returned to England alone. The Mosquito cruised along at 25,000 feet over Husum, Germany, when jumped by two Me 109s. On the first pass they came in at 8 o'clock from 29,000 feet. Gray turned into the 109s. On the second attack he performed a turning left dive to 4,500 feet seeking refuge in a low-cloud layer. He descended further to 500 feet and continued in cloud cover until reaching England. This was the first successful operational daylight Mickey mission producing good photo coverage (the first was flown 12 December) and both men received DFC medals.

Tail warning radar for Mickey aircraft flying unprotected at night was canceled. Now that missions were flown in daylight with fighter escort, they deemed the equipment unnecessary. The first P-38 from Burtonwood arrived modified with Mickey radar. Additional work performed by 375 Service Squadron made the aircraft available for operational flights.

On Christmas morning, mother nature provided a home-like touch as snow coated everything, turning Watton into a white winter wonderland. Christmas dinner, turkey and trimmings were enjoyed by all; including several hundred children from nearby orphanages invited by the Red Cross to spend the holiday with the troops.

MM371 on 18 December 1944 nighttime landing accident.
Photo G. Scruggs

A scheduled Graypea went awry the 27th. Major Willis Locke/Captain Richard Wright (NS508) took off first, followed by Lts. Morton Hunt/Richard Davies (NS519), then Lt. Richard Kenny/Capt. Chesley (MM393). It was a bright clear morning with ground covered with snow and frost layered the runway. The crew chief previously sprayed Prestone on the frosted windshield.

After the first Mosquito roared down the runway, Morton Hunt pushed the throttles forward and followed him on takeoff. The prop wash from the preceding aircraft hurled a cloud of frost particles that immediately clung to Hunt's windshield; it became opaque, hindering forward vision except the lower left hand corner of the perspex. By then he approached 70 mph with the tail now in the air.

Looking out the left side Perspex, the runway edge approached closer as the aircraft slowly veered to port. After applying differential throttle, the Mosquito swerved and skidded down the runway, tracking with left wing forward. He tried differential throttles again and over-corrected for the aircraft now tracked right wing forward. The aircraft slid along until touching a patch of clear runway; the wheels obtained traction causing one landing gear to shear off. The aircraft fell on the right wing and ruptured the 100-gallon underwing tank, spewing fuel over the airframe, when suddenly a ball of fire billowed 50 feet high. The Mosquito slid to a stop engulfed in flames.

The navigator ripped the hatch from the canopy roof and jumped out. Hunt released his safety belt and started out the top. With parachute pack attached, he jumped off the left wing and trudged across frost-covered grass in heavy boots until overcome by exhaustion. Soon, fire trucks arrived, spraying the aircraft with foam.

Lt. Kenny/Captain Chesley (MM393) followed Hunt on takeoff, but a piece of ice blown back by Hunt's Mosquito lodged in the pitot tube. Upon reaching 90 mph, the air speed failed to register, so Kenny aborted takeoff. Major Locke continued alone without fighter escort and made rendezvous with the bomber formation, providing Chaff coverage over the Bonn area.

Efforts to salvage the Mickey radar-mapping program resulted in new tactics. Watton initiated daylight missions with P-51 fighters escorting teams of Mickey Mosquitos. Instead of a single aircraft, the 654th now dispatched two Mosquitos together to the same target. On 28 December, Lts. Robert Shoenhair/Joseph Ward (NS686) and Norm Magee/Robert Balser (NS709) flew to the oil refineries at Politz. Only three of eight Mustang escorts promised arrived. They were reluctant to fly deep into enemy territory and planned to escort the Mosquitos only three-quarters of the route. Thankfully, the P-51s continued to the target and back. H2X problems marred photos from NS709, but NS686 returned with excellent results. Dispatching teams of Mickey aircraft became routine.

The first Ground Control Approach (GCA) equipment in England arrived at Watton in late December for approach and landing in zero-zero weather. Pinetree understood the importance of weather information and Watton became the only Eighth Air Force station assigned GCA. The equipment remained stored pending arrival of operations and maintenance crews from the USA.

Capt. Michael Lorenz returned to Watton the end of December from special duty in the USA. Posted to the Azores the first week of October, Group HQ ordered his return on 7 November. They arranged an aircraft to transport him and navigator Capt. Jacob Hochman to Pinetree. Both men would crew a B-17 transporting Col. Elliott Roosevelt to the USA.

Upon arrival at Pinetree, Lorenz received notification takeoff would commence shortly and to inspect the aircraft. It was a B-17 stripped of armament. The radio compartment had all military equipment removed and folding bunks installed on each side. The radio operator and equipment relocated to the cockpit immediately behind the pilot. The bomb bay had one 500-gallon fueltank on the left side, while the right side contained a wooden platform for baggage. Lorenz placed his gear on board and checked the distribution of luggage for balance.

27 December 1944. A sad demise for NS519 that flew in the first Frantic shuttle mission to the USSR.

Two NCOs approached with cages containing a Bull Mastiff puppy and a second larger animal, and loaded both in the rear aircraft. A short time later the Fort took off in rain for Prestwick, Scotland. On arrival Lorenz consulted the navigator regarding the flight to Iceland. Concerned that a sudden snow squall upon arrival would prevent a landing, he requested the Tokyo tanks filled to capacity as a precaution for return to Scotland. Based on current Iceland weather data the Operations officer objected to this action, but Colonel Roosevelt overheard the conversation and intervened. After listening to Lorenz, he ordered the Tokyo tanks filled as a safety measure.

The Fort took off and arrived at Iceland at night in clear weather. Prior to approach Lorenz dumped the excess fuel and landed on the snow scattered runway. After refueling, the B-17 set off for Bangor, Maine. Upon landing the two Bull Mastiffs were removed for quarantine. The aircraft then proceeded to La Guardian Airport, New York, to off load the passengers. The aircrew obtained special leave and Lorenz traveled home to Ohio to his wife, family and Christmas.

On 26 December, Leona and Michael Lorenz traveled to Washington and two days later received a White House invitation to a luncheon. A taxi delivered them to the main gate where staff escorted them to the Blue Room. Here Colonel Roosevelt and his bride Fay Emerson greeted the guests with Mrs. Eleanor Roosevelt in attendance.

Two days later, Lorenz traveled to New York and reported at La Guardia Field for the flight to England. MPs now guarded the aircraft. Apparently it had been sabotaged in Burbank, California. While undergoing thorough inspection for the return flight, workmen discovered someone cut all the gasoline transfer hoses to the main tank, thus the need for increased security. The B-17 departed New York on December 31 for an uneventful return flight to England.

January 1945

Severe weather that began in November, increased in December, and heightened to chronic proportion in January with wet and cold conditions. Weather became more an obstacle to flying than the enemy; ice, snow, and cold made work on windy hardstands difficult. It also increased the danger of icing conditions at altitude and restricted visibility. New Years day proved jam-packed with air activity.

Foo Fighters were encountered early 1 January. Lts. John Green/ Warren Barber (NS626) flying a WR at 30,000 feet observed pairs of bright, yellowish-orange lights evenly spaced, which seemed floating in the air, always to the rear or on either side of the Mosquito. The lights were brilliant and at times faded, either in the distance or disappeared. They first appeared over the Zuider Zee at 0230, and as the aircraft approached Berlin at 0310, two more were observed. Flying closer to Berlin at 0330, three more came into view. No attacks were made, hence the crew was uncertain as to the identity of these lights.

Postwar, Warren Barber attended a SAC lecture presented by a former Intelligence officer who provided further detailed information of the above incident. Apparently, on that particular night, one AAF and two RAF crews reported identical observations within the same area and time period. The Germans also reported observing strange lights at the same time and locations but thought they were Allied aircraft. On that same day, military sources in Tokyo also noted an identical phenomena occurring over Japan.

Scientific investigation attempted to provided a rational explanation. The bombing of European cities caused massive firestorms, polluting the atmosphere with smoke and dust, thus increasing charged particle production and the occurrence of electrical activity phenomena, identified as Foo Fighters. Meteorologist reported the brilliant crimson sunsets over the continent, the most beautiful ever seen, again attributed to increased dust particles from burning European cities.

407 BS, 92 BG, B-17 297778 skidded off icy runway on 12 January 1945.
Photo G. Scruggs

The cold month began with 8 CCU filming an air-sea rescue for a training film. This included action scenes of returning disabled aircraft ditching in the North Sea and the rescue efforts that followed. Cameramen T/Sgt. Cecil Brown and S/Sgt. Bill Newman covered air-sea rescue operation at RAF Station Martlesham Heath, as Brown relates:

> "We filmed ground preparations for an air-sea rescue and then took off in a Walrus to photograph the actual event. On the first patrol we spotted a British Mosquito pilot floundering in the Channel. He didn't stay above the water long, and there was no chance of saving the poor fellow. He went down before a dinghy could be dropped. On several occasions, the Walrus pilots spotted airmen who ditched and dropped dinghies to them. The weather was extremely poor limiting our filming the operations."

Meanwhile, Lt. Richard Geary/S/Sgt. Ralph Thiry (NS595) made Boxted a base of operations to film actual rescue attempts. The 5 ERS covered the heavy bomber routes across the North Sea. Their P-47s continually patrolled the area for returning aircraft requiring assistance. Geary loitered over the North Sea maintaining radio contact with the P-47s for instructions detailing direction and location of any potential aircraft ditching.

On 1 January, the 398 BG sent a mission to Magdeburg, Germany. En route over the North Sea, a bomber piloted by l/Lt. Howard M. Pinner experienced trouble flying formation at 5,000 feet. A blown cylinder head and cowl flaps created an uncontrollable engine fire. The pilot alerted crewmembers, and everyone prepared for ditching. The plane, now 120 miles from land, hit the waves and broke in half just behind the ball turret. Within ten seconds the bomber with three tons of bombs aboard sank.

Upon ditching the copilot manually set the life rafts free. The tail gunner and engineer-gunner, managed to escape in the rough sea at the point were the plane broke in half. However, a waist gunner and navigator drowned after the plane struck the water. Six of seven crewmembers climbed into the dinghy. At this time the tail gunner floated 50 yards from the life raft. For 30 minutes the crew made every effort to paddle toward him, but he disappeared beneath the waves.

The Gibson Girl emergency radio popped out of the radio room during touchdown and was found floating nearby. The men paddled after it and unpacked the equipment. The radio operator began transmitting their position. For at least another hour and a half, the wet, cold, and sick survivors continued to send their position by radio. Finally, three P-47 Thunderbolts made an appearance approaching the dinghy at 500 feet.

For the next several hours, the P-47s and others, which relieved the first group, flew over dropping smoke flares to make certain of the dinghy's position. Two Lindhom rescue dinghies were then dropped by a P-47 to within a few yards of the stranded men. They paddled towards them and the ball-turret gunner climbed into one and copilot into the other. Food, water, cigarettes, and protective clothing were available in the two rescue rafts, and the crew took advantage of all items.

Meanwhile, the alerted Mosquito arrived on the scene, but Geary experienced great difficulty locating the rafts among the high swells. A P-47 flew low and dropped a smoke marker and circled the raft so Geary could spot it. The Mosquito circled the area keeping the stranded men in sight while Thiry filmed the scene below.

Just before dusk, two RAF Warwick aircraft arrived each carrying a Flying Ditchman lifeboat under its fuselage. The first bomber dropped a rescue boat but the parachute failed to deploy and it crashed in the sea, bursting into a million splinters. The second lifeboat released landed too far away for the men to reach.

By this time, the P-47s returned to base while the Mosquito continued circling. Later, a third Warwick arrived and dropped another lifeboat. The parachute blossomed and it touched the water intact. Upon impact, rockets attached to 50-yard static lines shot out in all directions. The men secured a line to reach and climbed safely into the rescue craft.

The lifeboat came equipped with every possible necessity including two engines, a compass, foodstuffs, and clothing. After reading the instructions, the engineer-gunner started the engine and turned for England, but they were traveling north towards Norway instead of west.

The Mosquito, now short on fuel, left the scene. Unable to contact the men by radio, Geary flew low over the lifeboat in a westerly direction hoping they would take this as a cue, and he soon disappeared from sight. A British air-sea rescue launch reached the lifeboat and took the survivors aboard, greeting them with a "Happy New Year Yanks!" Bacon and eggs were served the survivors and all their immediate needs met. The following day, Geary and Thiry visited the survivors in the hospital. The bomber crew expressed their appreciation for a job well done.[64]

On New Year's day, the Luftwaffe launched Operation Bodenplatte attacking Allied airfields in Holland and Belgium. The day also heralded the final Aphrodite mission. The target: Hereford, Germany. This involved two task forces, each consisting of a pathfinder B-17, one Baby, a Mother, one Mosquito, and P-38 PR ship. After takeoff, control transfer and bailout successfully accomplished, the B-17 robots were guided across the North Sea to Germany. Heavy cloud cover obscured Hereford and the secondary target at Oldenburg had 8/10ths clouds. Two Mosquitos left Watton: Capt. James Matuska/T/Sgt. Cecil Brown (NS569) and Major Clarence Short/Sgt. Mark Ortega (NS619) to rendezvous with each task force. Ortega recalls:

> "We took off from Watton and flew to Knettishall where we hovered over the field waiting for the Mother and Baby plane to takeoff. We followed our Baby after takeoff and then lost it in heavy clouds over the North Sea. We kept searching until we reached the Frisian Island and then turned back for England."

This robot missed the target at Oldenburg and was dumped into the residential area of a nearby town as Brown relates:

> "We stayed on the tail of our Baby plane until the North Sea, then we lost it in heavy clouds. Near the Frisian Island we sighted it again and from there on in stuck close to it. We followed it to the target at Oldenburg, Germany. I took 400 feet of film at low altitude, in addition to flight shots of the Baby heading for the target and the strike itself, and then streaked like hell for home on the deck."

The drone blended with the snowy terrain and became difficult to detect and control. It crashed and exploded three miles south of the target.

An interesting outgrowth of the Mickey program involved a suggestion by the Radar Intelligence Section for an H2X remote reception system. The radar section conducted experiments to determine if H2X scope images in flight could be transmitted and presented to an observer on the ground. Instead of taking scope pictures by camera, a television camera mounted over the scope would transmit the images to a ground station. The line-of-sight range to ground from 30,000 feet was 100 miles, and even greater range could be achieved utilizing a relay ship at similar altitude and distance. JANIE, the code name given to this project, presented many possible tactical uses. The officers submitted a report on 10 January to Wright Field where parallel experiments were being conducted.

Previously, Capt. Michael Lorenz of 652 Squadron flew Colonel Elliott Roosevelt to the USA in December by B-17 aircraft. Accompanying the Colonel were two dogs brought home from England. On 11 January, Seaman lst Class Leon LeRoy hurried home to Antioch, California, on emergency leave to comfort his ill and recently widowed mother. LeRoy, along with a Seabee, and an Army T/Sgt. on their way to ailing spouses were removed from an ATC plane at Memphis, Tennessee, because they held third priority rating and had to provide room for a first priority rating passenger.

In their place ATC placed a huge crate on board the transport. Inside sat Blaze Hero, a pedigreed English bull mastiff owned by Colonel Roosevelt. Apparently, before returning to England, Roosevelt asked an ATC friend to fly the mastiff to his wife, cinema actress Faye Emerson in Hollywood, if an empty plane happened to be going that way.

Someone on the White House staff arranged with ATC for delivery of the dog, and Blaze received a "White House priority" or first priority rating. This comedy of errors resulted in the three servicemen being "bumped" from the aircraft and the incident, *L'Affaire Blaze* made the front page of newspapers.

An American fighter attacked Lts. Jerry Roberts/Charles Veldon (NS650) on 12 January while returning from a mission to southern Germany. After crossing into France, they descended to lower altitude. In the distance Veldon observed a P-51 circle around to position itself behind them. The Mustang began firing. Roberts took evasive action

64 Allen Ostrom, 398 BG Remembrances, Seattle, WA:Vanguard Press, 1989: 35.

with steep turns, climbs and jinking about the sky. Attempts to call off the fighter by radio were to no avail; it maintained its aggressiveness with continued tracers. The only option remaining was to signal colors of the day by Very Pistol. The hours of the day were coded by different color flares. Veldon attempted opening a small window to the slipstream but thwarted by a safety catch. Roberts yelled, "Fire the damn thing." After much consternation Veldon broke open the window and shot the flares. The P-51 pilot broke off the attack.

Mosquito navigator, Lt. Frank Catalano obtained a pass and visited his former unit, 407 BS, 92 BG. On the 12th his friends flew him to Watton in B-17G 42-97778. Upon touchdown the undercarriage link assembly failed causing a left landing gear collapse and ground loop.

On 13 January, Lts. Richard M. Kenny/Arnold V. Kuehn (NS712) flew a WR over Germany. According to Kenny:

> "At 1516 while south of Berlin at 30,000 feet, I turned west and then decided to make a 360-degree orbit to obtain a better visual inspection of the Berlin area. As I made the turn, we spotted two Me 262s approximately four to five miles behind and below the Mosquito. There was a low cloud layer extending westward from the Hanover area. I could not outrun the jet aircraft, so I prepared to make a dash for the clouds. Just before starting my dive, I made another turn for a quick check of the rear. To my surprise, there was an Me 262 on my tail with his four nose guns flashing. The jet quickly closed in at 500 mph, so I immediately applied full throttle and dove for the cloud layer. During the steepest part of the dive, we had a true ground speed well over 450 mph. The jet continued following close behind repeatedly firing.
>
> "I rolled the Mosquito on its back as if going into a split-ess maneuver. Instead of heading straight down, I rolled over off my back, then jinking, turning from one side to the other, and then climbing. I avoided flying straight or making any turns to prevent him from obtaining a deflection shot. The jet did not fire any tracer rounds during his alternate passes, closing in from 400 to 500 yards before firing. The jet kept firing below me and the cannon shells were exploding far in front of the Mosquito, as if they were time-fused rounds. The '262 made repeated attacks from all quarters, but I evaded his every move by weaving and performing extremely violent corkscrew turns. The attack began at 30,000 feet and wound down to 12,000 feet before the jet broke off his pursuit, either because of ammunition shortage or low fuel.
>
> "The Me 262 then pulled alongside on my right at a distance. The Luftwaffe pilot waved his hand, then turned around and headed for the Berlin area. The second '262 never made a pass and always remained off to the side, several hundred yards away. The encountered lasted from 1516 to 1525 hours with the jet continually on my tail firing at me.
>
> "The cloud tops near Hanover were at 12,000 feet and I continued the flight to England flying just above the clouds. Watton was closed due to inclement weather so I landed at Bradwell Bay at 1745. The RAF crews servicing the Mosquito claimed the wrong type of spark plugs were installed in the engines. That is one mission I will never forget."

An Me 262 attacked 1/Lt. Richard Kenny on 12 January 1944.
Photo R. Kenny

The following day, Lt. George M. Vanderleest ground looped MM364 while landing on an icy Watton runway.

January proved an unsettling month for Colonel Roosevelt as "the heavens opened up and I found myself involved in more wars than just that against the Nazis."[65] Besides the *L'Affaire Blaze*, an enigmatic event occurred. Without prior warning, Colonel Roosevelt was relieved of command on 17 January and Brig.Gen. Charles Y. Banfill appointed Commanding General of 325 PW. General Banfill previously directed the ground Intelligence school at Camp Richie, Maryland. On arrival he performed an intensive tour and personally inspected all sections. After several days he mysteriously departed to become Director of Intelligence, Eighth Air Force on General Doolittle's staff. Colonel Roosevelt resumed Wing command.

A week later, the Wing was subjected to another thorough inspection by Lt.Col. J.M. Leavens, Air Inspector, HQ Eighth Air Force. He repeated the previous review performed by General Banfill, visited every section in the Wing including Watton, interviewed personnel as to duties and qualification, and reviewed work performed by the various sections. Many Watton flying crew recall panicking to conceal personal liquor supplies and other victuals from the Azores. Just as mysteriously as he arrived, a week later the Air Inspector completed his investigation and quickly returned to Pinetree. What happened?

In his book, *The Recollections of Colonel Retread*, Col. Carlton G. Ketchum has no kind words for Colonel Roosevelt. Ketchum, an ardent and principal Republican Party fundraiser in Pittsburgh prior to the war, bitterly opposed both the Democratic Party and the Roosevelt administration's domestic and foreign policies, and lambasted both whenever the opportunity presented itself. Ketchum states, "Roosevelt was heartily detested by General Doolittle."[66] His book portrays Elliott as, a great party man who never paid his bills, leaving this to his junior officers; who criticized the British unfairly; who unjustly refused a promotion to Ketchum's personal friend; and that General Doolittle loathed and despised Colonel Roosevelt, he requested General Spaatz return the president's son to the U.S. for reassignment.

In early December, Colonel Roosevelt's name appeared on a promotion list for brigadier general. His countermanding HQ Eighth Air Force orders relating to the F-8s, and the negative canine publicity may have infuriated Doolittle. Also, Ketchum's opinions at command level may have profoundly influenced this antagonism. It would appear Doolittle sought General Banfill's and the Air Inspector's survey seeking evidence for grounds to either force Elliott's dismissal or cancel his pending promotion. Unable to unearth anything actionable, Doolittle had no choice but allow Roosevelt's elevation to brigadier general. Colonel Roosevelt previously nominated the 25 BG (Rcn) for a

65 Elliott Roosevelt, As He Saw It, New York: Duel, Sloan and Pearce, 1946: 229.

66 Ketchum, Op. Cit., 242.

unit citation, but bitter personal differences between him and Doolittle obstructed this endorsement.

January saw the worst weather experienced in England for many years with snow and windstorms sweeping in succession across the island, making flying hazardous and schedules difficult to meet. From time to time, 652 BS aircraft diverted to airfields across the British Isles and only a fortunate crew returned to base after flying a mission. When weather closed in for long periods over Watton, it became necessary to have diverted 652nd aircraft takeoff from other bases to accomplish a mission. This required the crews to continue flying after little sleep and with unchecked aircraft. Problems arose at other bases. An Operations officer refused to allow Capt. Michael Lorenz to takeoff for a WR in atrocious conditions. When Lorenz flashed his green card for unrestricted flying in any weather, the officer stared dumbfounded and allowed the crew to continue the mission. The squadron met the schedule of four missions each day in most cases, resulting in a constant flow of meteorological data necessary for projected AAF missions in the ETO.

When inclement weather diminished CCU aerial work, the unit decided to document Eighth Air Force ground activities by dispatching personnel to 1, 2 and 3 AD. Filming included, footage of scenes showing snowplows clearing runways, spreading salt; snow swept from the wings of aircraft; and mechanics handicapped by heavy gloves working on engines or warming up engines. Also scenes of takeoff from snow covered runways and dawn shots of heavy bombers circling over the field. Sgts. Thiry, Beach and Janowitz filmed in color the Liberator "Witchcraft" completing its 100th consecutive combat mission. Included were scenes of air and groundcrews receiving awards from Major General Kepner and Brigadier General Peck.

Sgts. Larry Barbier, James Zipperer and Dick Elliott covered the 3 AD from RAF Woodbridge. Returning combat damaged bombers either couldn't reach their home fields or adverse weather conditions forced diversion to Woodbridge for emergency landings. The camera operators covered different sections of the field filming B-17s and B-24s approaching with feathered props, badly damaged or partially missing tails and wings. One aircraft landed with the ball-turret missing and flak-damaged stabilizer.

Crewchief clearing snow and ice from Mosquito stenciled, Lt.Col. L.W. Gray and CC T/Sgt M. Cousins. *Photo G. Scruggs*

The 654 BS radar section installed H2X equipment in a modified longnose P-38 that became operational in January. Additional longnose P-38s were expected from Burtonwood. And on 20 January, Lt. Joseph Polovick (MM342) landed after a training flight and when applying brakes, slid sideways off the icy runway causing the right undercarriage to collapse.

Intelligence dispatched two PRU missions on the 21st with Politz and Merseburg as targets. They obtained excellent pictures of the Merseburg Synthetic Oil plant, but missed Politz entirely. The Third Reich heavily protected the highly prized oil producing plants. Nearly 300 antiaircraft guns guarded the Politz refinery, the same number defending the entire Berlin area.

Mosquito "Q" moved inside blister hangar for maintenance, January 1945. *Photo. G. Scruggs*

P-38L 44-23880 with H2X in nose, and under surfaces painted glossy black for night operations. All missions flown with fighter escort during the day. *Photo R. Freeman*

Major Hugh Bozarth on left with 1/Lt. Shaffer E. Horne in wintry conditions. *Photo G. Scruggs*

Lt. Richard Geary flew the 21 January mission to Politz Oil Refinery at Stettin with Lt. Floyd Mann as navigator. They had been on standby for this particular mission waiting for the weather to clear. The operations room had an enormous map that covered 25 feet or better of one wall. The missions for the day, the next day, or when weather permitted, were represented by colored yarn. A different color for each mission stretched from Watton to the target area. The yarn for the Politz mission went all the way from one end of the map to the other. Geary recalls air crew members asking, "Who the hell is going to fly that mission?"

It was a cold winter morning when an orderly awakened Geary at 0400. The weather cleared and the mission proceeded as scheduled. Geary went to the flightline and then to the parachute room to meet his navigator, Floyd Mann.

A thin layer of snow covered Watton as they took off at 0920 in NS569. Prior arrangements scheduled a rendezvous at 0925 with four 20 FG P-51s at 18,000 feet over Cromer. They would provide escort to Stettin and return.

The Mosquito met the fighter escort as planned, but now heavily loaded with 1,000 gallons of fuel, flew at a severe speed disadvantage. Geary attempted to maintain economical cruising speed but outpaced the P-51s and was forced to throttle back to continue flying formation with them. The Mustangs had long-range droptanks and were also fully loaded. Once involved with enemy action, they would jettison their tanks, and therefore, were attempting to conserve and obtain maximum range from their fuel supply. This exacerbated the problem. It was a very long flight to Stettin, and on three occasions Geary throttled back and did not receive the mileage planned.

The formation started out on a tough and difficult daylight mission. They flew across the North Sea, around the Frisian Islands, past Heligoland and over the neck of Denmark. While flying near Kiel at 1048 they encountered heavy flak as predicated, accurate for their altitude of 25,000 feet but not direction. A P-51 piloted by Lt. C.L. Huey developed engine trouble and returned to base.

After crossing Denmark, the four-plane formation flew over the Baltic Sea to avoid further flak areas. The sky appeared clear blue (CAVU) with unlimited visibility for miles around. Geary could see the long sinuous outline of the Swedish coast to the north. One Mustang flew 50 yards off on each wing, and the third lagged 100 yards behind and slightly higher. The P-51 pilots were Lts. Einhaus, Reynolds and King.

1/Lt. Richard Geary involved in the 21 January Me 262 attacks.
Photo R. Geary

The formation flew along the German coastline to the Elbe River then turned southward towards Stettin and the Politz Oil Refinery. Geary observed a large number of enemy aircraft several miles away to his left; approximately 70 escorted Ju 52M3s flew parallel to the American formation. As Geary recalls:

"A young P-51 pilot with a southern drawl broke radio silence and blurted, 'Are we going to jump them?'

"The flight leader replied, 'No. Our obligation is to look after Big Boy.'

"The young pilot responded, 'If they jump us, we can sure give them hell.'

"This display of bravado in such a dangerous situation was comforting. I wondered if the Germans on the ground heard the conversation. If so, did they marvel that someone was 'cocky' enough to take on 70 airplanes?

"At 1135 the alerted enemy defensive positions fired flak at us like you wouldn't believe. Their pattern included Barrage Flak, normally reserved for bomber formations, as well as Predicted Flak. In Barrage Flak, the antiaircraft guns fired at one time in a pattern. In Predicated Flak, antiaircraft guns aimed at and specifically followed a flight. The intense flak was accurate at 27,000 feet. I was diving and corkscrewing at close to 400 mph and the flight leader was yelling over the radio that the Predicted Flak was right on my tail. Bursts trailed me by 150 feet or less. I dropped from 27,000 to 24,000 feet before getting some relief.

"As we approached the refinery complex through all this flak, I instructed Mann to enter the nose and use the bombsight to take the necessary photographs. He discovered that a portable oxygen bottle required to enter the nose for photography was missing. To make the best of our situation, I attempted to take photographs myself. I turned on the intervalometer that automatically started taking photos at timed intervals, then attempted to lineup the Mosquito with the refinery. I intermittently dipped the nose to note my position in relation to the target until it disappeared from my sight. Unfortunately, the pictures stopped at the door to the refinery. We did obtain coverage of Ganserin-Janonitz, northeast of Politz and they served some value as targets of opportunity.

"I started to climb back up to altitude leaving behind the heavy flak that followed us. At 1150 the young Mustang pilot with the drawl shouted, 'Look at the SOB climb.'

"I looked out to my left side and saw this object streaking up from the ground. There it was, an Me 262 climbing like a 'bat out of hell.' This was the first jet I had seen. The fighters maneuvered in position to protect me. They wanted to position themselves with one Mustang below, one behind and one above me. I did not like this situation and attempted to fly below all three P-51s, using them as a shield.

"The Me 262 appeared head-on and began orbiting to get on my tail. I did not make a run for it, but remained with the fighters so they would have a chance at the jet. I looked back and all I could see was a small dot coming up fast. As I straightened my head again, the flight leader yelled, 'Break 28,' my call sign.

"I immediately placed the Mosquito in a steep bank and almost on its back. The flight leader yelled again with an urgency in his voice as if within any second I would be blown out of the sky. The tone of his voice excited my navigator who also yelled, 'Break Dick, break.'

"I was doing close to 400 mph in a left breaking dive, a customary maneuver. What else was there to do? I had no chance to look back. In a flurry of desperation, I slammed on opposite rudder and aileron. The Mosquito cartwheeled 180 degrees across the sky in the opposite direction. I don't know what kind of maneuver this was, and it is a miracle the aircraft did not disintegrate. God must have been on my side. I didn't even have my lap belt on. Dust flew up from the floor, emergency maps came off the wall and loose material floated in the cockpit. The Me 262 hurtled directly over me, seemingly a few feet from the cockpit canopy. There was just one big flash of silver chrome as the uncamouflaged jet shot by. He had me in his sights, but my unexpected action put us on a collision course. Instead of shooting at me, the jet pilot had to use all his talents to avoid a midair collision. That both the German pilot and I lived through the encounter, I credit to his reflexes.

"I lost visual contact with the jet and Mustangs at that moment but remained in radio contact with the escort pilots. I leveled off, pushed the throttles wide open and headed for home. After experiencing the superior speed of the jet, I questioned if I would make it back. All manner of options went through my mind. Then I realized, I would never make it home with the throttles wide open.

"I was now east of Berlin, flying northwest at 27,000 feet when some strange looking objects appeared in the distance ahead. I could see four black specks leaving intermittent contrails as they climbed swiftly toward me. After surviving the first attack, I dreaded being part of any further engagement. I called the fighters and they assured me they would be along quickly. I didn't know how far back they were. The four objects streaked closer and closer, head-on, four Me 262 fighters. They swiftly flew past on my right at a distance of perhaps 50 yards. I didn't make a break, for I was almost certain one of them would have tailed me. Assuming they would transfer attention to the fighters, I radioed the Mustangs to warn them. They acknowledged my call and that was the last I heard from my fighter escort.

"An immense blue dome of clear sky overhead enveloped snow-covered Germany. The atmosphere was crystal clear and immaculate, and ours the only contrail in the sky. Such weather was unbelievable! Fortunately, a towering range of clouds appeared as we approached Belgium. Two single contrails, we assumed to be fighters, approached us but we lost them in the clouds."

Now low on fuel, Mann provided a course for the shortest safe distance across the North Sea to England. Geary, now at 24,000 feet, turned towards the Schelde Estuary area of Belgium to reach Allied lines for safety. He radioed a Mayday and received a vector to Calais. The fuel gauge almost read empty, but he maintained altitude crossing the English Channel to Cromer. Throttling back even further, he banked for Watton and landed at 1410 with less than five minutes fuel remaining. The mission lasted five hours and 50 minutes. Both men expressed appreciation to be safely home.

According to the 77 FS, 20 FG debriefing report, the Me 262 combat engagements ensued from 1140 to 1200. One of four Me 262s attacked the formation over Politz, where the Mustangs chased the attacker and engaged three others in a dogfight. Meanwhile, the Mosquito now heading west on a withdrawal course encountered another four Me 262s and radioed a warning to the Mustangs. Lowell Einhous, the P-51 escort flight leader recalls the second engagement:

> "While climbing for altitude, we encountered the other four 262s flying our type of formation at our approximate altitude. We clashed with the jets in several 360-degree turns, firing at them on several occasions. Apparently the firing was without effect, and the jets broke off the engagement. While returning home north of Berlin, a single Me 262 flew parallel to us while four others flew further south but none attacked. We experienced trouble maintaining speed with the Mosquito because of our droptanks. The distance covered required that we carry extra fuel. The Mosquito pilot says the mission was five hours and 50 minutes. The P-51 escort was airborne considerably longer than that. We were also short of fuel because of our engagement with the (8) 262s and because of trying to stay with the Mosquito. Though we encountered (13) 262s, we did not shoot anything down that day. We tried but to no avail. Even so, someone from Watton called later and congratulated us for a job well done and for the safe return of the Mosquito crew."

On the 21 January, Lts. Jerry M. Roberts and Ralph E. Fisher (NS509) left Watton on a WR to Germany. They received a prior briefing to cover the Danish peninsula and northern Germany. The flight droned on performing weather observation duties, when suddenly and without warning, fiery red tracers encircled the Mosquito.

"Mustang," shouted Fisher, by which time Roberts had already cranked the control wheel, placing the Mosquito in a left diving turn.

"Yellow-belly Mustang," Fisher again yelled into the intercom.

Roberts glanced back attempting to locate the other aircraft's position. As he tightened the turn, red tracers moved across from his left side striking the port wing and engine. He then felt penetrating rounds hammer against the armor plating behind his seat as the deflection shot bore home. Both left engine and wing burst into flames almost simultaneously as fire and smoke obscured the canopy. He flung the Mosquito in the opposite direction for better visibility, then feathered the port engine and shouted for Fisher to bailout.

Ralph sat slumped over at the navigator's position. Roberts reached over and pulled his limp body to an erect position. Apparently the rounds struck the armor plate behind Roberts at an angle and fragments ricocheted off. They tore loose Fisher's oxygen mask revealing a blood drenched face. He failed to respond to shouts or physical stimulation.

Upon releasing the control column, the Mosquito proceeded into a lazy left hand spin resulting in flames and smoke again washing over the cockpit canopy making escape through the top escape hatch hazardous. But Fisher's body blocked clear access to the bottom hatch. Again Roberts pulled Fisher back but this time saw blood spurting from a neck wound; he was still alive but unconscious. Roberts quickly pulled up a scarf around Ralph's neck and covered the wound to prevent further blood loss. He then removed the plotting board from Fisher's lap. The Mosquito continued losing altitude as the swirling ground appeared ever nearer.

The spin aided movement towards the floor. Somehow Roberts retrieved a chestpack and snapped it on Fisher's harness, then managed to remove the inner floor cover and stomped on the bottom door until it flew off. It proved an arduous task struggling to lower Fisher's legs out the open hatch, then his hips, but the chestpack impeded further movement. Quickly Roberts yanked the packs D-ring, placed his foot on Ralph's shoulder and forcefully shoved him through the hatchway. Positioning himself in the opening, Roberts struggled to squeeze his harnessed frame through and safely exited the aircraft.

Almost instantly there was a jolting shock as the canopy deployed followed almost immediately by a second jarring encounter when the initial parachute swing slammed him to the ground. Then everything dimmed to a black out. After slowly regaining consciousness, he was unable to move. With great difficulty he cautiously managed to release the parachute harness, then lay there short of breath in physical agony.

The descent and crash was witnessed by local farmers who provided a makeshift litter and carried Roberts to a group of houses for help. They found Fisher still alive, but unfortunately he died before receiving medical attention. Roberts became a POW and later transferred to Dulag Luft, Oberursel, near Frankfurt. After the war, the military interred 1/Lt Ralph E. Fisher in the Ardennes American Cemetery and Memorial, Neupre (Neuville-en-Condroz), Belgium.

The night of 22/23 January, Lts. Richard Geary/Paul Lovett (NS556) went airborne on a night photo mission to Gransbergen, Germany. They men flew through heavy overcast and approached the European coast. Geary glanced out the side Perspex and thought he saw something. He directed a flashlight on the port engine and unexpectedly saw glycol pouring out the top cowling forming a horrendous foamy plume. Quite startled by the unexpected turn of events, he quickly feathered the prop and shutdown the engine then opened the bomb bay doors to salvo the PFBs into the North Sea. He obtained a reciprocal course from his navigator and turned for England. Then the starboard engine temperature began to climb. He radioed for a vector to the nearest base but was told, "Hold on old boy, hold on."

As they passed the English coast, he received a radio report that every base in the entire British Isles closed down. The only option available was to bailout. He asked Lovett for another reciprocal heading to the coast to set the controls and bailout where the coast should be, and hopefully land in England. Concerned with landing in the icy sea, Lovett refused to bailout, suggesting they again attempt locating somewhere to land instead.

Then, a soft glow abruptly appeared in the murky darkness below. Geary immediately recognized it as a FIDO field (**F**og **I**nvestigation and **D**ispersal **O**peration). It meant a large RAF bomber force was returning from a mission and hopefully, they were not in the landing pattern. He pushed the column forward and went down. During descent, the gleaming light slowly increased in intensity.

The FIDO field is surrounded by a series of pipes with spouts burning thousands of gallons of fuel, greatly illuminating the runway. The flames heated the air, lifting fog and mist clear off the ground, but the rapidly rising, hot air mass created eddies resulting in extremely rough air turbulence when approaching the field. Geary turned into the crosswind leg at 300 feet, then approached downwind, and that is when he lost his only engine. He had no choice but attempt a carrier-type landing. Both undercarriage and flaps were down.

He passed through heavy mist at 100 feet altitude as white snow flecks flashed by. Suddenly, a huge tree loomed ahead. If he pulled up and stalled, the aircraft would fall on one wing, and with plenty of fuel on board, the resulting fireball would incinerate both crew to a cinder. He pulled back ever so slightly, allowing the wheels to drag through the top branches. The Mosquito lost speed rapidly and crashed heavily in a snow-covered field prior to the runway. When Lovett, the navigator opened his eyes, a large opening appeared on the port cockpit side with the pilot missing. Uninjured, he crawled out.

The heavy landing cracked apart the cockpit fuselage and the impact threw Geary out a huge hole. He became entangled in electrical wiring and lay half buried face down in the snow, unconscious. Lovett attempted to revive him. Giving him up for dead, he began walking away just as Geary regained consciousness and hollered. Lovett quickly returned to help untangle Geary from the mess of wires. Geary told the navigator they should have bailed out, but Lovett responded, "Hell, I'd rather come down like this anytime."

Both men began walking from the wreckage towards a nearby caravan signaling an approaching aircraft. Soon personnel appeared and a vehicle arrived to transport both men to the base hospital. Next day, squadron commander Major Robert Houle visited Geary at the infirmary. He recalls Geary informing him the brutal impact caused short-term memory loss of immediate events prior to the crash. When Geary regained consciousness in a delirious state, his first thought upon seeing brilliant flames all around him was he died and had gone to hell!

The crash landing occurred at Foulsham at 0030 and became Geary's last mission. Hospitalized with jaw and head injuries, he received reassignment to Meridian, Mississippi, as an A-20 night-fighter flying instructor.

A Graypea mission launched the 28th to support a 1 AD mission to Cologne. The mission received a fighter escort. Upon return to Watton, Lts. Robert James and Woodrow Loden (NS594) touched down in a strong crosswind when the aircraft began to slide to the right. James attempted using brakes and rudder to correct this movement on the ice-covered runway, but the wheels provided no traction. The crosswind forced the aircraft sideways into a high snowbank, breaking off the tail section.

To Geary's amazement the life raft inflated on 'J' NS556 at RAF Foulsham, 23 January 1945. *Photo R. Geary*

On 28 January 1945, NS594 slid off an icy runway into a snow bank, snapping off the tail section. *Photo. G. Scruggs*

The main B-26 detachment sent to France on 30 November returned to Watton 28 January. The armorers remained behind to dispose of the PFBs. Kenneth V Lundberg recalls this duty:

"In the fall of 1944, with ground troops moving deeper into the continent, delivery of target photos took longer and longer, and concern grew that timelines became a problem. In November, Col. Roosevelt decided to move a contingent to northern France nearer to 12th Corp location of Intelligence headquarters for the Allied thrust into Germany. The 7 PRG provided a squadron, the 27 PRS flying F-5s and Spitfires, and the 25th with B-26s, formed a temporary unit. They established a base at Denain-Prouvey near Valenciennes, six miles from the Belgium border near Mons. The Watton contingent consisted of 11 officers – all aircrew – and 30 enlisted men. The commanding officer for the 25 BG (Rcn) personnel was Capt. Victor S. Doroski.

"Captain Doroski was from Riverhead, on Long Island, New York, and a veteran pilot by the time he joined the 25 BG (Rcn) in August 1944. He flew P-38s, B-26s and the de Havilland Mosquito. His reputation as a skilled pilot earned him a commendation from Eighth Air Force for being the first to pilot Lt.Cmdr. Steve Simpson on his OSS missions.

"On November 30, the Watton unit flew to Denain, the officers flight crews on the B-26s assigned to the unit and the groundcrew on B-17s borrowed from their sister squadron, the 652nd. The unit received one hour's notice and traveled with minimum personal baggage and equipment. Their order read, 'to set up operations pursuant to order of the CO.' This translated into providing photo Intelligence for the ground troops in 12th Corp on as fast a turnaround as possible and to follow the advance of the ground troops, if necessary.

"The results are chronicled in an article written by Hubert Childress, CO of the 27th many years later in The Journal of the 7 PRG, February 1991: 'The 27th was selected to move to France as it was deemed by Group and Wing commanders to be the best squadron in the Group ... The purpose of the move was to offer additional reconnaissance capability to the 12th Army Group. To that end, within hours of our arrival, I flew to Luxembourg, met with an Army Colonel of 12th Army Group Operations, explained that I had a squadron of F-5s and a flight of B-26s at Valenciennes, and offered to accept any and all army requests for day or night reconnaissance. The Colonel was very receptive to the offer and assured me he would contact me very soon ... As the Colonel and I walked out of his office, we ran into Colonel James Hall, the original CO of the 13 PRS and the 7 PRG. At the time, Hall was the Director of Reconnaissance, HQ Ninth Air Force, i.e., the person responsible for providing all direct recon support for U.S. Army forces on the continent. That was a most untimely meeting as the Army Colonel introduced me to Hall and related my very kind offer of recon support. Needless to say, I never heard from anyone in the 12th Army Group, and what Hall perceived as Elliott Roosevelt's attempt to expand his area of activity onto the continent was nipped in the bud. The B-26s never flew a mission from Denain.'

"That rebuff and the severe winter of December/January 1944-45 kept the special unit at Denain in a state of inactivity. A tent village was set up for the enlisted men, anchored by the one standing building on the former Luftwaffe fighter base – a motor vehicle garage. The officers were billeted in the city of Valenciennes in an old, but mildly comfortable 'chateau.' Here a photo lab and a communications center was set-up – neither ever used in the cause of the war effort.

"Nonetheless, the experience was a memorable one for the GI's involved. Just the requirements of survival in a primitive setting kept the men busy. Anticipation of missions 'on the next day' kept the unit at readiness and morale was high. Captain Doroski, who was popular with both officers and men, commanded with an instinct for balance between discipline and freedom. This paid off during the days of the German Ardennes offensive. For two weeks, December 16 to the end of the year, each day provided a series of uncertainties – 'break through', 'sighted German paratroopers', 'field at full alert', 'reported infiltrators in American uniform', 'false and real German overflights.' Through it all, the esprit de corps of the unit was excellent, and response to leadership, superb.

Tent tarpaulins protect B-26 from snow and ice at Denain-Prouvey near Valenciennes, France.
Photo T. Tate

On left is Roy Pearson with Tom Tate, B-26 crewmembers at Denain-Prouvey. *Photo T. Tate*

Water-filled bomb crater with groundcrew quartered in background tents.
Photo. G. Scruggs

Armorers at Denain-Prouvey, France. L-R. Front: Frank Boylan, James Walsh, Robert Thomas. Rear: Larry Levy, Jack Sheen, James Tennille. *Photo J. Sheen*

"Those of us close to the situation give credit to Capt. Doroski for this. We felt confident in his leadership – and our destiny. The day after we got back in January, one of my tent mates, Marion Sprague, the cook, looked me up to inform me that he had volunteered (to Doroski) our willingness to go on another special mission because Watton was too boring. I don't remember giving him permission to volunteer me, but the whole matter became academic many days later. I wrote to my parents on 13 February, 'the commanding officer of our unit in France is missing in action. We have lost an excellent officer and a good friend.'"

The German Army transported men and equipment into the Ardennes region under cover of darkness. One may speculate that had night photo B-26s become operational immediately, they might have detected enemy troop movements early enough to forewarn the impending Ardennes Offensive and saved thousands of American lives. It is difficult to imagine any officer in command would squander his limited night photography capability by denying its operational implementation because of professional or political bias.

The Allies launched widespread air attacks against German railway installations on 29 January. At Watton, Lt. Harry E. Peterson (NS595) took off on a transition flight. After gathering speed, the aircraft swerved left off the runway and struck an obstruction, breaking off the tail wheel.

From 26 December to 31 January, Mickey operations experienced success due to fighter support, allowing both navigator and pilot additional time to concentrate on the radar equipment and needed adjustments. Full-scale operations during January resulted in disappointing figures. Of 36 missions flown, the greatest number airborne in any one month, only four were successful and one partly. Mission analysis revealed: three failures to navigational errors; four from personnel failures; seven with unacceptable photos; one abort due to photo equipment failure; two mechanical aborts; and four aborts because of H2X equipment malfunction; and others involved multiple factors.

Various reasons were advanced for the lack of success. 1. Meteorological conditions affecting operation of the H2X set. 2. Fluctuating power supply resulting in a temperamental electrical system. 3. Lack of space in the Mosquito for the necessary equipment. 4. Consequent inability of navigator to operate the set efficiently. Less apparent was the lack of thorough training to handle the highly complicated equipment. Navigation required full concentration by the navigator/operator, and in addition, H2X operation by someone not fully skilled with the intricate mechanism, rendered a successful mission problematical. As partial remedy, the squadron initiated a rigorous training program managing Mickey equipment on training flights over the UK.

At months end USSTAF concluded from an increased PR aircraft loss rate, the Luftwaffe placed high priority using jet aircraft to intercept and shoot these down. To counter this threat, three alternative actions were investigated. 1. Provide fighter escort for PR missions. 2. Install .50 caliber external gun packs on Lockheed F-5 aircraft for self-protection. 3. Introduction of early production Lockheed P-80A jet aircraft known as the F-14 to fulfill PR, but not for weather reconnaissance which required a trained observer. The F-14 delivery date was November 1945. In lieu of the jet's arrival, P-51s would provide escort coverage on F-5 missions. The Mosquitos would continue to rely on speed and performance at high altitude to evade enemy attacks.

Bad weather restricted CCU filming the *Mosquito Story* to only six days during January. A total of 1,370 feet was exposed showing Mosquitos taking off and blowing up a cloud of snow; interiors showing pilot and navigator at their normal flight duties; close-up of the Gee Box in operation; control tower receiving and sending messages; the Mosquito flying and buzzing the field on return from a weather mission. Lts. Roger Gilbert, Louis Rogers and John Grassel flew Mosquitos during filming.

February 1945

Early February saw the greatest air blitz ever to strike the Third Reich. One thousand U.S. bombers attacked Berlin while Allied ground armies on the continent scrutinized the weather reports from day-to-day. The 652 BS was the largest single factor contributing to long-range forecasting in the success of these military operations. They continued flying four missions each day and several extras on the old Epicure route, which lies roughly between the Sharon and Allah routes.

Minor accidents took their toll 1 February. B-17G serial 44-6575 piloted by 1/Lt. Donald E. Peterson prepared for takeoff at 2045 hours in light drizzle with a strong 32 mph crosswind. He applied differential throttle to align the Fort with the runway. After a 300-yard run the aircraft drifted to the side and both throttle and rudder proved ineffective to correct this movement. The aircraft left the runway causing the tail wheel to collapse. The Fort returned to the dispersal area where Peterson's crew transferred to 43-38573 and took off to continue the WR mission.

Lts. Allan B. Bateman/H. Floyd Mann (NS559) flew a night photo mission to Pforzheim. On return they landed at Manston without the airspeed indicator or Gee. The next day they left for Watton. Coming in behind another Mosquito and a B-24, Bateman made a three-point landing and was at taxiing speed halfway down the runway. To avoid closing quickly on the Mosquito ahead, he applied full flaps and pulled off the runway to avoid hitting it. After rolling a distance on the shoulder, the left undercarriage collapsed.

On 3 February, 8 CCU sent four cameramen with two BGs, and a BBC correspondent with a third group, to Berlin, while other cameramen covered ground activities at the three bases. Fair weather permitted visual bombing of rail junctions and the southwestern and southeastern commercial districts. The combined footage resulted in a news film, "Inter-Allied Cooperation: Eighth Air Force Strategic Heavies Tactically Bomb Berlin for Soviets."[67]

The early morning of 4 February, Lts. Wallace B. Rouse/Joseph M. Levy (NS730) returned from a WR mission and diverted to RAF Exeter. Later that afternoon, they set off for Watton. During the first 700 yard run, black smoke poured from the left engine accompanied by a severe vibration. The aircraft lacked sufficient power to become airborne and traveling too fast to safely stop. Rouse applied throttles to bounce over a fence and down a gentle slope towards a creek and wooded hill. He swung left to avoid these, but now confronted rows of houses. He dug the left wing into an embankment, collapsing the aircraft on its belly and disintegrating in flames to a halt in the middle of a street. Both crew scrambled to safety.

Misburg and Hamburg were targets 8 February for two Joker night photo ships. The Deurog Synthetic Oil Refinery at Misburg had 10/10ths clouds, so Hanover was photographed as a target of opportunity by the first Mosquito. The second aircraft conveyed Capts. Victor S. Doroski/ Jacob Hochman (NS583). Both men were airborne at 0158 and last contact with the crew occurred at 0255 by continental ground station Nuthouse. When they failed to return, Watton contacted 12 Group, but the RAF maintained no plots on this aircraft. Later, a Luftwaffe report delivered through the International Red Cross reported: "A Mosquito crashed near Oberschonhage, 6 km east of Detmold. Fire destroyed 99% of the aircraft. Hochman and an unknown body were buried in a cemetery at Leistrup-Meiersfeldon." Apparently a partially hung PFB ignited in the bomb bay and the ensuing fiery explosion immolated both officers. Doroski previously flew secret OSS Joan-Eleanor missions and Hochman navigated Col. Roosevelt's B-17 to the USA in December.

Crash remains of Mosquito NS730, 4 February 1945. *Photo W. Borges*

67 File 520.923. Oct. 44-Sept. 45, Combat Camera Units, USAFHRA.

On the11th, Lts. John Carter/William Loden (MM344) flew a WR to Germany. At 1540 while at 24,000 feet an electrical fire with sparks ensued from the instrument panel. Loden, the navigator, climbed into the nose and reported smoke issuing from a junction box. Then a generator went out and Carter switched off all electrical equipment and turned back for England. But 30 minutes later smoke again poured from the nose area. The navigator attempted using a fire extinguisher but passed out from its toxic exhaust. Carter placed an oxygen mask on Loden's face reviving him immediately. Carter then spotted an airfield and landed safely at Davidstow Moore. Repairs were affected before return to Watton.

Bad weather grounded all bombers in England on 13 February but not Mosquito WR missions. Finally, Col. Elliott Roosevelt received promotion to brigadier general this day. His meteoric rise included: Captain, September 1940; Major, March 1942; Lt. Colonel, summer 1942, and Colonel, March 1943. Rapid advancement was not uncommon in World War II. Cinema actor James Stewart's ascended from 2/Lt. to Colonel in four years during an enlistment period briefer than Roosevelt's.

The Graypea Mosquito allotment increased. The 25 BG (Rcn) received instructions to provide four Graypea and one command aircraft upon request to each AD commander. Any Mosquito could be utilized for command purposes.

In one of several coordinated attacks by Eighth Air Force and RAF bomber command, 1,377 Forts and Liberators were dispatched in four task forces against Dresden and Chemnitz. A Graypea flight supported 1 AD to Dresden. En route, three Fw 190s flew at 27,000 feet and 2,000 yards distance, did not attack. Lts. Richard E. Petersen/Major Richard Wright (NS626) turned back near Dresden with fuel problems. An AAF radar station misidentified him as an Me 410 and initiated intercept proceedings. Petersen immediately called Baggage giving his position north of Strasbourg at 21,000 feet. Baggage requested he orbit, and with identification confirmed, the fighters called off. Petersen landed at Brussels to refuel.

That evening, three night photo Mosquitos flew to Amsterdam as primary and Zwolle, Deventer and Moppel as secondary targets. Lts. Kingdon R. Knapp/Robert Spaight (NS554) found the primary covered with cloud, so Knapp flew west of Krefeld, Holland, for a target of opportunity without clouds for a photo run. As Spaight recalls:

> "The pilot actuated the intervalometer set to release one flash bomb every 38 seconds. I gave Knapp our position line and he switched on the intervalometer, starting the photo run. One away … two away … after the ninth flash bomb released, we waited for the tenth.
>
> "Suddenly, Knapp reached over to salvo the remaining bomb load. A split second after clearing the aircraft there was a brilliant blinding light beneath us and a horrendous explosion that abruptly jolted and violently threw the aircraft about. I was taken completely by surprise and didn't have time to realize what had happened. The mental shock of what occurred caused my heart to beat heavily.
>
> "Upon gaining control, we gathered our thoughts and quietly returned to Watton. After landing we climbed out and inspected the aircraft. No structural damage was evident, but heat generated from the exploding PFBs scorched off the entire paint covering the Mosquito lower surfaces. Somehow, Knapp knew the 10th flash bomb had partially hung up and salvoed the load. How he deduced something was wrong, I don't know. I have always said, 'He saved my life.'"

Capt. Victor S. Doroski, KIA 8 February 1945, flying a Joker mission. *Photo R. Walker*

On 17 February, Pinetree assigned six additional Graypea Mosquitos to both 1 and 2 AD, and one for command purposes to each AD. Five aircraft modified at 2 SAD plus 11 more assigned this task totaled 16 Mosquitos.

Lts. Roger W. Gilbert/John K. Jackson (RF985) along with Mosquito RF992 and B-26 #226 flew to Alconbury on the 19th. Haze and foggy conditions obscured the runway. Gilbert experienced difficulty aligning properly, and on his third approach, touched down midway down the runway. The plane ran off the end, through a barbed-wire fence, across a highway and on to a field where rough ground collapsed the landing gear.

The first good weather in weeks appeared 20 February. Over a thousand bombers attacked German aircraft factories at Brunswick, Oschersleben, Bernberg and Leipzig. The attack continued on Monday, and on Tuesday. Eighth Air Force joined Fifteenth Air Force bombers in the first coordinated attack of USSTAF in Europe. On Friday, more than 2,000 bombers raided Regensburg, Augsburg, Furth and Stuttgart.

19 February 1945 accident of RF985 at Alconbury. *Photo W. Borges*

Trap shooting members at Watton. L-R. Front row: 1/Lts. Ray Spoerl, James W. Handren, Capt. Louis A. Rogers, Unknown, Capt. Allan B. Bateman, Major James M. McNulty Jr. Rear row: Capt. Richard Kenny, 1/Lts. Robert Spaight, Charles Finley, and Capt. J. Keith Werthman. *Photo J. McNulty*

A spate of accidents marred the 20th. During a training flight, Lt. Joseph A. Polovick (MM342) ground looped during landing, collapsing the right undercarriage. Lt. Warren C. Davis (MM345) returning from a local flight veered off the runway and collapsed the left undercarriage gear. That afternoon, he was again airborne (NS508) for another training flight. He over-corrected and swerved off the runway, washing out the aircraft.

The following day, Capt. Robert P. Howle Jr. ferried MM388 to Alconbury. After touchdown he rolled one-third down the runway when the left tire blew causing the aircraft to vibrate violently. The left undercarriage collapsed causing a 180-degree left turn before coming to rest.

The night of 21 February, Lts. Peter Dustman/Len Erickson (NS554) flew a night photo mission to Kamen, Germany. During the photo run, an RAF night-fighter intercepted them. As Dustman recalls:

> "We were close to Essen during an RAF raid on that city. Flying straight and level at 24,000, I started the intervalometer to drop the flash bombs at intervals. It takes 30 to 35 seconds at that altitude for a bomb to flash. The first bomb released failed to flash. Then I heard over the radio an Englishman say, 'Mosquito wiggle your wings.'
>
> "I knew it was an RAF night-fighter on our tail wanting us to acknowledge we were a Mosquito and not a German aircraft. Then I heard the English voice break in, 'Mossie, rock your wings.'
>
> "Erickson said, 'Let's get a couple of more pictures anyway.'
>
> "A voice then said, 'This Mossie will not rock his wings.'
>
> "A faint voice over the radio replied, 'Then shoot the bastard down.'
>
> "I quickly rocked the wings, turned, and got out of there. We did not get a single picture. It was an RAF Mossie night-fighter that was on to us. The Northrop P-61 lacked sufficient speed and could not climb as fast, so we did not fear them. It was the RAF night-fighters we worried about. This particular night-fighter was sent to protect the RAF bomber stream, and it was a mistake sending us to an area they were bombing."

February marked the final chapter in the Mickey program; 27 missions were airborne during the month before the order suspending further flights arrived. An analysis revealed 55 approaches flown, 34 accomplished and only four approaches accepted. The Mosquito electrical system could not handle the additional load imposed by H2X equipment. With the scanner switched on, low temperature at high altitudes congealed oil in the motors and the resulting current surge blew fuses, and in some cases burned the cables. Lack of space hampered the navigator, preventing efficient equipment operation.

The Mickey P-38 solved the power supply problem at the cost of reducing navigation efficiency. The navigator sat confined in the extended nose operating both H2X and camera. Lack of space prevented maintaining a navigational log and cramped him on long missions. After only four P-38 Mickey missions, it became apparent neither aircraft proved suitable for H2X work. An ideal plane had to combine speed, range and altitude, sufficient space for equipment, and enable the operator to adjust the set in flight; also, a power supply capable of providing smooth current and regulation. After phasing out the Mickey program, it became necessary to transfer many radar technicians to 482 BG and others to the 50 Fighter Control Squadron, RAF Station Hinston.

The 24th marked the transfer of all Graypea missions from 653 to 654 Squadron. Overburden with weather gathering duties, the shift of responsibility helped alleviate the workload. And with Mickey project cancellation, provided the 654th a new role.

Four instead of three aircraft were dispatched on Graypea screening missions. Initially, the fourth aircraft served as spare and returned to base when no abortions occurred. In other instances the spare completed the sortie and dispensed Chaff along with the other three aircraft. This new arrangement helped insure mission success. More Mosquito aircraft underwent modification to Chaff ships in contemplation of Graypea missions supporting all three ADs.

20 February, "Z" MM345 "The Beast" ground loop during a training flight. *Photo G. Scruggs*

After the MM345 20 February accident, the pilot ground looped NS508 "A" that same day. *Photo G. Scruggs*

Having lost France, the Germans launched V-2 rockets from Holland. On 27 February, Lts. Allan Bateman/John Cone (NS752) flying a WR mission witnessed three V-2s launched from The Hague at 0118, 0121, and 0137. Early that morning, Lt. Morton Hunt of 654 BS flew a Redtail mission:

"I flew to Horsham-St. Faith and picked up General W.R. Peck of the 96 Wing. I was to chauffeur the general so he could observe his Group formation en route and over the target. The operations officer informed me to my delight that we would have no less than four P-51 escorts guarding him. It was the only time in all my missions that I flew with a P-51 escort. The route that day was from Horsham to the Frisian Islands to Leipzig. Upon rendezvous with the bomber formation, I viewed a sight I had never seen before. An aerial armada of heavy bombers in the sky stretching as far as the eye could see, mile after mile in excellent visibility. It was an impressive sight.

654 Squadron members, L-R. Navigator1/Lt. John Morrison and pilot 1/Lt. Morton Hunt. *Photo J. Morrison*

"The general instructed me where he wanted to fly and I cruised along past the formations at 325 mph easily. We crawled up the line of bombers as they lumbered along to the target. Presumably, the bomber formations had been instructed to be aware of a red tailed aircraft cruising past or through the bomber formations. We observed machine-gun turrets aim at and follow us down the line. I recall General Peck, who was very cool and commanding, uttered several times very testily, 'Keep further away … that's too close, move farther off.' So I veered and flew somewhat further away from the bombers. At all times, reassuringly above and slightly behind or off to the sides flew the four P-51s. They escorted us as far as their fuel supply would allow to Leipzig, then peeled off and left for home. The mission lasted 6:20 hours which is a long time to be sitting in a Mossie. On return we diverted to Church Fenton, then later I returned General Peck to Horsham-St. Faith."

Not all Redtail missions received escorts. On 27 February, Lt. John Green collected a Colonel Stewart in NS708 from Hethel. En route to the bomber stream at 1145, a lone P-51 voluntarily provided support by escorting the Mosquito throughout the flight over enemy territory. They experienced heavy flak in the Leipzig area, accurate for height and moderately accurate for position. Green reported his passenger was actor Col. James Stewart. However, Stewart replied to a letter denying he ever flew in a Mosquito? Major Willis Locke recalls visiting Hethel previously to pickup actor Col. James Stewart, but Pinetree canceled that particular AD mission.

Also on the 27th, Lt. Thomas L. Merrill quickly taxied B-17 43-38709 on to the main runway at Lagens Field, Azores. A left brake failure caused the aircraft to ground loop 180 degrees resulting in the left wingtip scraping the runway shoulder embankment.

The communication section modified all new Mosquitos replacing the British radios with American VHF sets, along with both Gee and Loran radio navigation aides with installation of trailing wire antenna.

Zthe Joker missions also faced complex technical problems. Of twenty-three night PRs flown by 654th, one successfully covered the primary target, eight secured good photos of targets of opportunity while the remainder were failures. Technical difficulties varied. Time fusing with bombs igniting in the photograph; synchronization of camera with flash illumination proved insurmountable obstacles; and overcast clouds at the target ruined most photo results.

March 1945

The Loran training program established in November failed to produce a gratifying grasp of its navigational function in flight. To rectify the problem a team led by senior Group navigator, Major John Walch, designed a unique trainer by combining Loran with a Link recorder. They converted a standard Loran trainer, and using the Link recorder, sent a continuously moving signal from the transmitter, simulating moving symbols appearing on the Loran scope similar to a plane in flight. They constructed a special map on which the Link recorder or "crab" operated. The map was marked off with hyperbolas representing two Loran stations. A reading taken on two stations gave the exact position of the crab. The hyperbolic lines were superimposed on a Mercator projection, enabling the navigator to mark his fixes and determine a position. A check of the navigator's accuracy became readily available by comparing his map with the flight of the crab. All navigators recognized the practicality of the trainer.

In March, Watton established Loran classes utilizing the new trainer. Navigators and radar men from all ADs arrived to receive instruction. Soon training requests overwhelmed available resources. Therefore, the 25 BG (Rcn) formulated plans for development and assignment of similar trainers to BGs for navigator training.

The Loran section devised an additional aid. Inadequate space in the Mosquito cockpit compelled the navigator to operate the Loran over his left shoulder, normally beneath a shroud in complete darkness. To overcome this handicap, the section devised a blackout booth simulating exact conditions confronting the navigator in flight. With experience acquired in this contrivance, the navigator determined for himself the easiest manner of operating the equipment.

Wrecked Mosquito being scavenged by groundcrew for components with B-17 48507 parked nearby. *Photo L. Erickson*

Mosquito carcasses parked in a storage area and utilized as a parts department.
Photo. L. Erickson

Mosquito graveyard of airframes scavenged for parts. Note in background a camouflaged Mk.III trainer. *Photo L. Erickson.*

The communication section installed dual-VHF radio equipment in five Mosquitos to aid tactical communications between Redtail aircraft and Combat Wings of the ADs. Its first tactical employment occurred 2 March when Lt.Col. John Larkin (NS708) flew to Horsham St. Faith to pick up General Johnson for a mission to Chemnitz. Flak bursts at 25,000 feet shook the aircraft. Next day, two Redtails flew similar missions: Lt. Harold James took Col. Brown (NS708) while Capt. Robert Shoenhair carried General Peck (NS744), both observing a 2 AD bombing mission.

Foo Fighters appeared on 5 March. Lts. Theodore B. Smith/Arnold Kuehn (NS752) at 2146 flying a WR over Germany at 23,000 feet observed a large orange ball approximately 5,000 feet below the Mosquito and 20 miles in front. The ball seemed to hang in the air for five seconds. Again at 2135 a similar ball appeared to move horizontally at the same altitude as the Mosquito at a distance of 20 miles and remained visible for 15 seconds.

Watton experienced another series of accidents. On 7 March, Lts. James Evans/Robert Spaight (NS651) were taking off when the aircraft began to swing. Both brake and throttle failed to prevent going off the runway where the left wheel snagged in a drainage culvert collapsing that landing gear. Next day, Lts. Robert James/Terrence Hall (NS782) returned from a night WR and on touchdown the left landing gear began to collapse. James maintained speed and took off. He circled for 45 minutes attempting to lock the undercarriage in the down position. The engineering officer advised the pilot to land with both wheels down and to retract the right landing gear if the aircraft started sinking. Upon touchdown, the left landing gear began to collapse. James endeavored to retract the right undercarriage, but the Mosquito ground looped, completely demolishing the aircraft. The following day, Lt. Robert B. Hastie (NS595) took off on a transition flight when the aircraft swerved right causing the left undercarriage to collapse.

On 9 March, HQ 325 PW received notification AM production of Mosquito Mk.XVI will cease in favor of Mark 35. Further allocation after July 1945 will be Mk.34 Mosquitos. Also on this day, Major Hugh Bozarth/Lt. Harry Anglum (NS634) flew a night WR. At 10,000 feet near the North Frisian Islands the crew encountered three Foo Fighters – two were low and the other at 15,000 ft. They gave an appearance of white parachute flares, persisting for 15 seconds above a solid cloud at a distance of 25 miles. They eventually faded out like flares.

On return, Bozarth called Largetype ground control to clear the coast and his message acknowledged. Two minutes later, another Largetype called saying, "Hello Maypole 16, have you a message for me," repeating this three times. The transmission was weaker and in a masculine voice, while the first was feminine, louder and stronger. Bozarth did not respond to the call. Two minutes later he called Largetype and the original station came in. Nothing further was heard from the second, possibly enemy station.

654 Squadron NS651 "Woodpecker's Delight" accident on 5 March 1945. *Photo W. Borges.*

Increased Luftwaffe aircraft activity, notably Me 262s were encountered but proved less dangerous than friendly fire from bomber formations. On several missions, the Graypea screening force was fired upon by gunners in the heavies. Fortunately, no casualties were sustained. The inability of BG gunners to recognize friendly Mosquitos resulted in low morale and operational efficiency.

During March, Group devised a Graypea formation to provide the lead bomb groups with maximum screening cover. Prior to dispensing, the Mosquitos formed line-abreast, spread out approximately 150 to 300 yards. As they proceed with the "S-ing" run from IP to the target, the screening leader called the heading every 15 seconds, enabling the entire screening force to "S" in unison.

Mosquito V-Victor NS782 "Angela" accident 8 March at 0345 hours. *Photo W. Borges*

48 mission markings on 653 Squadron Mosquito. L-R. S/Sgt. W.F. Barnett, crew chief; 1/Lt Warren Borges, pilot; 1/Lt Wendell Biggers, navigator; and Sgt. Steve Spohr. *Photo W. Borges*

NS595 ground loop on 9 March 1945. *Photo W. Borges*

20 February 1945, 652 BS crew, L-R. T/Sgt. Reynolds, engineer; 1/Lt. Connors, copilot; 1/Lt. Robert Menken, pilot; 1/Lt. Martinez, navigator; Sgt. Eckert, gunner; Unknown radio operator; and Sgt. Cox, gunner. *Photo. R. Menken*

On the 10th, a Graypea mission supported 1 AD. The Mosquito formation flew over the Zuider Zee at 27,000 feet en route to the target when the interphone system failed on RF988, but Lt.Col. Alvin Podwojski/Lt. Len Erickson continued onward using handsignals or scribbled messages.

Erickson, working at his navigator's table, noticed shadow movements in the cockpit. He looked upward and observed the Mosquito's nose pointed high in a near stall attitude. He quickly glanced at the pilot and found him sitting motionless with his head slumped and both hands on the control wheel, unconscious from anoxia. Erickson reached over, pushed the control wheel forward to prevent a stall and placed RF988 in a shallow dive. He informed the other Mosquitos of his situation and they continued to the target. Erickson maintained descent for 10 minutes until reaching 12,000 feet when the pilot regained consciousness.

Podwojski went on his emergency oxygen and climbed to 25,000 feet. Attempting to rejoin the Mosquito formation, he flew parallel to the bomber stream and overtook a group of 36 bombers having the horizontal yellow tail stripe of the 4 CW, 3 AD. Approximately 500 yards separated the aircraft when at 1249, arcing streams of red tracers were directed at the Mosquito. Podwojski racked the Mosquito over to show the U.S. insignia on the wing and the aggression ceased.

Meanwhile, the three Mosquitos reached the target and encountered heavy flak bursting behind their formation as they released Chaff. RF988 flew over the target four minutes later to discharge its load and return to base.

The Graypea mission became somewhat complicated beginning 11 March. One Mosquito had a 36-inch focal lens camera installed in the nose to obtain strike attack photos for damage assessment. Weather permitting photos of the target area were taken prior to the bombing run. After discharging Chaff, one Mosquito circled around and followed the last bomber element to obtain post-strike photos over the target. The pictures were immediately dispatched to Wing HQ by special armed courier.

That evening, Lts. Richard Petersen/Richard Davies (NS752) encountered an unusual phenomenon. While flying a WR, the crew observed a yellowish-orange glow trailing the Mosquito at 13 miles distance. This glow followed them for 15 minutes, slowly closing in, and when three miles behind, it suddenly faded away. After crossing over from the enemy lines at 6,000 feet, they encountered two unidentified night-fighters. Evasive action was taken losing both in clouds before they could slip into position to fire. The men cleared with ground control Gondola for permission to cross the continental coast at 10,000 feet, whereupon they were coned by Allied searchlights and received intense 20 mm antiaircraft fire over Antwerp. The aircraft escaped unscathed.

A tragic ground accident occurred the 12th. Cpl. Richard N. Bartlett rode a bicycle with one hand holding on to the engine nacelle of a towed Mosquito. He lost balance, toppled to the pavement, and the landing gear wheel ran over him.

Manhattan Project officials developing the U.S. atomic bomb issued a special request to Eighth Air Force. They speculated German scientists were preoccupied with similar interests, developing nuclear material in Oranienburg, 15 mile north of Berlin. Located in the projected future Soviet zone of occupation after the war, the officials wanted the plant destroyed. On 15 March, Eighth Air Force dispatched two large bomber forces. To disguise the attacks true intent, one formation destroyed the Oranienburg plant while the other bombed German Army headquarters of *General* Heinz Guderian in nearby Zossen.[68]

Screening 1 AD bombers attacking Oranienburg were: Lt.Col. Alvin Podwojski/Lt. Floyd Mann (NS743); Lts. John Grassel/Joseph Ward (NS785); Clarence Korthase/Lucien Peters (NS512); and Joseph Polovick/Bernard Blaum (RF986). Screening the 2 AD raid on Wunsdorf-Zossen were Major Willis Locke/Lt. Louis Toth (RF992); Lts. Lyndon Lakeman/Robert Spaight (RF979); and Charles Finley/Robert Balser (RF983). As Spaight recalls:

> "I flew with Lakeman to a German airfield called Wunsdorf-Zossen, located 20 miles south of Berlin. Our Intelligence believed Hitler would visit the airport that particular day, and 2 AD would bomb the place. Our three Mosquitos dropped Chaff at Zossen, then detached ourselves from the bomber formations and headed southeast to the secondary target to photograph synthetic oil refineries at Merseburg, a distant suburb of Leipzig. At Merseburg, German smoke generators obscured the target area with a dense smoke layer – you couldn't see a thing! The Mosquito formation flew directly over the plant and our lead navigator continued flying past Merseburg. I informed Lakeman that we had already flown past the target and to turn around and head for home. Lakeman said we would follow the flight leader and maintain radio silence. We continued flying south and eventually, I told him we were far off course heading for Munich.
>
> "Finally, someone spotted a large city on our left and thought it was Leipzig so our formation turned towards it. Lakeman had me call the lead navigator and inform him the city was Nuremberg. After concluding I was correct, our formation turned for home. Having consumed considerable fuel on this grand tour, we all landed at Brussels for refueling and remained overnight."

Watton listed an entire crew from 652 Squadron officially KIA on 17 March after failing to return from an Atlantic Allah WR mission. All were 492 BG members attached to 652 BS flying B-17G 448601 The last radio contact received by St. Eval at 0010 requested a QFF. Later, a petrol fire was observed six miles off the coast in the vicinity of the track for the mission. An exhaustive search by 19 Group RAF retrieved aircraft parts identified as a B-17 and the flying jacket and letters belong to Lt. Anthony D. Pellegrini, navigator of the missing plane. KIA were l/Lt. Charles L. Sortomme, 2/Lts. Calman T. Estrach, Robert N. Clark, Anthony D. Pellegrini, S/Sgt. Robert C. Allison, Sgt. John C. Gerlott, Cpls. James G. Chressanthis and Arthur G. Eickhoff.

Mosquitos provided Graypea support for a lst AD mission to Berlin the 18th. They were at 25,000 feet over Berlin when an Me 262 approached the Mosquito formation at an angle. Lts. Kenneth Kenyon/Floyd Mann (NS743) immediately turned into the jet and it peeled off in a dive. The jet flew a 360-degree full circle and climbed under him, passing 100 yards off the left wing, then disappeared without making further contact.

March 20 proved an event-filled day launching four Mosquitos on a Graypea screening to Hamburg: Lts. Norman Magee/Len Erickson (RF988); Charles Finley/Robert Balser (RF999); Roger Gilbert/Ray Spoerl (RF992); and Joseph Polovick/Bernard Blaum (RF996). The Mosquito formation made visual and VHF contact with the bomber formations some distance from the target. At 1554, flying a four abreast formation, Polovick, Finley, Magee and Gilbert began discharging Chaff over Stade, Germany, while flying above the bombers at 26,000 feet. According to Gilbert:

> "We were not quite line-abreast and I was on the right and a little lower than the lead ship of Lt. Magee to my left. At this time I remember hearing three radio transmissions about Luftwaffe jet aircraft in our area. On the last warning from Magee, the situation became tense. Ray Spoerl who sat slightly behind me, unbuckled his safety belt, turned around and poked his head into the observation bubble in the canopy roof to inspect the rear. He reported shortly thereafter, an Me 262 appeared coming at our aircraft from six o'clock high. A few seconds later he shouted the jet was firing at us and to break. I immediately broke as hard left as I could to turn inside the jet and get out of his line of fire. After completing 45 degrees in the bank, 30 mm shells from the jet ripped into the aircraft. Four rounds struck the instrument panel while others shot out the radio and punctured our dinghy. Another burst tore off four feet from the outer left wing – I was startled, watching pieces of the wing fly off in the turn. This momentarily jammed the aileron. I had elevator and rudder control but no ailerons – they were in full, left turn position and jammed.
>
> "The damaged wing enabled me to make a tighter left turn. The Me 262 flashed by closely on my right. When I initiated the sudden sharp break, centrifugal force immediately pulled and pinned Spoerl to the floor. The cannon fire punched a hole right through the observation bubble he peered out to check behind our aircraft. I glanced down at him sprawled on the floor and thought he had been shot. But as the centrifugal force decreased, he attempted to recover to his position next to me. He was all right. I became greatly concerned with being unable to recover from this very tight spiral with a damaged left wing. We were rapidly descending with 240 indicated air speed and pulling a lot of 'G's. Finally, I broke the ailerons loose and they worked partially in one direction only. I then applied differential throttle by retarding the right throttle lever while advancing the left. I was able to recover from the spiral at around 20,000 feet.

68 Richard Davis, Carl A. Spaatz and the Air War In Europe, Washington, DC: Center for Air Force History, GPO, 1993: 579.

1/Lt. Roger Gilbert examines the top right wing damage which appears larger than when viewed from the bottom. *Photo R. Gilbert*

Roger Gilbert on left and navigator Raymond G. Spoerl on right.
Photo G. Gilbert

> "The radio was shot out, eliminating communication with anyone. I turned and banked the Mosquito, returning to England. Both engines operated smoothly and with the higher engine power settings, I maintained aircraft control. Once the Mosquito was trimmed properly I did not experience any further control problems until air speed dropped to 170 mph; this caused the aircraft fall off toward the left. As we proceeded towards the North Sea thoughts of having a punctured dinghy worried us."

Meanwhile, Polovick/Blaum completed discharging Chaff when they heard Magee's voice over the radio, "There are Me's all over the sky, you're on your own!" The crew spotted several jets then heard Finley shouting excitedly over the radio, "Joe, an Me on your tail at six o'clock low." Polovick broke left and down for several thousand feet. After recovery he flew a return heading northwest for the North Sea. Once clear of the target with no German fighters present, he joined Finley and climbed to 25,000 feet over northern Germany. When 15 or 20 miles off the coast over the North Sea, they changed course due west and flew a loose formation heading home.

When Polovick switched from underwing tanks to the main tank, one engine sputtered and stopped from a fuel air lock. Finley flew alongside calling Mayday and their position. Polovick then experienced a second air lock and radioed, "Y-Yoke, both engines out." He banked to reach land 25 miles away and glided towards Fohr Island on the German coast. The aircraft lost altitude quicker than anticipated. Once at 1,500 feet Blaum bailed out over the island and floated to earth. Polovick trimmed the aircraft and jumped from 500 feet. Both men landed near the crash site and captured by armed German farmers. The Luftwaffe transferred the crew to Stalag Luft One, Barth, Germany.

After witnessing the safe bailout, Finely/Balser continued to England. The other crew of Magee/Erickson feathered a prop after the starboard engine ran rough and returned safely.

Gilbert flew his damaged Mosquito to Watton:

> "Without a radio, we had to somehow communicate with the control tower. I decided to fly low past the tower so they could see that part of the wing was shot off for visual confirmation of our present problem. So I made the run across the field at 220 or 230 mph and stuck that damaged wing right in their windows so they would not miss seeing it. They alerted the crash and fire equipment. I was unable to keep the wings level below 170 mph and was forced to land at this speed ... upon touchdown I slapped the wheels right on the runway edge, pulled back the power, extend full flaps and the Mosquito safely rolled along. We then taxied into the dispersal area. Normally on return only the crew chief met the aircraft along with ground transportation to pick up the crew. But now there were a number of men including the maintenance officer and operations staff waiting to inspect the damaged aircraft.
>
> "Repairing the Mosquito is a story in itself. An English repair party arrived from the de Havilland factory. I visited the hangar and watched the civilians saw the left wing off with a giant saw – a large lumber saw, similar to those used for cutting down trees. The workers sawed right through the spar! They brought in a replacement wing, glued it on with splices, covered another piece of plywood over and placed a band around the splice joint, fasten with all kinds of screws.

Gilbert points to bullet hole with another visible in canopy dome from Me 262 attack. *Photo R. Gilbert*

"I met Bob Howle, my squadron commander and said, 'My God, Major, I'm not going to fly that damn plane. They cut the wing off and all they did was glue another one back on. That thing will fall off. I'm not going to fly it.' He asked what I wanted and I told him a different airplane. He issued me a requisition form. They had a group of new Mosquitos parked in a storage area to replace those we lost. I received a new Mosquito, had it painted in regulation markings and flew it. The maintenance officer, Capt. Robert Shoenhair, when he heard my refusal to fly the rewinged Mosquito replied, 'Hell I'll fly it.' And he did – he flew it the rest of the war!"

During March a number of RAF airfields experienced night attacks from Luftwaffe intruders dispatched to shoot down returning bombers preparing to land. Besides RAF airfields at Matlask, Swanton, Morely, West Raynham and Little Snoring, Watton also came under attack from night-fighters.[69] Late evening 20/21 March, Lt. Robert Biesiot returned to Watton in B-17 #533 from an Atlantic WR. During final approach the control tower shouted on the radio, "Stop and drop immediately, a Junker is behind you." Biesiot shutoff the landing-lights, chopped the throttles and raised flaps. The Fort dropped rapidly and came down in a field 150 yards short of approach as red tracers streamed overhead and ricocheted off the concrete runway in front. The Fort hit hard and bounced into the air. Biesiot applied full throttles to maintain flying speed and bounced safely on the runway edge. The Ju 88 disappeared in the darkness.

More accidents. On 22 March, Lt. Kenneth McGriffin (NS619) practiced night landings when the aircraft bounced on touchdown, then swerved left off the runway collapsing the undercarriage. Next day, Lts. Morton M. Hunt/Allan Morrow (NS552) took off at 0306 on a Joker night photo mission. The aircraft began to veer left, went off the runway and ground looped.

Night landing accident on 22 March, 654 Squadron, NS619. *Photo W. Borges*

Night accident on 23 March of NS552 Joker aircraft. *Photo W. Borges*

69 Simon W. Parry, Intruders Over Britain, Air Research Publications, 1987:160

Several tragedies ended the month. The first occurred 24 March on a special mission over Germany. In February 1945, Lts. Carrol B. Stubblefield and James B. Richmond as navigator, received orders to participate in a top-secret mission. They selected Mosquito NS711 in RAF markings from storage and flew to RAF Defford, Great Malvern, near the Welsh border of England. They reported to the American British Laboratory (ABL), part of Telecommunications Research Establishment at Malvern having scientists and engineers involved in electronic and radar research.

The Allies captured an intact Luftwaffe identification, friend or foe (IFF) set, the Pfleuger 25. American scientist Dr. Eugene Fubini led a team of men and women from Harvard University, develop a device to locate Luftwaffe aircraft operating German IFF. They designed special electronic equipment to pulse a similar IFF signal. Once a Luftwaffe aircraft receives and processes the signal for a particular length of time, it then transmits to the ground station, confirming it as a friend. Special equipment installed in the Mosquito would calibrate electronically the approximate distance and direction of the IFF source.

Engineers positioned a special antenna with vertical and horizontal dipoles the wavelength of the Pfleuger 25 frequency, behind the nose Perspex. They also installed in the cockpit, behind the navigator's position, cathode ray tube (CRT) equipment to measure the IFF signal. The Mosquito performed tests pulsing a Pfleuger 25 station on a hill near Malvern. This allowed ABL engineers to determine the feasibility and parameters for measuring distance and direction to insure the device functioned as designed. ABL formulated plans for the Mosquito, escorted by Mustangs, to fly over enemy territory. The crew would emit pulses to entice a Luftwaffe aircraft IFF to respond, then vector the Mustangs to the enemy for combat engagement.

Col. David Shilling, 56 Fighter Group, flew Richmond in a modified two-seat P-51 to an airfield in Holland near the German border. The site provided radar coverage over enemy territory and arranged for code words such as Nuthouse and Rosebud, etc. At Defford the British applied American markings to NS711 but did NOT paint the tail section red.

On 24 March the Mosquito left Defford at 1500 to unite with eight P-51 Mustangs over Watton on an escorted mission over enemy occupied Europe. The group flew to Antwerp, Aachen, the Remagen bridgehead, to Limberg, Siegen, then headed northeast towards the Kassel area. The Mosquito continued weaving at 18,500 feet to allow the escorts to keep pace with him. The Mustangs flew at 19,000 feet and trailed 500 to 1,000 yards behind.

At 1700 hours Nuthouse reported bogies heading west and others going east. Those flying west were identified as eight P-47s at 17,000 feet. Four made a 270-degree turn and flew in formation with the P-51s. The Mustang pilots attempted calling them on C channel without success. The Mosquito weaved, then banked on a northeasterly course, informing Nuthouse of return to his original heading.

As the Mosquito continued to weave, two P-47s flying with the Mustangs turned inside and closed down on the Mosquito. One P-51 pilot radioed a warning the Thunderbolts were attacking, as the Mosquito pilot replied, "I thought they were you."

At RAF Defford, L-R. Civilian American engineer, 1/Lt. Carroll Stubblefield, American secretary, 1/Lt James Richmond. *Photo J. Richmond*

High school photo of Carroll Stubblefield, senior class of 1941.

Immediately, the P-51 group leader dropped his wing tanks and said, "I'm on the P-47s." Meanwhile, the two Thunderbolts started firing and obtained hits on the tail and both engines, setting the left on fire. The Mosquito then broke into three pieces and fell to earth. A parachute appeared at 10,000 feet from one of the tumbling pieces. Lt. Carrol Stubblefield was killed in this action.

The P-51 group leader positioned himself numerous times behind the P-47s, but the Thunderbolts rocked their wings violently, so he held off firing. After the Mosquito fell from the sky, two Mustangs of the group announced they would follow the Thunderbolts to their home base near Le Culot, Belgium. The P-47s had yellow around the nose and a yellow tail, bearing the markings 7UV and M7U (36 Fighter Group).

According to Richmond:

"Our call sign was Rosebud. We had eight Mustangs escorting us and went searching for bogies to test this concept. We were flying on our particular course emitting pulses but did not pickup any IFF signals. The Mosquito is very cramped. The set I was working on is located behind the pilot and I'm turned around about 135 degrees looking at the CRT. I was sitting nearly backwards under a shroud to view the display screen. The CRT had a scale on the screen showing blips and measures on a base line giving us distance. My attention is focused looking at the screen, and preoccupied with this task, I'm oblivious as to what was happening around me. The first thing I knew, I heard Stub say, 'God damn, we're being chased.'

NS711 at RAF Defford prior to application of U.S. insignia. The tail section was NOT painted red. *Photo J. Richmond*

"I said, 'What do you mean?'

"He replied, 'Someone is shooting at us' and he started screaming into the radio (to the Mustang escort), 'Get them off, get them off.'

"When I pulled my head out from beneath the shroud, there were streams of tracers passing by the cockpit all around us. At this moment the port engine caught fire, and then the starboard engine was afire with flames. Stub feathered that prop and shouted to me, 'Get the hell out of here Rich. GO.'

"I immediately kicked out the bottom trap door and dropped myself down into the opening. I was partially out the hatch when the aircraft exploded and I was blown clear. Before I knew it, I found myself tumbling and falling through the air. I managed to pull the D-ring and the parachute deployed. I immediately looked around and could see three pieces of the Mosquito falling to earth. I watched and watched but never saw a parachute open as the remains of our aircraft plummeted to earth."

Richmond landed at Brilon near Magdeburg, Germany. During final descent he crashed through high trees injuring his right leg. Armed civilians observing the event assisted him to the village where he remained in custody for 48 hours before transport to Pinnaburg near Hamburg. The Luftwaffe transferred him to Stalag Luft One at Barth on the Baltic. On 1 May Soviet troops liberated the facility and POWS repatriated to Camp Lucky Strike, Le Havre, France. A few days later, S-2 escorted Richmond to SHAEF HQ Paris for interrogation by a major general regarding the 24 March mission and the exact disposition of the secret equipment.

For the third time in March, the Altenbeken viaduct became target for night photo missions on 23 March, as two night sorties flew to assess activity on the viaduct. The same target received similar attention the 24th with four aircraft producing large-scale photos from 12,000 feet. According to SHAEF documents:

> "In ... March, the enemy withdrew across the Rhine using bridge and ferries, during which period the river ... was covered with stratus at about 4,000 feet. These positions were strongly defended by flak which combined with the low-cloud ceiling made daylight PR almost suicidal. The only alternative – night photography. Stretches of the river were photographed to cover important crossings and the result of the recces showed bridges and ferries in use and provided an estimate of the volume of traffic. The Mosquitos allotted to this task were aided by a late rising moon and the photos obtained were of good quality and permitted a certain amount of detailed interpretation. Such an occasion gave night photography an opportunity to prove its usefulness."[70]

Also on 24 March, Capt. Keith Werthman/Lt. Edward Hodges (LR530) taxied out along the perimeter track in preparation for takeoff on a checkout flight. The Mosquito had flown an hour earlier by another pilot practicing landings. Hodges completed a turn on the perimeter track when the brakes failed. Werthman used differential throttle to move the aircraft back on track, then cut the switches and stopped both engines. Unfortunately, the perimeter track had a downgrade and with no brakes, the Mosquito rolled off the track towards a parked aircraft, clipping the wingtip of C-47 #871.

25 March 1945, 1/Lt. John Green standing next to a Mosquito that accidentally rolled into C-47 wing. *Photo J. Green*

70 SHAEF, RG331, Entry 266, Box 53, NARA, College Park, MD

Shortly after midnight 24/25 March, Lts. Louis Rogers/Charles Veldon (NS578) flew a WR to Germany. En route at 0149 they observed an explosion with a blinding flash at 30,000 feet presumed a V-2. Early that morning, Lts. Warren Borges/Robert Gordon (NS774) took off in a crosswind for a test flight. Approaching flying speed, the aircraft swerved left, but right brake, left throttle and rudder control failed to prevent a ground loop, collapsing the right undercarriage.

A tragedy occurred 25 March when Lts. Bernard J. Boucher and Louis Pessirilo (NS752) left at 1448 hours on a WR flight to the continent. American forces had crossed the Rhine River with Hitler's Third Reich on the verge of total defeat. The flight continued uneventful with no Luftwaffe aircraft seen. Cloudy weather covered Germany and Boucher took the Mosquito down to determine the exact altitude of cloud levels that were difficult to estimate. Unfortunately, this brought the aircraft within range of Flak batteries.

During descent flak struck the starboard engine that erupted into a raging inferno streaming flames and black smoke. Before Boucher could feather or shut it down, a second burst rocked the aircraft, severely damaging the tail section. This second interruption caused loss of directional control and the Mosquito started a spiral descent.

Both crewmembers prepared to abandon the aircraft. Unfortunately, Pessirilo's chest pack slid loose from between his legs and centrifugal force from the spiral forced it into the nose section. Lou bent down attempting to retrieve the pack but the spiral quickly progressed into a tight spin. Boucher jettisoned the top canopy hatch and started to climb up through it. Halfway out he observed Pessirilo struggling on the floor unsuccessfully. Now sitting on top and outside the canopy hatch, Boucher held on with both his legs inside the cockpit as the swirling aircraft and air currents pulled him outward. Boucher looked inside as Lou shook his head "No" he could not reach the pack and waved for Boucher to go. Instead Boucher gave a handsignal to come up then slapped his hand against his parachute harness. This indicated Pessirilo's only option to survive the impending disaster.

Pessirilo managed to overcome the centrifugal force and climbed up from the floor, reached the top canopy opening and began to squeeze through. Boucher released one leg from the hatch opening while holding the other tightly inside against the canopy. Pessirilo forced himself through with Boucher pulling from the outside. Having finally exited, he crawled out on to Boucher's body, slung both arms through and gripped tightly the parachute harness. Boucher tightly wrapped both arms around Lou's upper body and the one free leg around his lower section. He then released his other leg from the canopy and both men fell free from the aircraft.

Several days later, Boucher slowly regained memory to find himself a POW in a hospital suffering from severe concussion and amnesia of what transpired. As he gradually recovered, minute recollections and flashes of insight of previous events helped overcome some memory loss to reach the following conclusion. Upon releasing his leg from the canopy, Boucher struck and bounced off either the tail section or wing and lost consciousness. His arms and legs automatically went limp from around Pessirilo who assumed correctly that Boucher was unconscious. To save themselves Lou had to release one arm either fully or partially to pull the D-ring. Upon chute release and with Boucher's arms and legs limp, Pessirilo could not overcome the sudden jolt of canopy deployment and fell to his death. His action saved Boucher's life but cost him his own. A letter from Group HQ informed the Pessirilo family that secrecy of the mission "required parachutes not being worn."

Watton launched a four aircraft Graypea mission the 29th. Lts. Stephen Luce/Joseph Ward (NS512) were fired upon by the 91 Group leading the formation in the vicinity of the IP, with rounds striking both fuselage and tail. On another Graypea flown this day for 3 AD an unidentified formation of bombers fired at two Mosquito aircraft without scoring hits.

25 March 1945, NS774 "M" of 654 BS after takeoff in strong crosswind. *Photo W. Borges*

Ground Control Approach radar (GCA) equipment arrived in the ETO with ABL technicians introducing the system to field personnel, assisting with maintenance and operational problems.[71] The first GCA began operating at Etain, France, on 22 January where P-47 units and a P-61 night-fighter squadron, located 30 miles from the front lines, supported the Third Army. A second unit began service 25 February at Forennes, France.

On 6 March the third unit arrived at Watton and became operational the 16th. All pilots received a debriefing on GCA, shown a film, then observed an approach and radar operator at work. Group HQ instituted a training program and all pilots provided checkout flights using GCA equipment. Over 300 hundred approaches, including 27 emergencies with seven in zero-zero conditions, were achieved: 25 in March, 150 in April and 125 in May.

An example of an extreme emergency requiring GCA at Watton occurred when the base commander accomplished a guided approach under extremely appalling conditions. After the aircraft touched safely on the ground, zero-zero visibility prevented seeing the edges of the runway and thwarted taxing to the dispersal area without the aid of a Jeep leading the way.[72]

Col. George Goddard, father of AAF PR, developed the Sonne S-7 "moving film strip" camera, considered technically shutterless. The first combat use of the camera occurred 8 January 1945, when 1/Lt. DeYoung of 5 Photo Squadron in Italy flew an experimental run over the mouth of the Garigliano River in a P-38.

Mobile GCA unit with rotating antennas could be moved to runway by truck.
Photo USAFHRA

71 BBRL – British Branch Radiation Laboratory, MIT component

72 File 519.90631-13, ASP Report No. 12, Survey of GCA (AN/MPN-1) in the ETO, January-May 1945, USAFHRA.

On 12 March Major Willis D. Locke, Capt. Joseph J. Pelligrini and T/Sgt. Thomas M. Tate along with Frank McKee flew a B-26 to airstrip A-42 (Villacoublay) in France for duty with the Air Technical Service Section, USSTAF Europe. Lt. Peter D. Dustman/George F. Henion flew in Mosquito NS559 for S-7 placement. Technical equipment difficulties postponed installation and Dustman returned the aircraft to Watton the 15th. Sometime in April a Mosquito arrived at Villacoublay, along with a P-51 for S-7 camera installation. This particular Mustang returned to the USA in August 1945 on board the aircraft carrier HMS Reaper during delivery of captured Luftwaffe prop and jet aircraft.

Eventually ten P-38s, three P-51s and one Mosquito were assigned to 302 Transport Wing, Villacoublay for S-7 camera installation. According to then Major Harry C. McCool, a Doolittle Raider, SHAEF established a postwar bombing survey project. Aircraft flying at 300 feet altitude across designated targets with the Sonne camera were sufficiently low to photograph inside ruined factories and buildings. This allowed photo Intelligence personnel to access data more expeditiously and accurately for bomb damage assessment.

Technicians and operators inside a GCA van. *Photo USAFHRA*

April 1945

Operations declined in April with the rapid Allied advance into Germany. As more German territory became occupied, there were less major targets to bomb. Even so, operations brought both excitement and tragedy to Watton crews.

Internment in Sweden

Six Mosquitos flew a Graypea mission 3 April in support of 1 AD to Kiel. Lt.Col. Alvin Podwojski/Capt. Lionel Proulx (NS650) took off first at 1502 followed by the others. They circled the base allowing each aircraft to marshal, then climb en route to 25,000 feet. The flight met the bomber stream 35 miles southwest of Kiel, flying two elements of three aircraft each. Podwojski led the first and Capt. McNulty the second flight, one minute behind. During the mission, the starboard engine temperature of NS650 began to rise.

The bombers reported they were 12 minutes from the IP. So, the Graypea flight changed to six aircraft flying line-abreast, ahead and above the bomber stream. No flak interfered with the bombers though 12 to 20 bursts trailed the Mosquitos by 100 yards. After Chaff discharge the aircraft left the target and bomber stream behind. On the return flights they seldom maintain formation and flew individually.

Podwojski in NS650 banked and turned northwest. Proulx loosened his seatbelt to peer behind periodically for enemy aircraft. At that moment something struck the starboard engine. Proulx glanced back and spotted three Me 109s. He shouted a warning as the pilot cranked the control wheel and sought refuge in nearby clouds. The starboard engine oil and radiator temperature reached maximum, as oil pressure dropped dramatically; then the port Merlin began running rough. The navigator provided calculations to reach friendly lines or Sweden. The pilot notified the flight he planned heading for Sweden. The Mosquito banked and turned near the German coast due west of Flensburg. The navigator pulled out a chart with escape routes and immediately provided a course to Sweden.

P-51s from 359 FG escort Capt. Louis Rogers during Redtail mission, 30 March 1945 with General A. Olds in NS636. *Photo L. Rogers*

After feathering and shutting down the starboard Merlin, Podwojski jettisoned the wingtanks. Only the starboard underwing tank released initially, but after completing a steep bank to check for enemy fighters, did the port tank fall free. With 8/10 to 10/10ths cloud, the Mosquito skipped from cloud to cloud pursed by Luftwaffe fighters. Flying on instruments they managed to reach Sweden at 8,000 feet while maintaining a slow descent. Royal Swedish Air Force Seversky P-35 fighters intercepted and escorted the Mosquito to an airfield near Malmo called Bultofta.

On the downwind leg to approach the runway, the port engine failed, forcing Proulx to handpump the undercarriage down for a dead-stick landing. Upon touchdown, fire vehicles equipped with foam turret guns sped parallel with nozzles trained on the aircraft. The Mosquito slowly rolled to the runway end and stopped. The crew wasted no time exiting the bottom hatch. Swedish Air Force personnel brandishing weapons soon surrounded both men. They allowed inspection of several flak holes in the bottom, starboard Merlin cowling.

The military escorted the crew to a command building for interrogation and confiscated all flying equipment and sidearms. After providing name, rank and serial number, followed a tasty meal. A Swedish officer escorted both men on a train to Stockholm, then another train for Falun. They arrived at the internment camp after dark to clean, comfortable, but Spartan quarters. The crew later received orders transferring them to the U.S. Legation in Stockholm with accommodations at the Stockholm City Hotel awaiting transfer to the England. Count Bernadotte, the Swedish Red Cross director arranged an agreement for prisoner of war exchanges on a quid pro quo basis. Both men received transport to Bromma Airport under cover of darkness, loaded on board a black B-24 of Bernt Balchen's unit, and flown to England.

On April 4, Lt. Theodore B. Smith (NS635) flew his first Redtail mission and went to Bungay to pickup Col. Troy W. Crawford, CO 446 BG. Colonel Crawford would coordinate operations of the front-third formations of B-24 bombers targeted for Parchim Airfield west of Berlin. Heavy overcast at the primary forced diverting the bombers to the secondary target, Wesendorf Airfield. At 0930 Colonel Crawford radioed the lead bomber his Mosquito was joining the formation.

Impressed with the bomber formation, Crawford instructed the Mosquito pilot to fly position on the left flank to obtain photographs. At this time, the formation underwent attack by a Fw 190 and a dozen Me 262s. To seek protection, Smith flew closer to the AD bombers, when gunners of the lead squadron opened fire on the Mosquito. The right engine received strikes and moments later the left engine, resulting in complete loss of power. Unfamiliar with the Mosquito escape procedure, Smith shouted to Crawford instructions to unlock the floor hatch. Crawford then jumped followed by Smith. Both landed safely and became POWs.[73]

SHAEF emphasized the importance of locating all POW camps in Germany. Wing HQ issued a high priority dispatching PR missions to film all known and suspected camps while photo-interpreters evaluated the results.

NS635 shot down 4 April 1945 with 1/Lt. Ted B. Smith and Col. Troy Crawford, both POWs. *Photo PHB/H. Holme.*

73 Harold E. Jansen, History of the 446 BG (H), The Hague, Netherlands, 1990:261

T/Sgt L.P. Palmer against canopy, with pilot 1/Lt. Ted B. Smith, POW on 4 April 1945. Note Bleeding Heart or Broken Heart. *Photo H. Steinsson*

Late that evening, Lt. John F. Murphy (NS569) practiced night landings at Watton. After the third, he taxied along the perimeter track to prepare for another takeoff. The pneumatic brakes functioned on two turns of the track, but on the third they failed. The momentum carried the Mosquito towards a B-17 parked in the dispersal area. To avoid a collision, Murphy gunned the right engine to rapidly turn the aircraft. As the tail swung 180 degrees it narrowly missed the B-17. But unexpectedly, the Mosquito tail section slammed into a maintenance stand concealed by darkness, snapping off the rear fuselage.

Orburg and Nuremberg were targets 5 April and eight screening aircraft flying the mission released Chaff and took photos. Lts. Cornelius Engle/Albert D. Rasmussen (NS783) were fired at by four bombers of the third group while making a camera run across the target. No hits were scored. Next day, bomber command received orders restricting attacks against towns, only in direct support of ground troops. The precaution reduced ground forces exposure to friendly fire as Allied armies moved deeper into Germany.

A Graypea launched at 0730 on 6 April supporting 1 AD to Leipzig suffered a mishap. Lts. John A. Pruis/Claude C. Moore (RG113) were the final of four aircraft to takeoff. During liftoff, the right engine burst into flames and immediately shutdown with propeller feathered. With flames extinguished, the pilot applied left engine power to gain altitude. The aircraft held at 1,000 feet and flew over a nearby practice range to jettison the underwing tanks. He then entered the landing pattern and descended quickly on final approach. Flames resumed spewing as the right prop windmilled out of control. A rapid touchdown halfway down the runway left no room to safely stop. The crew initiated a ground loop; after releasing the overhead emergency hatch, the navigator retracted the undercarriage. Now skidding on its belly, the Mosquito swerved off the runway and spun on to the grass field before grounding to a halt. Both crew escaped out the top hatch. Emergency equipment and crews arrived to douse the flames.

Four Graypea Mosquitos targeted Zerst Airfield the 8th, flying at 22,000 feet. Between the rendezvous and IP, four P-51s closed for a head-on attack. One P-51 opened fire but fortune turned aside the bullets with no hits scored. The Mosquitos made rendezvous with the bombers as planned. They attempted to pull within recognition distance to identify the tail markings. Gunners in the lead formation fired streams of tracer bullets forcing the Mosquitos to turn away. Periodically thereafter, they were fired upon by the preceding groups while en route to the IP.

At the IP the Mosquitos received instructions to fly starboard of the formation ahead to allow a left turn. As the Mosquitos crossed over behind the B-17s, the latter opened fire with a continuous burst from the top-turret guns of the entire formation until out of range. Fortunately no hits were scored. The bombers firing were identified as the 305 Group, 409 Combat Wing.

On 4 April, the pilot swung NS569 and struck a maintenance stand in the darkness.
Photo W. Borges

Mechanical failures forced Lts. John M. Carter/Woodrow E. Loden (NS744) to abort a WR the night of 8/9 April. First, the airspeed indicator malfunctioned, then the left engine malfunctioned forcing its shutdown. Executing an emergency landing at Manston the aircraft descended rapidly for a single-engine approach at 0345. Unable to bring the aircraft to a safe stop, the pilot overshot the end, across a ditch and road, through a hedgerow into a plowed field where the undercarriage collapsed.

653 BS Light NS744 landing accident 8 April at RAF Manston.
Photo W. Borges

On Monday afternoon of 9 April, a Graypea mission took off for an eventful mission with tragic consequences. Lts. Lyndon Lakeman/ Paul Lovett (NS783) took off first at 1411 followed by John Pruis/ Claude Moore (NS792); Robert Foster/Louis Toth (NS743); and Major Robert Howle/Lt. William Miskho (NS793). The Mosquitos preceded a maximum effort mission by lst AD to the Dornier aircraft plant at Oberpfaffenhofen, a suburb 20 miles north of Munich in southeastern Germany. As Moore recalls:

"We rendezvoused with the bomber formation prior to the IP that was some 28 miles from the target. At the IP we climbed 2,000 feet above the bombers and began dispensing Chaff. I activated the cameras as we flew over the target area to obtain photo coverage before the bombing began. After passing the target, we orbited and followed behind the bombers to obtain damage assessment pictures of the destruction. Thoroughness was one of my characteristics and having unused film, I instructed Pruis to circle over the target again for more pictures. This gambit distanced us far behind the other three Mosquitos. We set our course on a 314-degree heading to make eventual contact with the Mosquitos and return to base. This was Pruis' third Mosquito mission with the 25 BG (Rcn). He previously completed a tour in B-24 bombers. As pilot, he was concerned with fuel consumption, especially considering the distance of our mission. Though fuel consumption would be close, all concerns for engine and aircraft performance were minor.

"Flying at 20,000 feet, we could see three aircraft in the distance on course toward England. We presumed these to be our Mosquitos. As we came closer, Pruis and I decided they were P-51s, but with French AF roundels. They banked to the left and I no longer gave them any thought.

"Some minutes later, I was noting observations on my navigational chart or log, when there was an explosion. With two engines running, the indication of an explosion was not so much a noise as a jolt. I realized we were in trouble when Pruis slumped over the wheel. I attempted to help but was unable to move him. The Mosquito began a right hand spin with the right engine in flames. With a right hand spin, the flames from the right, or starboard engine, blazed across both the canopy and bottom escape hatch. Black smoke poured from the right side of the nose between the cockpit and the nacelle. I did not know what happened. I had not seen any enemy aircraft in back of us nor in the sky.

"I never wore a parachute pack on my missions. A seatpack placed me too high in relation with the canopy and the worktable. A chestpack pushed me too far from the plotting table and was burdensome to work with. I stored the parachute pack on a small shelf behind the pilot. The centrifugal force from the spinning aircraft was horrendous. I experienced a terrible time reaching for my chute behind the pilot. It took the strength of both arms to reach the pack and pull it toward me. I then attempted to connect it to my harness but due to the centrifugal forces, managed to connect only one side. I suspect I wasn't being deliberate or thorough enough, but later learned I had shrapnel wounds in my left arm from aircraft cannon shells.

"If I exited the bottom hatch, which would be normal, I was afraid the flames would come into the cockpit. I wasn't able to help Pruis but if he should regain consciousness and was able to do something for himself, I did not want the flames coming into the cockpit. Thus, I released the canopy escape hatch and it flew off. Leaving the craft through the top is a high risk endeavor. The probabilities of being struck by the vertical stabilizer are fearful.

"I struggled through the hatch and somehow became entangled with the oxygen and communication cables inside the aircraft. I was stuck half-in and half-out the top opening. I had been struggling without my helmet and wore no gloves as tongues of flame from the right engine lapped at me. Fortunately, I still wore my RayBan sunglasses. Flames lashed my upper body, seared my face and scalp, flesh burned from my hands and wrists, blood covered my left arm, while the earth below swirled closer. I moved back into the cockpit to remove the tangled cable obstructions that prevented my escape. Once in, I again experienced trouble getting out.

"Now I was more weaker from the burns and wounds than in my first attempt to exit. The pain was driving through the tension to my senses. The centrifugal forces of the spinning craft were becoming more intense by the second. There was no sensation of up or down, just the vision of earth spinning closer on a collision course. I felt my first and only taste of despondency. In a few moments it would all be over and I would abide with the sense of peace.

"Despondency was followed by the thought that my family and friends expected more of me than passive submission. Pruis was dead or unconscious, but my limbs still responded to my will. My limpid reaction to the slaps of fate made me furious with myself. Rage thrust all frailties from my mind. With a profane utterance I made my utmost and final effort to climb through the hatch.

"The next thing I was aware of was floating in the air. As I bailed out, I glanced at the altimeter. It read one thousand feet. I interpolated that I left the aircraft at 800 feet. I am certain I was not fully conscious when I pulled the D-ring. I was jolted heavily to consciousness by the opening parachute. I swung violently in 360 degree circles as I descended, probably due to the right chest hook being unfastened. I attempted to pull the riser to get the chute buckle fastened, but with the burns and injuries to my arms and hands, I failed. It was then I noticed the bleeding from my left upper arm and forearm.

"As I descended, I was circled by fighter planes. They were P-51s with the French roundel markings. The Mustangs circled only minutes then left the vicinity. I did not see the aircraft that fired at us. I knew only that the shells entered the fuselage diagonally from the left rear side and, that Pruis and the starboard engine received the brunt of their effects. I also remember the three P-51s flying previously ahead of us banked left and had flown past us to our rear.

"I turned my attention to the ground where I could see the Mosquito wreckage. The fire was confined to the right engine and had not spread to the rest of the aircraft. Its wings, nose and the forward part of the fuselage was intact. At that moment a gust of wind swept my RayBan glasses from my face. I was fortunate they shielded my eyes from the flames as I exited the escape hatch.

"Below lay evergreen trees with tall spindly trunks and an abundance of weak, flimsy limbs. I attempted to guide the parachute to smaller fruit trees nearby but failed in the short drop. The parachute caught in the top of a tall evergreen leaving me hanging by the stringers 50 to 75 feet above the ground. Interestingly, information provided at briefings indicated we were deep in enemy territory. There were several armed German soldiers near the base of the tree where I was hanging. They were contemplating my plight but did not seem too concerned. I must have passed out for when consciousness returned the German soldiers had disappeared.

"I unfastened the harness of the chute with great difficulty. My hands were burned raw, the left arm was bleeding and my face felt like it was on fire. The First Aid pack attached to the parachute harness tore off as I plowed through the branches. I attempted to climb down the pine tree, but the limbs would not support my weight. Suddenly the branches were snapping, and I was falling. I fell straight down in an upright position and hit the ground.

"I attempted to stand, but the pain was intense. My back, left ankle, knee, and foot were injured. I thought of using a tree branch for a cane to reach help. We understood the Catholic Church had a very good underground establishment functioning in the southern part of Germany. I was determined to find such an organization. It was close to dusk when I crawled to a tree limb and attempted to stand. Physically it was hopeless. I looked at my wristwatch and except for a rim around the dial, the face was burned. It said 1745. The forest was beginning to darken. I needed medical attention quickly. I could hardly see. I decided to take my chances with the Germans instead of spending the night in the forest. I wondered about their disappearance. I always carried a whistle attached to the zipper pull tab on my jacket. This was an RAF custom that I had appropriated. I blew the whistle and heard the rustle of twigs and brush. I blew again and a soldier appeared holding his rifle at the ready across his body. We were told to avoid capture by the civilian population, youth corps, Gestapo, or the SS troops. Since the soldier wore a tattered outfit and I couldn't see very well, I concluded he was one of the German Youth Corps.

1/Lt. Claude C. Moore survived the fatal shoot-down of 9 April 1945.
Photo. R. Walker

"He then questioned, 'Yank' which irritated me.

"I replied, 'Ya. What the hell is it to you?'

"The soldier responded sympathetically, 'It's OK buddy, we're Yanks too.'

"He yelled to his seven companions that I was located and walked over to me. They joined us immediately. Apparently, the American soldiers had been watching me come down in the Mosquito. They gave me hell for not getting out sooner. They had been trailing a pocket of SS troops when they observed our aircraft in trouble. When they saw me parachute, they gave up the chase and diverted their attention to my welfare. So, I had seen German troops and did not imagine it.

"I cautioned the soldiers to please be careful. I thought my back was broken and wanted to avoid a severed spinal cord. They took off their shirts and jackets, gathered two large tree branches and made a litter. They carried me down the hill to two Jeeps. I was placed diagonally on the flat platform of one Jeep and held closely to prevent jostling on the bumpy ride. They transported me to a first-aid station located in a commandeered German home. The makeshift stretcher and I were passed through a window to medics inside. As the soldiers were leaving to resume their patrol, they reclothed themselves in the shirts and jackets used on my makeshift litter. This clothing was stained with blood from my wounds and the serum from my burns. I asked them for their names so I could thank them later in written form. They declined. They did not believe their act was beyond the call of duty, though I felt differently. This was the last thing I remember until awaking two days later in an American hospital in Reims, France, the effects of morphine I presume.

"Some days later, I received a slip of paper that had been placed in my flying suit. It said, 'Picked up near Waldkatzenbach, Germany, about 10 kilo west of Eberback by lst Sgt. Nelson Griffin and boys, Co. D. 1st Bn. 410 Inf. Regt., 103 Inf. Div., APO 470, C/O PM, New York City.'

"I was disturbed about having lost a pen and pencil set my brother had given me. It was the one I used for navigational duties with the aircraft. It was a personal item I valued highly. Later, when I was in the hospital someone handed me a personal effects bag and inside was the pen and pencil set. It was a surprise. Subconsciously, perhaps, in the turmoil of the spin, I had grasped the set and placed it in the pocket of my flight coveralls."

Lts. Robert W. James/Terence S. Hall (NS757) were dispatched on a training flight, Monday, 9 April. They flew formation at 1,500 feet with another Mosquito and passed over RAF Station Sherpherds Grove. At 1520 hours, witnesses observed the lead Mosquito drop its port wing sharply to turn, enter a shallow spin with both engines at full throttle. It recovered briefly but went into a right spin, then stalled at 400 feet and dived vertically to the ground in a ball of flame. Both crewmembers died. Because of complete destruction and death of the crew, it proved impossible to determine the exact cause of the accident.

A Graypea effort supported 1 AD to Oranienburg on 10 April. Mechanical problems forced one aircraft to return early. The three Mosquitos reached the IP eight minutes early but found the bombers some distance ahead proceeding over the target. When the target was reached, three bomb groups had already completed their attack. The Mosquitos therefore did not discharge their Chaff.

At Steinhurder Lake an Me 109 passed 200 yards to port on a reciprocal heading. It turned to attack, but the Mosquitos quickly entered clouds for protection. Later, five Me 262s appeared south of Oranienburg 2,000 feet directly above at 27,000 feet. One '262 attacked from four o'clock high, but the Mosquito flight made a sharp right turn and the jet skidded by without firing. Five P-51s followed close behind as the '262 streaked past.

Back at Watton, Lt. Andrew C. Horvath (LR530) returned from a transition flight. On touchdown, the left landing gear seemed to buckle, so he attempted to keep the aircraft straight using right brake, rudder and aileron. But when the tail wheel touched the runway, the left undercarriage completely gave way causing a ground loop left off the runway.

Graypea Mosquitos encountered friendly fire on 11 April supporting the lst AD to Freiham, Germany. The 351 BG fired a 30 second tracer barrage but no hits were scored. At Watton, Lts. Edward F. Hodges/Donald E. Beard (NS516) made a smooth landing after a transition flight. On touchdown, the left tire went flat causing the aircraft to vibrate and veer left halfway down the runway. It ground looped, wiping out the undercarriage and breaking the fuselage in two.

Newly painted RG154 of 654 BS (S) with crimson tail and spinners. *Photo L. McKinely*

It was midnight when Brig.Gen. Elliott Roosevelt sat listening to the final BBC news broadcast. He heard the announcement his father, the president, died in Warm Springs, Georgia, on 12 April. Roosevelt immediately summoned a staff meeting for 0300 and made preparations to return immediately for the funeral. He appointed Col. Leon Gray, CO 25 BG (Rcn), temporary Wing CO, and assigned Lt.Col. John R. Hoover temporary CO 25 BG (Rcn). Three days later an impressive memorial ceremony by the entire Group paid tribute to the late Commander in Chief.

Lts. James Handren and Robert Gordon both completed a combat tour with the 490 BG and volunteered for another tour flying Mosquitos. One day, a crew chief walked into the Operations room and asked for someone to test fly a Mosquito (MM387) that received a major engine overhaul. Pilot Gordon and navigator Handren took off on 13 April to check the port Merlin. On liftoff, the engine vibrated excessively and suffered a runaway prop. A burning junction box filled the nose with smoke. Gordon initiated a steep approach and landed fast, midway down the runway, went off the end where the undercarriage collapsed. It also severed cables for the night landing light system.

Lt.Col. John Larkin witnessed the accident and instructed Gordon, "Keep your gear on, we're going up again and I'll show you how to fly a Mossie!" That afternoon, with Larkin at the controls, they took off (NS512) and circled the airfield and approached for landing. The aircraft bounced, then took a slight crab and dropped ten feet at the same time. On touchdown, the shearing force collapsed the undercarriage and the Mosquito slid down the runway before stopping. Now grim faced, Larkin said only one thing to the crew chief, "Get me another plane!" Gordon went back up with Larkin and spent the afternoon practicing takeoffs and landings. Next day, Capts. Robert Shoenhair/William Miskho (RF988) returned from a Graypea. The tail wheel failed to extend, but they executed a safe landing on the grass without damage.

A general reorganization of 25 BG (Rcn) began 15 April. The plan eliminated the Arms and Service branch distinctions in the AAF and created the formation of an Air Service Group to handle all service functions. This was in preparation for 25 BG (Rcn) redeployment, less 652 Squadron. This resulted in the elimination of 375 Air Service Squadron, 301 Station Complement Squadron, 1070 QM Company Service Group (Avn), 1274 Military Police Company (Avn) and the 2134 Eng Avn FF Platoon. Activated to replace these units was the 652 HQ & Base Services Squadron, Air Service Group. This also included the 828 Air Engineering Squadron, 652 Air Material Squadron employing personnel from the disbanded units and all functioned temporarily within the 25 BG (Rcn).

Also on the 15th, Lts. Warren Borges/Wendell Biggers (NS745) returned from a WR to western Czechoslovakia and southern Germany. As Borges recalls:

"That evening my navigator Wendell Biggers and I were about 28,000 feet between two dark cloud levels – like in a closet black as ink – cruising close to 300 knots. Strictly on IFR, aircraft trimmed and quiet except for the reliable two Merlin engines purring along. Suddenly, I thought I saw a light reflection out of my left eye. Observing the left wing, I almost panicked – it looked like a wing fire. It was a huge ball of fire rolling about the wingtip, but not a fire, and disappeared after one or two minutes. I did not tell Biggers as one panic person in the cockpit was enough. The same phenomenon appeared on the right wing but just for a moment, then disappeared not to appear again. I constantly checked my flight instruments and kept a roving eye on each wing. I knew it was not St. Elmo's fire but something I never experience before."

On the 16th, a Graypea mission screened 1 AD with Regensburg the target. Chaff discharge began at 24,000 feet while making S-turns and climbing to 25,000 feet. Flak over the target burst high above the bombers. Three groups had already passed the target before flak began. On return to England, a P-47 attacked Lts. John Murphy/Frank Catalano (NS783) at 4,000 feet. Murphy went full throttle, but the P-47 continued pursuit for five minutes. Exploiting the Mosquito's superior speed he outdistanced the Thunderbolt with no hits scored.

Dresden was the target 17 April for two Graypea missions. The first supported 1 AD. Six to eight Me262s circling over the city attacked the bombers but were non-aggressive to the Mosquitos. The other mission supported 2 AD. The men observed two B-24s tumble in flames near the target after a midair collision. No parachutes appeared. At Watton, Lt. Warren C. Davis/M/Sgt. Walter Jorgensen (RG145) returned from a training flight. After a normal approach and touchdown the Mosquito ground looped, washing out the landing gear.

Two Redtail missions supported a 3 AD effort 18 April. Lt. Kenneth Werthman (MM337) picked up Col. Madsen from Snetterton Heath. Two P-51s jumped the Mosquito 40 miles east of Munich and Werthman took violent evasive action. One Mustang fired a short burst but stopped when his wingman recognized the red tail. The P-51s then acted as escort until forced to return because of low fuel. The other Redtail involved Lt. Wallace Rouse with Col. C.E. Marion (NS748) of 393 BG.

Friendly fire continued to plague Graypea missions. The solution: provide fighter escorts to protect the Mosquito flight against attacks from both friendly bombers and fighters. Meanwhile, Lt. John M. Carter (NS748) practiced circuits and bumps, made a normal approach and upon landing veered off the runway collapsing the undercarriage.

Fire department spray down MM387 after landing accident 13 April 1945. *Photo L. Erickson*

653 BS NS748 broken in half by 19 April accident. *Photo W. Borges*

The 1 AD celebrated Hitler's birthday on 20 April with an attack by 1,000 bombers on Berlin. During the Graypea support, eight P-51s provided the Mosquito flight protection from the bomber formations throughout the mission. Meanwhile, Lt. Lyall Barnhart (MM368) returning from a transition flight made a hard touchdown causing the aircraft to bounce. He attempted a three-point landing but lost control and veered off the runway, collapsing the undercarriage.

NS553 on Skywave mission to San Severo, Italy, 19 April 1945. Standing at nose is Capt. Robert D. Shoenhair. *Photo R. Shoenhair*

19 April at San Severo on Skywave flight. Navigator 1/Lt Robert M. Balser sits on tail. Note serial number and tail is painted red. At this stage of the war, the need for night photo missions decreased, therefore NS553 flew Skywave day missions with painted red tail applied for AAF identification purposes. *Photo R. Shoenhair*

From 21 through 24 April, three Skywaves were flown and 14 Graypea aircraft dispensed Chaff. U.S. and Soviet troops met at the Vistula River on the 25th. And 654 BS supported a special mission. This involved 8 CCU cameraman Sgt. Larry Barbier who departed Watton in the nose of a Droop Snoot P-38 to film the bombing at Royan, France. The Graypea Mosquitos screened the bombers subduing a stubborn German garrison the Allied armies bypassed and who denied use of Bordeaux's port by controlling the mouth of the Gironde Estuary.

This day marked the Eighth Air Force final heavy bombing mission when 1 AD visited the Skoda Works at Pilsen, Czechoslovakia. It was also the last Graypea mission. The long route required extra weather predication. Prior to midnight 24/25 April, Lt. Peter Dustman/Capt. William Miskho flew a Skywave to the Czech border. Then after midnight, two weather Mosquito aircraft dispatched at intervals with Capt. Allan Bateman/Lt. John Cone (NS742) and Lts. Harry Flook/John Nash (MM338) covered the route to and from Czechoslovakia. Early that morning Lts. Wallace Rouse/Joseph Levy (NS753) again covered the route to the Czech border. The final Graypea flight of four aircraft screened the attacking force at Pilsen.

As more German territory came under Allied control, the number of strategic targets to bomb decreased. Therefore on the 26th, strategic bombing ceased while tactical air supporting ground forces continued. AAF grounded all operational aircraft in the UK except at Watton. The 652nd maintained scheduled Atlantic WR flights while 653 covered the ETO gathering weather data for forecasting purposes. For many days these missions constituted the entire Eighth Air Force summary of daily operations.

During early April excellent weather adorned England, but during the closing days an unusual cold, turbulent, and moist air caused poor flying conditions. Ceiling and visibility became dangerously restricted during takeoff and landings. On 28 April, Capt. Stephen Donegan returned from an Atlantic WR in extremely adverse weather with one engine feathered and a second running rough. For three hours the entire B-17 crew stood ready for bailout. Severe icing stuck the throttles in a half open position. Donegan chose to proceed to Watton rather than a crash strip elsewhere. A second engine cutout immediately on touchdown but he managed to stop the Fort before running out of runway.

21 April photo shows 654 BS letters BA-N under left wing of N-Nan. *Photo L.* Erickson

Graypea mission with NS785, 21 April 1945 flown by 1/Lt. Peter Dustman/Capt. William Miskho taken from RG155. *Photo L. Erickson*

Rare photo of Graypea mission on 21 April 1945. 654 BS with following aircraft 1/Lt. Peter Dustman/Capt William Miskho (NS785); Lts. John Grassel/Len Erickson (RG155); John Murphy/Frank Catalano (NS793); George M. Vanderleest/Richard Tyhurst (RG154); James Evans/Robert Spaight (RG148); Take off time 0811 and landed 1248 hours. *Photo L. Erickson*

Another view of 21 April 1945 Graypea with "M" possibly returning to base from Munich. *Photo L. Erickson*

May 1945

Late morning of 5 May, Lt. Lyndon Lakeman/Sgt. George Senseny (NS590) returned from a Mosquito test flight. On approach, the right undercarriage failed to fully extend. Senseny hand pumped the landing gear down and Lakeman attempted a touch-and-go landing. When the aircraft settled to the runway, the undercarriage began to fold, so he took off again and remained airborne. After discussing the problem with the engineering officer, he initiated another landing. This time the right undercarriage finally locked down but the left landing gear collapsed causing a ground loop.

The 654th flew four Skywaves, two on 1 May and one each on 3 and 4 May. Lt. Harry N. Flook unknowingly flew the final World War II Bluestocking WR flight on 6 May. Next day, Capt. Richard Kenny taxied his Mosquito to the departure end for takeoff on a WR. The tower ordered his return to the hardstand because Germany surrendered. Then the celebrations began.

NS590 accident scene 5 May 1945. *Photo W. Borges*

V-E Day on 8 May terminated all operational Mosquito missions, but 652 BS continued flying scheduled Atlantic WR flights. A period of relaxation set in and passes liberalized, providing the men an opportunity to visit many historic places. In anticipation of redeployment, all personnel received vaccinations and overseas physical examinations. Wing HQ authorized Trolley flights over the continent providing groundcrew the opportunity to observe the devastating bomb damage wrought by Allied bombers. Both B-17 and Mosquito aircraft performed flights to France, Germany, and Belgium, on low-altitude sightseeing tours.

On 18 May, JPRC at Benson dispatched a letter of appreciation for work accomplished by 325 PW units:

> "The record attained by the 7 PRG and 25 BG (Rcn) is deserving of the highest praise. During the period this Committee was in session, from 1 June 1944 to 30 April 1945, these two groups flew 4,085 PR missions day, night, and H2X, involving some 11,000 operational flying hours. Approximately 72% of the missions were successful. Nowhere else has the spirit of inter-Allied cooperation been so prevalent as that which exists in reconnaissance. It is this fine spirit of cooperation that has made reconnaissance such a success in this theater and a great deal of this is due to you and your staff, and to units of your Command."

B-17 used on Trolley sightseeing tour of war-ravaged Germany. *Photo L. McKinely*

Cologne Cathedral, Rhineland, amidst bomb destroyed buildings. *Photo L. McKinely*

Trolley flight of B-17s fly formation en route to Germany, 2102970 "J" with 338833 "O" in background.
Photo L. McKinely

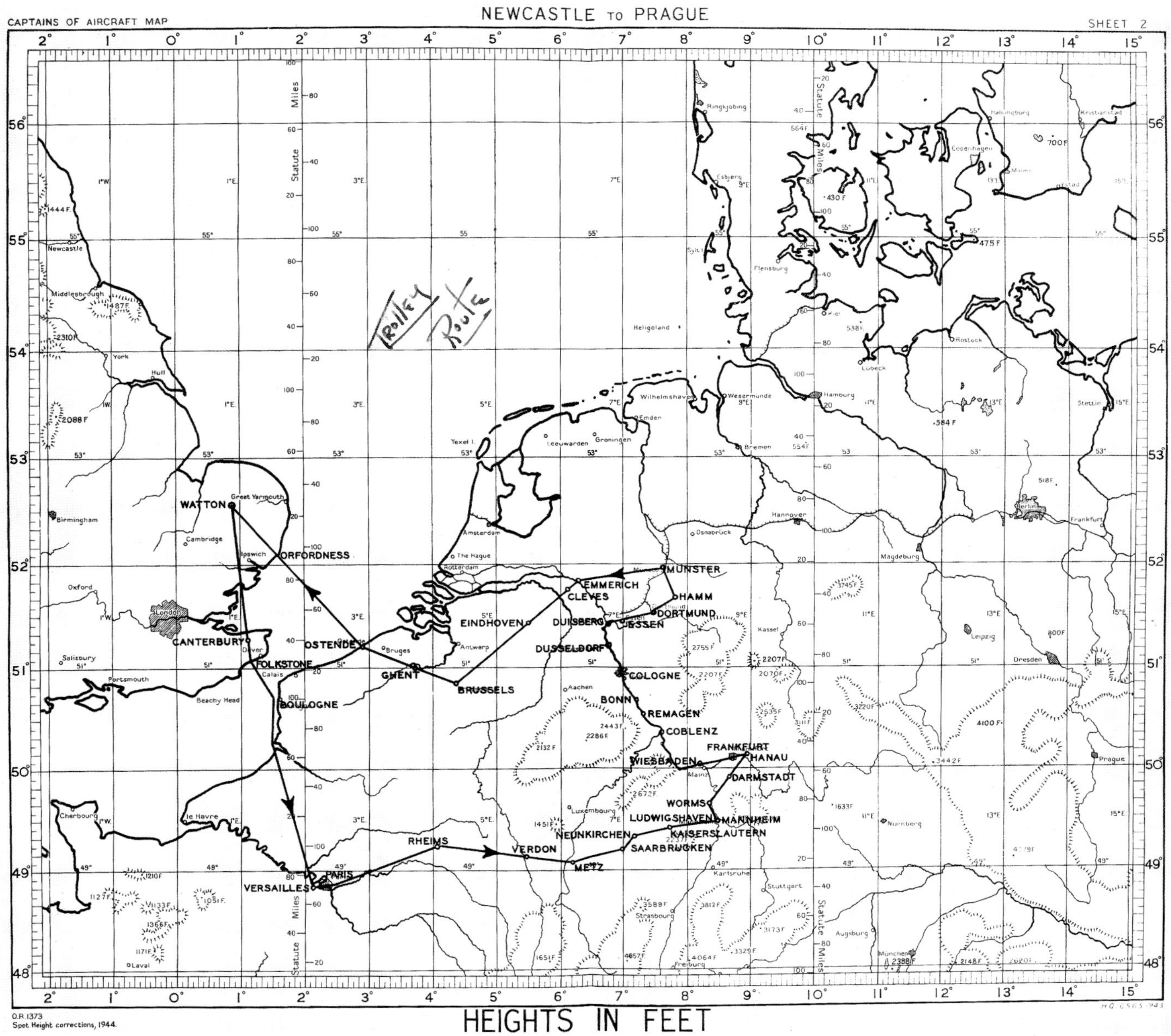

Mission map shows Trolley route taken across France and Germany. *Photo W. Borges*

Mosquito banks away from 652 BS B-17 on Trolley run. *Photo L. McKinely*

Lts. Harry Flook/Walter F. McKinney (NS758) flew a local test flight 28 May. At 3,000 feet, Flook switched from outer to inner fueltanks. The right engine developed a vapor lock so he feathered the prop and returned to Watton. On approach he attempted to unfeather the prop, but the engine failed to start. The Mosquito set down on the first third of the runway at an excessive speed. Braking action proved ineffective and the aircraft proceeded off the end, on to rough ground causing the left undercarriage to collapse.

"Southern Belle" NS758 accident on 28 May. *Photo W. Borges*

Capt. Richard M. Kenny and T/Sgt. Andy R. Earnest (NS708) flew a Trolley flight to Germany on 30 May. Near Bremen, Kenny switched from outer to inner fueltanks when the right engine stopped due to an air lock. Unable to clear the fuel line he attempted to feather the prop but a weak battery prevented this action. Flying at 2,500 feet, he released the underwing tanks when the left engine also quit. Then an airfield came in sight and he attempted a wheels down landing at 1500 hours. Kenny applied braking after touchdown, but unfortunately, the thick green grass, wet from previous rain hindered traction. The Mosquito slid a considerable distance before crashing into an antiaircraft emplacement. The ensuing ground loop snapped the Mosquito in half. Both men were shaken but unhurt.

NS708 on 30 May with fuselage broken in three parts after collision with a German gun emplacement. *Photo R. Kenny*

During June, the 452 Air Service Group continued its function of housekeeping, handling of personnel, and operation of the base security, firefighting, utilities, post-exchange, transportation, and flying control sections. On 19 June, 828 Air Engineering Squadron received an alert for transfer to USA. All sections completed their work orders then packed and prepared for shipment home. Their L-5, Oxfords and UC-64s transferred to the 653 BS who ferried 29 Mosquitos to various bases in the UK, leaving six Mosquitos, two Oxfords and one UC-64 for training and routine flights. The last two Oxford aircraft R6034 and V3628 were ferried to Portsmouth Aviation, Portsmouth, 21 June for scrapping.

653 BS NS739 with WX-F squadron identification letters located only under left wing, May 1945. *Photo. E Jones*

By months end, 76 Mosquitos transferred to the RAF including three Mk III trainers. The 652 BS continued flying Atlantic WR missions providing meteorological data to support Operation Home Run, the return of bombers to USA.

Meanwhile, on 18 June, Lt.Col. John R. Hoover and Maj. Ernest H. Patterson (MM388, BA-H) flew to Chalgrove. Previously, a colleague from this base performed a high speed, low-level beat up of Watton. Hoover now answered this challenge. At 2055 hours he buzzed the Chalgrove hospital living quarters three times. On the fourth low-level pass the Mosquito hit the water tower on top a Quonset hut rendering the left engine and propeller unserviceable. He immediately climbed to gain attitude for an emergency landing. The extremely short, high-speed approach failed to allow the right landing gear time to fully extend and lock down. The Mosquito bounced twice and upon third contact with the runway, the right undercarriage collapsed starting a violent ground loop. The aircraft left the runway, tearing off both propellers with reduction gears and severely damaging the right wing. The aircrew survived the ordeal unscathed.

The 652nd received transfer notification to 1 AD, Alconbury. The final mission flown from Watton took place12 July, and on the 13th, all further flights scheduled from Alconbury. Bad weather delayed movement of men, aircraft and material by air to their new home.

Meanwhile, Capt. Richard Kenny flew to Malmo, Sweden, to retrieve Mosquito NS650 previously forced down in April. He took a Swedish Air Force general for a ride in the Mosquito and impressed him with its performance and beat up the field. On 28 July, Kenny left Malmo for Watton but landed in Copenhagen, Denmark, with a coolant leak. The Mosquito required an engine change and later returned by another crew. The last day of July found the 652 BS preparing for Air Force Day.

Air Force Day on 1 August brought several hundred British civilians to Alconbury and the 652nd displayed a B-17. Also on this day, the Allah "A" flight flew a new track and the Sharon missions came to an official close when the wheels of the last operational flight touched the runway at 0227 that morning. Major Michael Lorenz, unit CO, flew a special mission to the Azores on 2 August. He returned the 4th after closing the operations detachment at Lagens and retrieved Lt. Robert Menken with 10 enlisted men to Alconbury.

After dropping the atomic bombs on Hiroshima and Nagasaki, Japan surrendered on 14 August, V-J Day thus ending World War II. Meanwhile the 652 BS remained at Alconbury but assigned to 3 AD on 25 August.

Mosquito MM345 "Z" of 653 BS. *Photo R. Astrella*

September opened with normal flying activity and rumors. During the month, 652 BS became involved in a large personnel reshuffle. It soon became depleted of men, forcing Major Lorenz to cease flight operations. The last mission landed at 2120 on 26 September bringing the monthly total to 47 flights. The month ended with a beer party held by the enlisted men to which officers were invited. The men from 861 Air Engineering Squadron attended and a good time spent drinking beer and swapping rumors, talking about the good and not so good time the men had together. It was as much a farewell party as possible. October opened with 652 BS preparing the Fortresses for overwater flying to USA.

The 652nd maintained status and further demobilization as a temporary holding unit on paper only, with assignments 1 September to 1 AD, then on 12 October to 3 AD. On 25 October, 652 BS transferred from Alconbury to Raydon less equipment and many personnel from various units. It remained here under Eighth Fighter Command until 17 December when the unit transferred on paper, less men to Camp Kilmer, New Jersey, and deactivated 19 December 1945.

On 5 August, 653 BS personnel moved to Chalgrove, Oxfordshire, assigned to 374 FS, 361 FG, staging for return to the USA on the USS *Nunton B. Baker* troop ship. The 653 BS then served as a holding unit on paper for personnel returning to USA on a point basis. On 5 September, its complement consisted of two officers and one enlisted man assigned to 7 PRG. By the 12th, unit strength increased to 48 officers and 229 enlisted men. But by 27 September, unit strength reduced to one officer and one enlisted man with no equipment. On 15 October, 653 BS transferred to 3 AD, Hitcham, Suffolk, briefly and again on 1 November to Eighth Fighter Command. On 17 December, the unit transferred on paper, less men, to Camp Kilmer, NJ, and ceased to exist on 19 December 1945.

After the surrender of Germany, USSTAF initiated plans to transfer ETO units to the Pacific war zone. HQ 25 BG (Rcn) and 654 BS personnel departed England on board the Queen Mary, arriving in New York Harbor, 2 August. Issued a 30-day leave, members visited family members prior to reporting for duty at Drew Field, Florida. Several HQ members arrived in early August to prepare transitioning personnel to Douglas A-26 light-bombers. Japan's surrender on 14 August canceled this initiative and the 654 BS disbanded 4 September, while 25 BG (Rcn) deactivated on 8 September 1945. The 654 BS allotment was re-designated 54 Reconnaissance Squadron (Long Range, Weather) on 4 September. Assigned to Third Air Force on 8 September, the unit flew B-29s from Guam on 27 November 1945.

652 BS B-17 48520 "Corn Fed Virgin." nose art. *Photo E. Jones*

And finally at Watton, the base reverted to British control, home to the RAF Radio Warfare Establishment, which became the Central Signals Establishment in September 1946.

In September, 8 CCU transferred to the continent. Prior to movement they cabled Pinetree seeking guidance on the disposal of documentary films made at Watton. They received instructions to send the reels to Culver City, California, or New York City, or to a British film company. The color documentaries include "Aphrodite Story, Mosquito Story, P-47 Air-Sea Rescue" to name a few. None exist today and are believed destroyed prior to CCU transfer.

In Germany, CCU became involved documenting and cataloging German military films and negative archives. Late in 1946 the unit filmed the Nuremberg War Crimes trials and also produced various postwar documentaries on German reconstruction under the Marshall Plan.

The vital importance of meteorological surveillance in forecasting weather trends continued postwar. In 1947, a conference of the International Civil Aviation Organization allocated a number of stations in the Atlantic to an international fleet of weather observation ships from the USA, UK, Canada, France, Belgium, Holland, Norway and Sweden. This established an ocean-going meteorological observation capability providing international cooperation in sharing weather information in the north Atlantic area. These vessels continued this service, supplemented by international commercial aviation flights reporting weather data, until space satellites took over the role in weather forecasting. Today, a modified weather scheme established by the World Meteorological Organization, a United Nations agency, combines satellite information with data from ground stations, commercial aircraft, and merchant vessels, to aid facilitating global weather forecasting.

The Secret Mosquito War Against Germany

"This chapter reveals the history of a most remarkable, clandestine radio system, for communicating with secret agents hidden in the European countries during World War II, including Hitler's Third Reich. It was developed and perfected by a most extraordinary Texan, Stephen H. Simpson, Jr. Being the pilot who flew these missions, vivid memories flood back through my mind as I read this history so meticulously researched and presented. I was "there" again in the cockpit, hearing those beautiful Merlin engines of the Mosquito, and the roar of the R-2800s in the A-26, skimming low level over Germany and evading the night-fighters at high altitude at night. This is an exceptional, accurate, revelation of those hectic days … long ago." Robert P. Walker, Lt.Col., USAF Retired

At three minutes before midnight on 1 September 1944, an ex-German national named Jack parachuted under a full moon from an RAF Hudson aircraft on to German soil near the town of Sogl. He became the first of 102 Office of Strategic Services (OSS) agents to infiltrate into Germany between that date and 6 May 1945.

The following chapter details one segment of the OSS secret war against Germany involving radio-equipped Joan-Eleanor (JE) Mosquito aircraft to secure Intelligence information from agents deep inside the Third Reich. A brief overview detailing the OSS air operations evolution and 801/492 BG "Carpetbagger" procurement in this endeavor, followed by the JE development history and 25 BG (Rcn) participation in the secret air war, is presented as a best-case scenario. Where conflict exists regarding the dates of events, logbook entries from the pilots involved is cited rather than data listed in OSS documents.

"V-2" nose art. Photo *E. Jones*

In the spring of 1942, General Wladyslaw Sikorski, head of the Polish government in exile and Commander in Chief of the Free Polish Force, arrived in Washington, DC. He petitioned President Roosevelt to obtain six long-range bombers to airdrop supplies and men inside Nazi-occupied Poland by his forces in England. The Secretary of War discussed this request with General William J. Donovan, OSS. After consulting HQ AAF and state department officials, government authorities decided such aircraft should be controlled by OSS to coordinate covert activities among the Allies. Lack of similar air support handicapped current OSS plans resulting in a desperate need for aircraft to fulfill future operational requirements.[74]

In preparation for D-Day and Allied operations in France, inserting agents into enemy-occupied Europe received top priority. The British operated Westland Lysander and Lockheed Hudson aircraft for pickup and delivering agents. The British counterpart of OSS, the Special Operations Executive (SOE) refused to release Lysanders for OSS duties due to the limited number of aircraft available.

As an option OSS studied parachuting agents behind enemy lines from Stinson L-5 and Norduyn UC-64 aircraft, and employing UC-64 aerial extraction of agents. Limited range and vulnerability to ground fire excluded both aircraft. Research programs at Wright Field investigated utilizing B-25 aircraft to both dispatch agents and retrieve messages by aerial pickup in the field.

In December 1943, OSS investigated acquiring helicopters for post-invasion operations. The craft did not require large landing areas and permitted agent delivery irrespective of moon period under conditions of poor visibility or low ceiling. Vertical extraction of agents achieved more Intelligence information than with W/T communications. In the UK, OSS examined two Royal Navy Sikorsky YR-4s and attempted to obtain three advanced YR-5 examples, allotted to Ninth Air Force for part-time operations. Unfortunately, YR-5 production allotment would not begin until 1 August 1944, well past the planned D-Day invasion.

Clandestine Air Forces

With AAF support, HQ OSS, Washington, set in motion developing two heavy-bomber organizations for delivery of men and supplies behind enemy lines in both the ETO and MTO. Aircraft acquisition progressed in North Africa. On 26 September 1943, OSS received three B-17s for covert duties. The Forts immediately underwent modifications to dispatch supplies and agents. Removal of the ball turret formed a 48-inch-diameter "Joe hole" for dropping agents known as "Joes." The guns received flash suppressors, engines had flamedampers installed, and the aircraft under surfaces painted glossy black.

OSS launched the first operation to southern France the night of 20 October 1943.[75] After flying over rough mountainous terrain, the Fort released ten containers and five packages at the pinpoint. On return, heavy flak caused substantial damage to two engines and loss of fuel. The pilot throttled back for maximum range and landed his Fort at 0800. By February 1944, further missions delivered over 50 agents and supplies into Vichy-controlled France.

In preparation for D-Day, OSS developed a second organization with additional squadrons to supply their secret army, the Maquis. Tasked with restricting movement of German reserves and dislocating enemy communication systems, the Maquis proved inadequately equipped with arms and explosive to maintain a continuous offensive campaign. To achieve maximum support required providing weapons and munitions to over 200,000 men of whom only 35,000 were amply armed. A total of 50,000 men required 3,300,000 pounds of supplies, allowing 65 pounds of equipment-per-man to provide organizational support for an Allied invasion. Equipping this secret army increased the demand for transport aircraft. Thus was born the Carpetbagger Project.

The first "Rockaway Babe" B-24 crashed 22 September 1944. "Rockaway Babe II" is a B-17 . *Photo E. Jones*

74 RG 226. Letter, William Donovan to Harry Hopkins, Secretary of War 14 Oct. 1942.

75 Monroe MacCloskey, Secret Air Mission, New York: Richards Rosen Press Inc., 1966: 31.

"Lilly Ann." *Photo E. Jones*

652 BS B-17 undergoes complete engine change. *Photo. C. Hackman*

652 BS Heavy Weather B-17 46577 "A" with no upper or lower turrets. Photo *C. Hackman*

In August 1943, Eighth Air Force Antisubmarine Command, operating B-24s in the Bay of Biscay disbanded and this duty assigned to U.S. Navy.[76] The dissolved 479 Antisubmarine Group comprising the 4, 6, 19 and 22 Antisubmarine Squadrons, released personnel from 4 and 22 for OSS duty. Attached to the 482 BG under the Carpetbagger Project, they became the foundation in developing both 36 and 406 Bombardment Squadrons respectively.[77] These units formed the nucleus around which developed the 801 BG (H)(P) on 28 March 1944. In May, two additional units, 788 and 850 Bombardment Squadrons joined the Group. The 801 BG formalized on 9 August 1944 as 492 BG "Carpetbaggers" providing logistical support to occupied countries of western Europe. The four squadrons eventually became the (36th) 856, (850th) 857, (406th) 858, and (788th) 859 Bomb Squadrons.[78]

Initial supply missions confronted problems. The Maquis became a beacon drawing entrepreneurs eager to gain access to supplies for a burgeoning black market. Agents returning from France reported supply canisters either looted at the drop zone, or entire containers completely disappeared. Increased screening and organizational control, along with stronger discipline within its ranks, helped the French underground overcome this difficulty.

Carpetbagger operations continued until September 1944 when the liberation of France eliminated the 492nd role supporting the French resistance. On 16 September, SHAEF ordered the Group relieved of OSS duties and delegated hauling fuel to Belgium for the advancing American army. At Harrington the B-24s underwent modifications with wood flooring installed in the bomb bay for fuel tank placement. The tanker program ran from 21 to 30 September.

USSTAF experienced difficulties redefining the future 492nd role. A conference convened 25 September at High Wycombe (Pinetree), HQ Eighth Air Force, by General Doolittle's staff. Attending were OSS and SOE officials, and 492 BG officers. When SHAEF delegated the 492nd fuel hauling duties, they acknowledged disregarding OSS operational requirements. Air support from SOE sources was adequate for projected OSS sorties to Holland, Denmark, Norway, and Germany, but mandated dependence on SOE. OSS planned dispatching agents and supplies to Slovak partisans in Moravia and Bohemian requiring 170 sorties over a six-week period. SOE could not fulfill this schedule from RAF Tempsford because of prior commitments.

On 15 October, Eighth Air Force retained 856 Squadron for OSS duties. Two of three remaining 492nd squadrons converted to fly night bombing missions for 2 AD. After discussions, Eighth Air Force agreed to eventually transfer 859 Squadron to Italy to support OSS airlift in that theater. Surplus aircrews received assignment to 25 BG (Rcn) at Watton, while several supplemented Bernt Balchen's Nordic OSS operations between Scotland and Sweden.

OSS bitterly opposed dismembering the 492 BG considering the difficulties and efforts expended developing the unit for Carpetbagger activities. General Donovan investigated transferring the organization to another theater for OSS work to preserve Group integrity. USSTAF was receptive to the proposal provided the War Department and theater commander concerned initiated a request for such a transfer. A 492nd survey party flew a B-24 to the CBI to investigate OSS requirements in that theater. Apparently OSS operations held a low priority, and General Joseph Stilwell, CO Tenth Air Force, requested one squadron from 492 BG for bombing purposes only. In fact Stilwell wanted to keep the visiting aircraft and return the crew to Harrington. Thus, the 492 BG remained in a state of uncertainty.

OSS met with Seventh Army Intelligence on 23 October. Previously, 200 agents infiltrated through Allied lines and enabled Seventh Army to know exact movements of every enemy unit from battalions down to small units. Now the Rhine River formed a physical barrier.

The army had no means to penetrate agents east of the waterway. The only option available required parachuting "tourist" agents 50 to 200 miles into enemy territory with instructions to follow a proscribed path back to American lines, reconnoitering en route specific points for enemy units and positions. The tourists would provide valuable tactical Intelligence to troop elements that overran them. The Seventh Army now requested OSS assistance. Slowly, the 492 BG expanded its role and missions.

Enter Steve Simpson

In 1928 Steve Simpson joined Radio Corporation of America (RCA) at Riverhead, Long Island, New York, as a student engineer. Here he performed research involving the technical development of radio receivers, television, and 16mm sound movies. In 1940 the U.S. government delegated the navy responsibility for international communications and the army, domestic transmissions. Summoned to RCA president's office, Simpson met with an admiral and captain seeking secure people with technical expertise in international communications. Several months later he accepted a navy commission as Lieutenant Commander, the equivalent of his reserve army rank of Major.

He worked on navy censorship until September 1940, then drafted to full-time active military duty. When the U.S. declared war against Japan, the government placed censorship under civil jurisdiction. In 1942 Simpson transferred to OSS to organize their Middle East communication system. They expected the region to become a center of military activity. The tide of war changed, and he remained in Washington to organize the worldwide OSS communication branch. This involved both personnel and facilities handling messages throughout the world, more importantly, "special" communications with points in occupied-enemy territory. The latter responsibility required his transfer to London on 16 October 1943.

In early 1944 Simpson attended dinner at a London restaurant with OSS staff members, discussing the best method in acquiring Intelligence information from Germany. The underground in Holland, Belgium, and France provided vital assistance to the Allies, but OSS had no such assets in Germany.

German engineers developed and built an excellent radio detection and accurate direction-finding system. They utilized three direction-finder (D/F) vans to triangulate a signal source. It normally required numerous agent attempts to establish contact before transmission of Intelligence information took place. In that time the Germans could easily D/F the signal location.

Simpson described the latest advances in electronic technology. IBM developed a wire recorder that electronically preserved on a moving wire, every word said during a conversation. Previously in 1943 he observed RAF experiments using a VHF radio to communicate with a high-flying aircraft. The RAF successfully employed similar equipment to garner Intelligence from agents inside occupied Europe.[79]

Simpson and colleague Lt.Col. Henry Shore of RCA, concluded the only successful form of communication for an agent in Germany was a compact radio carried inconspicuously on his person, allowing him to communicate undetected with a specially equipped aircraft. RCA had the technology available to develop such a device. When combined with a wire recorder inside a circling aircraft, the agent could convey more Intelligence information in ten minutes that normally took a week with wireless telegraphy.

Simpson's proposal reached OSS in Washington where General Donovan and his staff agreed to finance research and development of the special radio equipment. Simpson contacted Shore who put the idea into concrete form; Major Madison P. Rehm of RCA, Simpson's OSS assistant for three years provided further practical suggestions.

76 Warren A. Trest, Air Force Roles and Missions: A History, Air Force History and Museums Program, Washington, DC: GPO, 1998: 83.

77 The 406 Night Leaflet Squadron formed after the 406 Bomb Squadron designation was released from the 801 BG on 13 August 1944. The 406 NLS started operations around 1 September 1944 and flew from Cheddington until mid-March of 1945, then transferred to Harrington.

78 Robert W. Fish, They Flew By Night, 801/492 BG Association, 1992: 4-9.

79 Ron Clarke, "Cloak and Dagger," Air Enthusiast, (July/August) 2007: 27.

Electronics engineer DeWitt Goddard conducted research at the RCA Riverhead plant. One day director H.L. Peterson summoned him to his office where he met with Simpson. They sat down and Simpson detailed current OSS communication problems and obstacles they would later confront in Europe, once the war front moved into Germany. He informed Goddard the OSS wanted a radio device meeting certain specifications. 1. Communicate with a B-17 flying at 40,000 feet immune from enemy fighters and flak. 2. Contact an agent on the ground using a high frequency. 3. Be portable and easily concealed on the agent.

The British developed a communication apparatus called an "S" phone. It consisted of a waistbelt holding battery cells and radio receiver/transmitter, thus, not easily portable, too bulky, and impractical for OSS use. Could Goddard develop something similar but more simple? He thought he could.

Goddard's main concern involved designing a workable device without going into years of research. He built a small copper box in which he placed a high-frequency acorn tube creating a transmitter. He mounted the box on top a 50-foot pole with a dipole antenna and a modulator from a broadcast receiver. He then drove off from the RCA laboratory by car with receiving equipment to determine the distance he could detect a signal. Goddard received the radio transmission a distance of five miles from the Riverhead plant which he interpolated as equal to 40,000 feet in an airplane.

Next he decided a 250 megacycle frequency modulation (FM) signal was a VHF range the Germans were not using. Goddard constructed a small transmitter/receiver with a heterodyne circuit. He placed this in a small l.5 inch x 3 inch x 6 inch copper box having a ten inch folding antenna with transmission line attached.

In March 1944 he conducted air-to-ground tests from the Riverhead Airport. He selected a single-engine, four-passenger Stinson aircraft having a camera port on the rear cockpit floor to extend an antenna. He placed a radio from a Ford Trimotor on the rear seat, using the receiver's intermediate frequency (IF) stage as an amplifier; a six-volt battery served as power source. Three test flights conducted over eastern Long Island ended with forced landings due to engine trouble. Both the ground and air equipment exhibited promise for further development.

After several months work, he refined both ground and air radio units, and OSS provided a B-17 for high-altitude research. He conducted further tests from the Riverhead Airport with the Fort flying at 20,000 feet. Three test flights successfully demonstrated communication of radio signals 40 miles from Mantauk Point at the end of Long Island, to the airport.

The next step required building a more substantial apparatus because of OSS emphasis to accelerate development. The research program received Triple-A priority to authorize the requisition of strategic materials to expedite the project. After constructing a preliminary, full-scale model, OSS provided another B-17 for further test work.

The Freed Company of New York undertook manufacture of 12 aircraft units complete with wire recorder. But difficulties arose in locating an electronic manufacturer to build the ground or pocket unit. The Dictagraph Corporation of New York that manufactured hearing aids received a contract to produce the ground units and subcontracted the work to Signal-U Manufacturing Company of Youngstown, Ohio. Code name selected for the air unit was "Eleanor" for Mrs. Eleanor Goddard and the ground unit "Joan" after a WAC who worked for Simpson; hence, Joan-Eleanor or JE. OSS test flew the final JE product in a B-17 from Bolling Field. The equipment performed beyond expectation and was now ready for operational use.

OSS shipped the JE units to England on 16 July where Simpson immediately began evaluating the equipment in a B-17. The initial test flight from Bovingdon proved satisfactory.

On 18 July Simpson performed a brief five-minute demonstration conducted from the roof of London OSS at 71 Grosvenor Street. The following day he achieved excellent communications 35 miles from Bovingdon at 20,000 feet with good transmissions exchanged as far away as 53 miles.

Further operational evaluation encountered problems. Testing at RAF Sledge Green revealed signal distortion when the aircraft unit operated at high-gain frequency resulting in a range of only ten miles. Simpson took the equipment to the electronics research facility, American British Laboratory (ABL), Great Malvern, Wales, for further study. The British engineers labeled the problem an antenna design flaw and recommended using a four-foot square, corner-reflector antenna as the solution. This proved completely impractical; an agent could never conceal this on his person.

Simpson contacted OSS in Washington and arranged for Goddard's transfer to England to assist with the development problems. In August 1944 Goddard was inducted in to the U.S. Navy overnight, traveled by C-54 across the Atlantic and landed at Bovingdon. He visited ABL and examined the JE equipment and discovered an IF amplifier in the aircraft unit regenerating. This resulted in an asymmetrical IF amplifier band pass, thus limiting the signal range. Changing a damping resistor in the fourth IF amplifier rectified the problem.

Goddard test flew the JE equipment on 31 August confirming the problem was successfully solved. Elated with this news, Simpson provided General Donovan and his staff a demonstration on 1 September from the London OSS roof.

OSS issued orders to procure a B-17 or B-24 aircraft with full equipment and groundcrew as an introductory unit to place the JE technology into operational service. Eighth Air Force targeted the 492 BG for this duty because of past experience dropping agents and supplies into enemy territory from B-24 aircraft.

Chief of Operations (CoO) Eighth Air Force, OSS, and 492 BG officers met at Pinetree on 25 September to discuss assigning a B-24 this duty. Upon being told OSS planned to fly the bomber at least 100 miles behind the German frontlines, the CoO declared it impractical to operate B-24s on this mission profile; the aircraft lacked high-altitude performance required to fully utilize the JE equipment potential. Also, Germany developed new radar-directed antiaircraft guns. It would be suicide dispatching a Liberator near highly defended areas. Instead, CoO suggested a Mosquito aircraft as more appropriate for the role. It was fast, highly maneuverable, and possessed a high altitude capability. He promised to investigate obtaining an example for 406 Leaflet Squadron at Cheddington for exclusive OSS employment.

A 1930 wedding photo showing DeWitt Goddard. *Photo S. Simpson Jr.*

Next day, CoO informed OSS he deemed the de Havilland Mosquito aircraft more suitable for JE operations. On 27 September, Eighth Air Force arranged to assign two Mosquitos to 492 BG and convened a joint conference to discuss further experimental and operational flights using the aircraft.[80] OSS did not inform Simpson of 492 BG inclusion in the proposed operations. As he recalls:

> "All involved realized we had not come into this work for the love of flying, and that someone was going to get hurt. But, we knew that if we got only ten percent of the information planned, we would save many lives and possibly shorten the war by years. Nevertheless, the Eighth Air Force was adamant, stating we could use only Mosquito aircraft for which there was only one American operated base, at Watton."

On 30 September, Director of OSS Air Operations, Lt.Col. Charles C. Bowman, Simpson and Goddard visited the 25 BG (Rcn) at Watton. After detailing mission requirements, Group HQ showed the men two former Mickey (H2X) Mosquitos. Both had radar equipment removed and were recently withdrawn from squadron operations awaiting return to RAF. According to Simpson:

> "Goddard and I approached Watton with misgivings, and the first thing we learned was the Mosquitos assigned to us had been grounded because of undetermined trouble. The cause was soon discovered and changes were made by the entire Group. Our operations were so secret, they could not be divulged even to the Group commander."

Bowman discussed with Watton officials the difficulties inherent in having Mosquito aircraft operated by another unit, retraining of aircrews unfamiliar with the aircraft, and maintenance difficulties. While these discussions continued, Simpson and Goddard proceeded to examine the two Mosquitos. They discussed with an engineering officer the technical details of modifying the aircraft. Conversion work would require at least one week. Simpson now possessed aircraft for his project.

OSS Lt.Col. Bowman visited Harrington 1 October and spoke with the 492 BG. He relates the following:

> "I started off very tactfully and soon found that the quicker (492 BG) could get rid of the problem of operating the two Mosquitos, the happier they would be … entirely in favor of our suggestion the Mosquitos be operated by the 25 BG … that I call … Pinetree and recommend new orders be issued to this effect."

After discussions, 25 BG (Rcn) volunteered to perform the project. To become truly Mosquito proficient required many flying hours. They were apprehensive of allowing inexperienced crews fly the aircraft. Harrington officials reaffirmed this decision. The 25th would maintain operational control of the aircraft until future events warranted transfer to the 492nd.

On 5 October, Pinetree notified OSS they refused to allow 492 BG B-24 flights further than 50 miles inside Germany. Such operations could be adequately handled by SOE at RAF Tempsford, therefore there was no justification in maintaining the Harrington organization. But OSS wanted complete independence from the British and were adamant in supporting and maintaining the 492 BG role.

At Watton conversion work began on the two bulbous nose Mosquitos, NS676 and NS707. Both aircraft already possessed an aft operators' compartment and serve as prototypes to design and install the JE equipment. To lighten the aircraft required removal of all armor protection and IFF from the tail section, to help balance the added equipment in the JE operator compartment. Because Simpson wore a Navy uniform, everyone referred to the Mosquitos as the "Navy" aircraft. According to Goddard:

> "Watton previously grounded the Mickey Mosquitos because several exploded in the air killing the air crew. Apparently, sparks from the radar electrical equipment in the rear operator's compartment ignited fumes from a leaking bomb bay tank. We did not want our aircraft to blowup, so I designed a ventilating system. An external scoop was installed in the aft radio compartment to provide a forced draft of air into the fuselage. Another exit scoop in the bomb bay aided in the elimination of any gas fumes which may accumulate, thus preventing a possible explosion caused by sparks from the radar equipment.
>
> "Also, the aircraft was refueled on the ground at an angle. When flying level after takeoff, excess fuel spilled from the tanks. I had the overflow fuel vents relocated to the hinge area of the bomb bay doors from their normal position where the bottom door halves met. Unfortunately, the fuel leak problem was never fully solved."

Goddard suggested an understudy be assigned to thoroughly learn the JE unit intricacies in preparation for his eventual return to USA. The next night of 7 October there was a knock on the door of Simpson's apartment. When Goddard opened it, there stood a stranger in full uniform with a duffel bag over his shoulder, who asked, "What the hell am I doing here?" His name was Captain William Sawyer who transferred to England from an OSS detachment in France and reassigned to the project. Goddard recalls:

> "He was very knowledgeable about radios. He followed my advice and we had no differences of opinion. He was perfectly wonderful, and I was very happy to have him as my understudy."

Sawyer recalls his introduction to Watton:

> "When I first arrived in London, I heard about the V-1 'Buzz-Bombs' the Germans were sending over. During my two days there I did not see a single V-1. However, the first night in Watton, we all went out to the radio shop to inspect the JE gear. It was after supper and near dark. We heard what we thought was a plane approaching rather low. Steve and DeWitt remarked it did not seem to be aligned to the runway. We went outside and discovered it was a stray 'Buzz-Bomb' flying over the base at about 200 feet. It continued another three or four miles and hit the ground in a violent explosion. Fortunately, it came down in an open field. My first Buzz-Bomb! And 60 miles off course."

The 654 Squadron assigned pilot Lt. Richard Geary and navigator Lt. William Miskho to test fly and evaluate the Mosquito after equipment modification. On 17 and 21 September, Geary/Miskho test flew NS676, and on the 25th conducted the first flight with Simpson in the aft section. Goddard operated the ground set with Sawyer observing the proceedings. The aircrew had no access to JE conversation and obeyed Simpson instructions during orbit to achieve stronger signal reception. The test proved successful but during flight the aircraft reacted sluggishly to the pilot's control.[81]

On 26 October, Geary/Miskho took Sawyer for his first flight in the rear of Mosquito NS676. Sawyer recalls:

> "During my stay, the 654th assigned five pilots and three navigators to fly our planes, among their other flight duties. All pilots, though they were US personnel, had joined the RCAF early in the war and accumulated previous experience flying Mosquito aircraft. Two or three new JE operators arrived toward the end of my stay. One officer was Lt. George Fogarty, a conscientious addition to our group; others were Captain Victor Layton and Lt. Calhoun Ancrum."

Besides Geary and Miskho, the additional aircrews mentioned by Sawyer were pilots Capt. Victor Doroski, Lts. Robert P. Walker, Norman Magee, James Matuska, and navigator Roy C. Conyers. Several months later other squadron members joined the operations.

Lt. Richard Geary/Capt. Jessie Abbott, unit engineering officer, and Sawyer test flew Mosquito NS676 on 31 October. Sawyer recalls:

> "Shortly after my arrival at Watton, the pilots reported one of our Mosquitos did not handle as it should. When no problem could be found, a doctor made a flight with a pilot and took samples of oxygen in his mask, assuming the possibility that the pilot was suffering some vertigo problems due to lack of oxygen. The problem became resolved when it was discovered the two 50-gallon droptanks on the wings were full of fuel. They were supposed to be empty on our flights."

80 According to OSS documents, Mosquito NS584 and NS677 were selected for transfer to 492 BG. Yet, NS677 went MIA on 18 August 1944. NS677 may be a typo error for NS676.

81 RG 226: Steve Simpson, Lt. Cmdr, USN, Joan-Eleanor Log, cited throughout text.

Steve Simpson kneeling next to former Mickey Mosquito, bulbous nose NS 676, still with invasion stripe. Note porthole on starboard aft JE entrance hatch. Also extended Eleanor antenna from JE compartment. *Photo W. Sawyer*

Another view of NS676 with aft hatch open and extended Eleanor communication antenna. *Photo W. Sawyer*

Inside NS676. JE operator's compartment looking forward. In foreground is Eleanor antenna with handgrips for turning dipole for maximum signal. Top left is a compass, and on the bulkhead is the wire recorder. Lower left is dynamotor with battery and the TR-6 Eleanor radio is right of the antenna holder. *Photo W. Sawyer*

Goddard visited ABL at Great Malvern and learned the Germans currently operated sweep receivers to detect surreptitious communications. They had "Y" stations located 120 miles apart covering Germany and monitoring all radio frequency transmissions. These were small buildings fitted with numerous aerials and constituted part of the Radio Intercept Regiment (*Funkharschregiment*). He recommended every agent endeavor to determine the location of such a station in their vicinity and operate at least five miles away to avoid detection. In the event of a forced landing in enemy territory, Simpson installed explosive charges to destroy the JE radio equipment.

The lack of definite policy regarding the 492nd role resulted in confusion among all higher OSS/USSTAF/492nd echelons, hindering development plans as to size and scope of future operations. The initial location for inserting the first agent was not to exceed a distance of 600 miles from Watton. Therefore, OSS designated Holland and northern Germany as drop areas. But Pinetree refused B-24 flights to northern Germany because of heavy Flak defenses and Luftwaffe night-fighters. OSS planned JE missions from Italy, but Fifteenth Air Force rejected a request for facilities to house a 492 BG squadron for future operations, forcing OSS to abandon this proposal.

OSS selected potential JE agents fluent in English who previously lived in the target areas, had family members and friends there, and spoke the regional accent. They received practical instruction with the JE radio equipment. The ground Joan unit possessed a folding dipole antenna consisting of two elements, when placed in the horizontal position, constituted a half-wave dipole, serving as an efficient signal radiator. The antenna unplugged from the ground unit and easily folded conveniently for pocket storage. The agent directed the antenna in an end-on position for maximum transmission and reception. When receiving a null, he could determine the orbiting aircraft's position. VHF signal transmission became attenuated near dense foliage or in deep valleys. Simpson instructed the agents to avoid these areas during contact – they must operate in wide open spaces.

Simpson planned a unique method to contact agents. Each man received a straight-line extending from the Channel coast all the way into Germany. The Mosquito pilot did not know the agent's location, but knew he would report along the line, so they went searching for him. The pilot received a course and flew along the line with the JE operator calling to the agent. When he replied, the operator issued instruction guiding the pilot in the direction to receive maximum signal contact. The aircraft generally flew 20 miles away to some other area of the sky within radio range, flying a wide circular pattern or a long and wide figure-eight configuration near the pinpoint, or a variation of both to receive maximum agent radio signal.

The Allied armies' rapid advance changed the fluid nature of the frontline and not all areas along the line offered secure open space for contact. The Channel line proved impractical. OSS selected to fly JE aircraft directly to the agent's pinpoint – less time consuming to locate him.

The First JE Agent

OSS established the TYL project and assigned agent Anton Schrader, known as Bobbie, to parachute in to northern Holland. Goddard instructed Bobbie on JE radio operations, and several training flights between 6-8 November provided practical application using the equipment. Simpson scheduled a nighttime training session for Bobbie on 9 November. Goddard recalls supervising ground instructions:

> "Arrangements were made to send a coded message on the 10 o'clock BBC news giving time of contact. Bobbie had a cigarette carton-size radio receiver to tune-in the BBC for our coded message.
>
> "A night test was performed from the OSS communication station at Hurley Bottom, where the OSS sent and received coded messages from other operational sections and secret agents in Europe. We went out at 10 o'clock in the pitch black night to make contact. It was a very cold September night. I wore an electrically-heated flying suit attached to a battery box that kept me comfortably warm. Other members, ill-prepared for the weather, shivered in the cold evening air.
>
> "I built a special hand unit with a half-dozen jacks for additional headphones, allowing several people to listen to the radio conversation simultaneously. We accompanied Bobbie to the middle of a field. Simpson flew overhead in the Mosquito (with Doroski/Miskho). We stood around for several minutes and soon Simpson made contact with Bobbie, and they started talking to each other. All of a sudden Simpson said, 'Hey, wait a minute. All hell is breaking loose here. I'll get back to you.'
>
> "There were a few moments of silence, then he said, 'There are tracers going by the airplane and we're getting out of here. Good Bye.'
>
> "That was the last I head from him."

Joseph Persico in his book states:

> "Simpson impatiently tore open a hatch and stared below. Beneath him there unfolded a gorgeous fireworks. The plane shuddered with each burst. He knew he would have to straighten out the pilot. Simpson seized the intercom. 'Were in a storm Captain, You'd better get us out of here."
>
> "Commander that's no storm. You're being shot at!" [82]

Photo shows Steve Simpson in aft compartment at the JE operator's position in NS676 dated 12 November 1944. *Photo D. Goddard*

Goddard continues his recollections:

> "The following morning I found out what the problem was. It seems Steve was blissfully cruising along at 20-25,000 feet in the Mosquito when they tangled with a flight of RAF bombers heading for Germany on a night-bombing mission. Of course those gentlemen would not stand for a strange airplane in their midst. So they reported the intruder to the antiaircraft batteries. Fortunately the Mosquito was not hit."

Simpson conducted a second session at midday from the London OSS rooftop. Here, OSS Intelligence service reported the Germans operated four communication-receiver facilities. Three used by the military were capable of tuning to 250 megacycles, while the fourth could reach 1,000 megacycles. Goddard visited Great Malvern and modified the JE equipment, moving the operating frequency to 260 megacycles, placing the JE signal out of range for three of the four German stations.

The night of 10 November, with a barely waning moon, OSS scheduled a mission to insert Bobbie. A 492 BG Liberator left Harrington and flew low over the North Sea to Holland. Identifying the city of Ulrum, the lumbering bomber descended through clouds to 1,000 feet and Bobbie jumped "blind" at 2315. OSS scheduled a JE mission the following night, but adverse weather conditions at Watton made takeoff hazardous, so they canceled the flight.

For administrative purposes, HQ 25 BG (Rcn) nicknamed all OSS sorties Redstocking when categorizing and reporting monthly mission census reports to Wing HQ. Even though 25 BG (Rcn) assigned both Mosquitos to OSS, Watton airmen flew the aircraft. Therefore, Group

82 Joesph Perscio, Piercing the Reich, New York: Barnes & Noble Books, 1997: 164.

HQ wanted to receive mission credit. Consequently, they labeled all OSS flights Redstocking for credit purposes only and not to disguise the mission intent.

On 14 November, a Mosquito took off past midnight for Holland with Simpson but failed to establish contact with Bobbie. On return, the elevator controls jammed and the pilot experienced considerable difficulty controlling the aircraft. After a safe landing the engineering officers determined a wooden covering placed over the elevator and rudder control cables in the aft compartment to protect the JE operator, interfered with normal function. The wooden enclosure underwent modification and no further problem of this nature occurred.

The following evening of the 15th, Sawyer flew to Holland, but after repeat attempts, he too failed to establish contact, so the pilot turned for England. On return, inclement weather closed Watton. As Sawyer recalls:

> "We were diverted to Tangmere in southern England. Due to my location in the tail of the aircraft, I could not see outside. On first contact with the runway, we hit very hard and bounced. To me, it seemed like an eternity before we hit the second time. However, we did land successfully, but while awaiting debriefing in the tower building, the pilot mentioned something about two rows of runway lights on each side of the runway. When told there was only one row on each side, he promptly changed the subject. Obviously he had been suffering from a slight case of vertigo from oxygen lack. The RAF mechanics who serviced our plane after landing, advised us we had a cracked right undercarriage strut. The plane had to be repaired before it could return to Watton."

Undercarriage replacement sidelined NS676 for at least a week, requiring urgent efforts to bring NS707 into immediate service.

The lack of TYL contact worried OSS. They interviewed the 492nd delivery crew:

> "A few days went by and nothing had been heard from him. We were then called in, and we went over the whole mission, pointing out exactly where he had been dropped. After several more days went by, and still no word had been received from him, we were called in to London to go over the whole thing again."[83]

On 22 November, Mosquito NS707 became operational and went airborne with Simpson successfully establishing a 20 minute contact. Bobbie lost many provisions and pleads for replacements. He provides Intelligence, reporting German units flooding large areas to form a physical barrier; assessment of Allied bombing against rail and canal transportation of war material. Poor signal reception force confirming the next rendezvous for 26 November and break radio contact. The Mosquito returned to Watton with knowledge Bobbie was safe and well, successfully operating inside German-occupied territory. During return, the pilot's seat came loose resulting in difficulty flying the aircraft.

Another Mosquito contact on the 27/28 November resulted in an immediate response. Bobbie inquired when to expect additional agents, issues a list for provisions, and promises a reception committee with proper signal lights for next delivery. After sign-off, the aircraft diverted to Bradwell Bay due to inclement weather at Watton.

DeWitt Goddard left the OSS JE operations 30 November for return to USA:

> "My boss, H.L. Peterson was a shrewd fellow and a terrific engineer. Before leaving the USA for England he told me, 'The OSS will want anyone knowledgeable about radios, like you, to stay in England. Your rightful place is back here in the laboratory, not over there. Anyone can drive a truck, but not too many people can design and build one.' Those were his words. So, I was previously warned the OSS would attempt to keep me in England. Later, I learned Steve Simpson wrote to my boss and asked to have me reassigned permanently to the OSS project; they urgently required someone with my engineering background. But Simpson never received an answer to his letter, and I returned to the RCA laboratory."

With dispatch of Goddard the JE project experienced a great technical loss. OSS had two aircraft sets in the UK and 20 ground units. OSS in Washington gave top priority requesting immediate dispatch of additional JE radio equipment.

December 1944

At midnight, 1 December, Simpson flew in a Mosquito to Bobbie. He gave notification several agents will arrive by air and confirms signals for the delivery. Bobbie stipulates that both agents and supplies must arrive together. Simpson instructs him to monitor the BBC for an arrival date. On return, inclement weather force diverting to Bradwell Bay.

Since 654 Squadron maintained OSS aircraft for nighttime operations, they applied the identical paint scheme carried by night photo Joker Mosquitos to OSS aircraft: glossy black covering all lower surfaces, aft fuselage, and tail section; top wings and upper forward fuselage remained PR blue. The JE pilot also used the Joker radio call sign, Smalleek.

The 492 BG attempt a supply delivery mission on 5/6 December. Unfortunately, solid cloudcover with rain and snow completely obscure the drop area. Even from 600 feet the crew failed to see the ground. The Liberator aborted and returned to base.

The task of delivering supplies to Bobbie continued to confound OSS. Sawyer flew by Mosquito to Holland on the 10th. He arranged reception signal lights for visitors and packages, and if inclement weather, signal for packages only. Simpson then visited Bobbie on 12/13 December. Aborting the past few missions caused Bobbie's friends to lose confidence. Clear weather prevailed and he heard the aircraft arrive one night and gave the correct signal lights, but the plane never returned. Apparently, the reception members failed to maintain the flashlights lit long enough. It was imperative supplies be delivered within the next few days, otherwise, wait until the next moon period. Simpson promises delivery tonight or tomorrow. Bobbie will prepare a reception tonight to receive packages. His batteries weaken and contact is lost.

Simpson concluded that Bobbie's credibility with the underground movement would seriously be impaired. He formulated plans to deliver supplies by Mosquito. If successful, future employment by this method would facilitate missions deep inside Germany beyond the operational zone of Carpetbagger B-24s. Watton provided a night photo Mosquito (NS559) with bomb racks, permitting delivery of four to six 100-pound containers. Two canisters arrived from Harrington and were immediately loaded on board the aircraft.

83 Ben Parnell, Carpetbaggers, Austin, TX: Eakin Press, 1987: 115.

The delivery of supplies required a concerted effort to coordinate the operations of two Mosquitos; one circling high above establishing JE communication with Bobbie, while the other flew low to deliver the containers. The high aircraft acted as controller, relaying instructions from Bobbie to the supply Mosquito on where to make the airdrop.

On 15 December, a JE Mosquito with Sawyer roared off the runway at 0117 followed by the delivery Mosquito carrying the two containers. The JE Mosquito successfully established contact with Bobbie. Unfortunately, the drop zone was located near the ocean and fog now covered the area, limiting surface visibility. On the ground, one light malfunctioned leaving three to signal the circling aircraft. The high Mosquito endeavored to relay Bobbie's approach instructions to the low-flying aircraft, but all attempts to establish radio communications between both Mosquitos failed. The impenetrable fog prevented the delivery Mosquito from viewing any signal lights on the ground. Sawyer notified Bobbie to return next night at 0100 for another delivery.

Simpson led the 16 December supply mission. The delivery Mosquito left after midnight with the JE Mosquito following behind. The JE aircraft established contact with Bobbie, but again experiences VHF communication failure with the delivery aircraft, even though different antennas had been installed. The delivery Mosquito circled the drop area swooping down low but failed to detect any ground signal lights. Simpson attempted to arrange a return visit when Bobbie's batteries slowly lose power and contact is lost. Both aircraft returned to Watton.

OSS confronted a dilemma. The B-24s' slow speed, poor maneuverability, and vulnerability to light flak at low altitude, and night-fighters, made delivery missions to Holland and Germany hazardous. The previous two unsuccessful Mosquito resupply attempts demonstrated a requirement for a different aircraft-type to fulfill this mission. They examined specifications of available U.S. aircraft and determined the Douglas A-26 Invader met JE mission requirements. It could easily penetrate Flak defenses and evade enemy night-fighters with sufficient speed and maneuverability to insure a reasonable chance of reaching the target. It also afforded the option of combining aerial delivery of agent/supplies and communicating with agents by JE radio.

On 18 December, Simpson and OSS officials visited Pinetree and successfully secure two A-26s on loan for evaluation. Several sat parked at Bovingdon. Simpson and pilot Lt. Robert Walker arrived at the airbase and escorted to parked A-26B 43-39329. Walker located the entrance hatch and climbed in. After inspecting the cockpit, he determined there was no problem flying the aircraft. The engines wound, then coughed, and sputtered to life, and both men took off immediately.

On approach to Watton the nose wheel failed to extend. The pilot circled the field while Simpson made repeated but futile attempts to lower the nose gear. An engineering officer was summoned to the tower and provided instructions on hydraulic operations and possible options. With this knowledge it did not take Simpson long to employ his mechanical genius to correct the problem and extend the wheel. The aircraft landed safely and taxied to the ramp. The second aircraft A-26C 43-22500 arrived later that day and work began modifying both aircraft for dispatching agents. Future OSS plans for inserting two-man teams into northern Germany relied on USSTAF allocating A-26s on a permanent basis. 1/Lt. Lt. George H. Fogarty and Capt. Victor J. Layton arrive on base and join the Watton OSS operations.

Mosquito NS559 flown in supply delivery attempt to OSS agent Bobbie in Holland. Photo taken at A-83 Denain-Proust Aerodrome, Valenciennes, France, April 1945. *Photo R. Hiatt*

OSS selected Stuttgart as the next target for JE agent insertion. In preparation, NS707 flew a training flight over London on the 31st with agent TROY on the London OSS roof.

Simpson achieved access to all Watton departments and sections enabling the JE project to reach fruition. He initiated the Joan-Eleanor Log Book detailing a critique of daily events. It served as a communication device for personnel, listing daily activities to keep members informed on work accomplished or required. Simpson also documented against any barrier placed in his path. It details issues encountered with both AAF and London OSS meeting his agenda.

January 1945

The new year began with A-26 test flights to evaluate hydraulic system problems. Apparently Pinetree grounded 60 A-26s at BADA for similar reasons. They underwent replacement of Bendix regulators and selector valves with Vickers-built units.

Tonight, 3 January, Simpson scheduled a combined JE contact/supply delivery to Bobbie. He left after midnight in an A-26 and flew to northern Holland at 1,000 feet. The pilot slowly circled the area, made several low passes, but no ground lights were seen and no JE radio contact established. The A-26 returned to Watton.

On 5/6 January, a JE Mosquito with Sawyer contacted Bobbie, but experiences radio communication difficulties. Bobbie requests a return visit in two days during the full moon period. Sawyer asks him to save his batteries; a new receiver set will arrive the following day.

After midnight the 7th, Walker/Conyers/Layton and Simpson thundered off the field in an A-26C followed by a JE Mosquito. Walker flew on the deck across the North Sea while the Mosquito climbed high above to 25,000 feet. Upon contacting Bobbie, the Mosquito would relay by radio, his instructions to the A-26.

At mid-channel during the climb, the Eleanor unit malfunctioned and the Mosquito pilot aborted his mission. The A-26 continued over the North Sea encountering light Flak at 1,000 feet near the Frisian Isles. The aircraft flew low over the rendezvous area, now completely enshrouded in darkness. The pilot flew so low the A-26 antenna broke off preventing further radio contact. Slowly he circled, making several more passes over the pinpoint, but still nothing – no lights at all. Simpson instructed him to head back. He banked and turned for England.

On final approach to Watton at 0500, Layton peered through the belly-turret periscope and confirmed only the nose wheel appeared fully extended; the right main undercarriage lowered only partially and the left remained locked in the nacelle. When the pilot raised the right undercarriage, it too remained locked inside. Both nacelle doors refused to crack open; the latch assembles would not unlock. The emergency hydraulic system proved inoperable. Violent pullout dives and rocking the aircraft failed to break the main gear free. All routine auxiliary and emergency procedures were tried numerous times, but only the nose gear extended. He circled the field while Simpson expended every effort to extend the main gear, but to no avail. The tower summoned the engineering officer for consultation, but without success.

1/Lt. Robert P. Walker at the Mosquito controls. *Photo R.P. Walker*

1/Lt. Roy Conyers sitting at Mosquito navigator's position.
Photo R.P. Walker

With only 15-minutes fuel remaining, pilot Walker released the bomb bay supply canisters in preparation for a wheels-up landing. He slowly brought the A-26 to the runway, feathered both props in fine pitch and used full-high RPM so the prop blades would bend easier. Upon touchdown at 0645, the A-26 slid along the concrete surface on its belly, trailing sparks and smoke to a safe stop. The aircraft sustained bent props, engine, lower fuselage and turret damage.

Tonight, 9/10 January, Simpson flew by Mosquito to contact Bobbie. When asked why the aircraft failed to establish communication the previous rendezvous, Simpson explained an equipment malfunction prevented contact. If radio contact fails and he hears the aircraft's motors, Simpson requests Bobbie turn on the signal lights so they may determine his exact position. They arrange for another supply delivery attempt.

Failure to deliver supplies resulted in OSS delegating 492 BG this task. Solid clouds enshrouded the target area at 3,000 feet. The pilot attempted flying beneath the undercast but encountered 10/10ths cloud below 400 feet severely restricting visibility in the drop area. The B-24 aborted the mission and returned to Harrington.

Simpson exchanged A-26B 43-9329 for A-26C 42-2524 at Warton. The Douglas Aircraft Company technical representative provided lectures on hydraulic and fuel systems. Hydraulic troubles plagued all A-26 aircraft arriving in England, a problem the manufacture experienced difficulty rectifying. Simpson claimed he recommended so many changes that Douglas Aircraft instituted, he virtually redesigned the entire A-26 hydraulic system.

Accident scene photo showing A-26 43-22500 that bellylanded on 7 January 1945 by 1/Lt. Robert P. Walker. *Photo R.P. Walker*

Preparations to insert an OSS agent in or near the Berlin area progressed. Simpson discussed with senior Group navigator, Major John Walch, the uncertainties of flying deep into Germany at very low level. Walch intended achieving greater course accuracy by designing a special experimental Loran plotting board to measure four LOPs from both UK and MTO chains.

OSS initiated plans on the 21st to dispatch missions to southwestern Germany during the January moon period. Lyon, France, became a base for operations. With a waxing moon above, the first B-24 mission combined several agent deliveries in one Liberator sortie. This consisted of JE agent TROY, and two two-man teams PITT and MIMI. TROY safely parachuted near Stuttgart. The MIMI insertion failed because a leg bag became detached and accidentally fell through the Joe hole, almost pulling the agent out with it. Having a deployed eight-foot parachute on the end, it proved more than the dispatchers and other agents could handle. They cut it loose and the MIMI team returned to base after dispatch of PITT. OSS revamped the MIMI mission and discontinued using the leg bag.

The 492nd attempt a delivery to Bobbie on the 21st. On arrival the signal lights went on, but a misunderstanding by the reception committee, extinguished these just as the B-24 flew over. The aircraft made three separate runs and finally aborted and returned to England. The Liberator passed over an enemy convoy at 80 feet and received damaging fire, narrowly escaping without casualties.

On 22/23 January, Sawyer flew to Holland by Mosquito to visit Bobbie. Contact initially proved good, but Bobbie's batteries quit after 10 minutes. The previous night a B-24 arrived 50 minutes late. The reception committee heard the aircraft and ran back to switch on the signal lights. Sawyer responds, the aircraft observed lights for only a brief period and requests Bobbie's men leave the lights on. Due to adverse weather forecasts, tomorrow is the last delivery opportunity until the next moon period. Future contacts would prove impossible until fresh batteries replenish Bobbie's exhausted supply.

Rare photo of OSS A-26 taxing at Watton. *Photo G. Scruggs*

Poor weather for several days worsened the 26th with heavy snow cleared from the Watton runways, leaving slippery surface areas. Simpson checked with Intelligence to plan an operation to visit new agent TROY. A Mosquito reached the Stuttgart area the 29/30th and remained an hour within range but failed to obtain a reply from TROY.

Layton recalls this mission:

"It was during the latter part of January 1945, when I was given orders to fly these missions as an operator. I had been picked because I spoke the languages of our agents behind the lines, local nationals who had volunteered to do some work for us in their occupied countries. There was another reason; I served with the Maquis and had been liberated by the Allies in northern France, therefore, knew the dangers, pressures, and gnawing fears which always come with the job.

"It was like a return journey into the womb. The opening to the JE operators' station in the rear was so narrow that I could barely squeeze through. The new flight suit and the stiff webbing of the parachute harness made me feel as if I were a trussed turkey. The crew chief, a very helpful fellow who had braced himself firmly on the hardstand, was gleefully pushing on my rump. Inside, the glow of my flashlight reflected from the exposed control cables, pushrods and assorted valves, reinforced my initial anatomic impression. In front of a tube-and-canvas seat there was a small steering wheel. Beyond it on a rough, plywood shelf squatted 'Joan-Eleanor.' This was my first operational date with her, and I knew that we would become quite intimate during the long hours I would spend with her in this flying radio shack.

"The airborne unit was large, quite elaborate and powerful because it used the aircraft's electrical system. This radio was tunable and had conventional dials and controls. The mike input was compatible with the lip or throat microphone used by the plane's crew. In this instance, a connection was provided to record all communications on an early recorder, one which used thin magnetized steel wire. The recorder was mounted next to the radio. The antenna, consisting of several elements was directional, and attached to a long steel tube which extended below the fuselage. It was retractable during takeoff and landing. The small steering wheel at the upper end served to turn the antenna in flight and to orient it for the best reception. In addition to the electronics, the radio operator had in front of him a full display of basic flight instruments, a magnetic compass, connections for his oxygen mask, and an outlet for his heated flying suit. The whole thing had been jury-rigged in what normally was an inspection compartment in the forward part of the tail just behind the belly compartment, which now was packed with range extension tanks.

"The idea behind the equipment, rest upon a peculiar propagation characteristic of VHF in which the signal reflection from the ground forms an inverted cone within which two-way communications are possible. The role of the airborne station was to locate and fly in this propagation cone at the highest possible altitude (where the orbit would be largest). The rotating antenna could be adjusted for the best signal, indicating where the propagation cone was located in space. This directional information was then passed to the flight crew in the cockpit, so they could keep the plane on station. During periods of traffic between the ground and the aircraft there was practically no chance the location of the transmitter could be DF by triangulation, and while it was conceivable that the signal from the aircraft could be picked up by a listening station on the same frequency, there was not much the enemy could do about it. We were outside his capability to knock us out. All communication was recorded on wire and the spools were sent by special courier to headquarters right after a mission for analysis and target designation. But on this my first operational flight, no contact was made with TROY and we returned to Watton."

February 1945

Snowy winter conditions prevailed at Watton the 1/2nd. A Mosquito mission to TROY took off with Walker/Miskho and Layton, when the right undercarriage failed to retract. After numerous unsuccessful attempts to raise this structure, Walker decided to proceed as scheduled. Layton began calling 60 miles from the pinpoint. Reduced airspeed from the extended wheel delayed arrival to the Stuttgart region. The Mosquito circled the area for 24 minutes but failed to establish contact. Fifty years later, when asked why he continued the mission with the undercarriage down, Walker replied:

"The agent placed himself in great peril volunteering for the mission in German territory. Therefore, it was incumbent upon the Mosquito crew to disregard their personal safety and establish contact to assist the agent survive his present situation."

On 3 February, the 492nd attempt delivering supplies to Bobbie. En route, the B-24 received numerous hits from 20 mm antiaircraft gunfire, and no signal lights appeared in the drop area.

That evening a Mosquito staged from Dijon, France, to rendezvous with TROY. During climb to altitude, at 20,000 feet the port engine ran rough and cut out violently. Pilot Walker informed the crew to prepare for bailout. Because Dijon possessed inadequate night-landing facilities, he elected to fly to Manston and executed a single-engine landing.

A mission to TROY on the 7th experienced navigation instrument failure. Uncertain as to whether they reached the pinpoint, the crew remained 35 minutes but fail to establish contact.

And late evening the 8/9th, Sawyer flew to Holland by Mosquito and reached Bobbie past midnight. They confirm supply delivery next week but Bobbie's batteries become weak and his signal disappears.

The evening of 12 February a Mosquito took off for Holland, when the right undercarriage failed to retract. Undeterred, the pilot again elected to continue the mission. Now behind schedule, the navigator suggests a direct route over British AAA areas to shorten their flight time. Arriving on schedule, Simpson establishes contact with Bobbie, who issues confirmation for delivering supplies only; he desperately requires batteries and provides coordinates to another drop area.

Ken Lundberg, a 654 Squadron radio technician provides his assessment of Simpson's personality:

"We stood shivering on the windswept hardstand listening intently for the familiar, steady drone of N for Nan. The night was typically English; cold and damp with sheets of mist occasionally soaking all who were forced to be out at the unearthly hour of 2 am. A gust of wind carried in the first horsepowered whisper of a bi-motored plane. As it approached from the east, the hum increased – to a drone – to a roar, and finally to a crescendo of sound which shook the ground beneath us.

"N for Nan, as human as its occupants, welcomed the touch of sold earth, raced up the ramp to its hardstand, settled back on its haunches, sounded a final roar of defiance, coughed and with an air of smug satisfaction became silent. This scene had been enacted many times before, but as we sprang into action, we knew the routine would not continue; for Lt. Commander Simpson was aboard.

"Silence prevailed but a moment before a clamor arose within the plane. One of the fellows hurried to unlock the Commander from his coffin-like section in the fuselage. Without interrupting his tirade against weather, planes, and things in general, the Commander tumbled to the ground and was on his feet in an instant. His appearance was familiar to us, but to others would have proved a surprise. In the faint blackout lights of our command car, this giant hulk of a man attired in flying togs gave the appearance of a Martian warrior. His uniform was as indeterminable as his personality. Royal Air Force boots, Army shirt and Air Force heated suit underneath a Navy jacket.

"As we drove to the Operation Shack, he kept up a steady flow of conversation asking questions without waiting for answers. He had the prestige of a man who did dangerous things courageously – while enjoying it. Our eyes blinked as we entered the Operations Shack, and although no one called attention, we noticed that everyone unconsciously 'snapped to' as Simpson walked in. He was a Navy Officer at an Eighth Air Force base, but his authority came from personality, not rank. His actions were quick and decisive, his voice brisk and curt, and his manner reflected a strong self-confidence. His face was stern but not cold, and his eyes snapped as he talked. The lines on his face and his silvered hair were not from age; those long bitter cold nights over the continent left their mark.

"Commander Simpson was an enlisted man's officer. Many a conversation in the enlisted mens' barracks started with his latest escapade, and the secrecy of his mission only enhanced his aura. His utter disregard for formal procedure and red tape caused grave concern in some quarters, but no one dared reproach him.

After leaving JE operations, OSS inserted Victor Layton as secret agent into Austria. *Photo V. Layton*

"Commander Simpson was colorful in many ways, but his main claim to distinction was his refusals to be assimilated into the uniformed mass; his ability to keep his individuality.

"We knew very little about Commander Steve Simpson and that is what it was supposed to be. He used our Mosquitos for special missions. From conversation in our work 'banter', we gathered that he made regular contact with 'friends' on the continent behind the lines. We assumed by his mystery that he was OSS on special assignment from the Navy to the Air Force. He then disappeared as quietly as he came." [84]

On 13/14 February, a Mosquito flew to Stuttgart and successfully contacted TROY. The agent replies having previously waited four times at the assigned pinpoint. The JE operator issues instructions to listen to BBC at 1300 instead of 1100 hours, and provides a pinpoint for next contact. Finally! Tonight the 492 BG successfully complete the first supply delivery to agent Bobbie after numerous abortive attempts since mid-December. This success paved the way for dispatching agents scheduled for TYL II project within the next moon period.

Calhoun Ancrum, an OSS language specialist, arrived at Watton. His father served in the U.S. diplomatic corps in Moscow, Berlin, and Paris. Calhoun spent his formative years raised and schooled in these capital cities, and thus fluent in Russian, German and French languages. OSS planned returning Layton behind enemy lines in Austria. Ancrum underwent JE training as his replacement.

The impending Allied assault across the Rhine River made imperative the bulk of agent deliveries be achieved during the March-April moon period. Therefore, OSS consolidated the May-June insertions with March-April, creating a hectic operations schedule. This commitment required more A-26 aircraft. To overcome this handicap, OSS approached SHAEF, who in turn instructed USSTAF to heed this request. USSTAF assured OSS full cooperation if the Berlin insertion mission is successful. Therefore, Sawyer and Fogarty initiate concerted efforts to complete aircraft modifications to allow A-26 flights to approved pinpoints near Berlin. OSS received permission to insert agents northwest of Berlin and one north of the Ruhr; JE flights over the Ruhr and Berlin, not more than one flight every five days. USSTAF prohibited agent insertions east of the city due to rapid Soviet advances.

On 17/18 February, a weather Mosquito flew over Holland preparatory to a scheduled evening JE mission. Ad hoc airmen became involved when Col. Leon Gray/Lt. Robert Morrow flew Simpson to Bobbie arriving at midnight. Bobbie received new batteries and expressed appreciation for the previous supply packages delivery. He requests another delivery in several days, packages only – no friends. After receiving his Intelligence information, the Mosquito signed off and turned for England.

The February-March moon period began 19 February and became complicated by an unbroken seven-day period of inclement weather in England and Germany. OSS scrubbed many missions during the closing moon period creating a backlog of agent insertions.

Simpson scheduled a training flight over London the 19th with the HAMMER team on the London OSS roof. Thick clouds covered the afternoon sky. Lt. Norman Magee piloting the Mosquito narrowly escaped colliding with a B-24 by a scant ten feet when flying through clouds. Layton, in the aft compartment, recalls experiencing severe turbulence as the Mosquito missed the Liberator.

On 20/21 February, Simpson flew to contact Bobbie. When requesting a drop location, Bobbie interjected he wanted no visitors, only packages, then communicates his Intelligence information. Simpson arranged the next meeting, then broke contact. The following 22/23rd, Layton flew to Stuttgart to contact TROY. Searchlights probed the distance horizon as the aircraft arrived on schedule at midnight. He called for half an hour but TROY failed to respond.

OSS formulated plans to perform experimental jump tests for adapting the A-26 to deliver agents. On the 24th, A-26 43-22524 arrived at British Airborne Experimental Section, RAF Station, Beaulieu, Hampshire, with Walker, Miskho, Simpson and Fogarty. The men immediately begin constructing a platform covering the forward bomb bay. This allowed the agents to sit facing the tail and jump through the open rear bay. According to Fogarty:

"I did the design and construction of the dropping platform in the first A-26. Subsequent ones were copies. While jumps from the B-24s of the 492nd were from a circular Joe hole, the arrangement in the A-26 was much quicker and dirtier. About three-quarters of the forward bomb bay was floored over with plywood. At its extreme rear edge, about 18 inches of the floor was hinged to drop downward. Until drop time, the platform was held in place by pins which could be pulled by a cable held, not by the pilot, but by the navigator in the nose. He also controlled a set of signal lights located in the rear bomb bay.

"When time for the jump approached, the dispatcher dropped into the bomb bay to act as jumpmaster. Presumably, if the shelf failed to drop our daring agents, the jumpmaster pushed them off. The system was so crude and simple, it was difficult to see it failing, and I don't believe it ever did. As far as the agents objecting to their prerogatives as paratroopers being taken away, this is just plain silly. These agents were not trained paratroopers, and in any case, a low level, static line parachute jump at night would have hardly given them any chance to exercise any expertise. The whole idea was for the fully trained navigator to recognize the proper spot and release the agent like a load of bombs."

A UC-64 photographic aircraft with parachute experts took off and flew alongside the Invader. When two dummies were discharged, both parachutes opened aft of the A-26 bomb bay with canopies scraping along the fuselage beneath the tail section. Upon landing the men proceeded to saw off 16 inches from the rear platform and lengthen the 13-foot static line with a five-foot strap. During a second test the chutes again dragged along the bottom fuselage, snapping off the IFF antenna. They increased the static line another 6 feet, and two expert parachutists volunteered to perform a live jump. Flying 155 mph at 1,500 feet, both made excellent exits and safely landed. When their feet hung down from the platform, the slipstream exerted considerable pressure, tending to pull the jumpers out. Spoilers installed at the head of the bomb bay alleviated the slipstream pressure.

Lt. Robert Walker approached the end of his combat tour. Simpson asked him to recommend a replacement pilot. He nominated 1/Lt. Kingdon R. Knapp, ex-RCAF who received a DFC from King George VI of England at Buckingham Palace. Though only five-foot, five-inches tall and weighing 135 pounds, his flying skills more than made up for his small stature. Knapp teamed with navigator Lt. John K. Jackson and both men formed the backbone of JE flight operations. These officers later brought the OSS JE project through a turbulent and difficult period to a successful conclusion at wars end. Knapp would fly JE Mosquito missions while ex-RCAF veteran, Lt. Oliver Emmel joined the project to pilot the A-26.

On 26th/27th February, Knapp flew his first JE mission, with Mishko and Layton to visit TROY, but the agent failed to respond to repeated radio calls.

Late evening the 27th, Simpson initiated a dress rehearsal for the HAMMER mission. Pilot Lt. Robert Walker, navigators Lt. William Miskho and Major John Walch, and dispatcher/gunner Sgt. Derr (USMC/OSS) started off in the A-26 and followed a proscribed course over England. They flew at 1,000 feet altitude with a waning moon at 35 to 50 degree angle as expected on the Berlin flight. Pilotage from the nose proved excellent with no difficulty identifying checkpoints, etc. The practice flight proved beneficial to all concerned allowing crewmembers to practice as a coordinated team.

The Berlin mission required flying a predetermined course through several weather fronts, over antiaircraft defenses and German airfields at levels from 100 to 500 feet; and locate the pinpoint. Both Walch and Miskho researched data, then developed the whole flight plan basing navigation on blind flying instruments, absolute altimeter pilotage, checking topographical features throughout, and utilizing Gee and Loran navigation instruments. Meanwhile, Walch – creator of both Gee and Loran trainers – continued perfecting an experimental navigation board for plotting tracks simultaneously from two Loran sources. Both men then pre-flew the entire course on the Loran and Gee trainers so their maps would contain information expected from the A-26 navigation equipment.

84 Bulletin Board, 11:3 (May 1997): 13.

Tonight, 28 Feb/1 March, a 492nd Liberator dispatched from Lyon, France, delivered JE agent ANZIO near Stuttgart, Germany. Meanwhile, Knapp/Jackson and Sawyer flew to Holland and at midnight establish contact with Bobbie. Sawyer apologized for not delivering the packages due to weather conditions. Bobbie answered back, "Yes, Bill, I understand the weather was bad, but it was BLOODY awful to stand there in the cold. Will you come next time? Over."

Bobbie gave his prearranged "danger signal", a swear word, notifying he was under duress. Sawyer provides reference coordinates for location of the next supply rendezvous.

Bobbie responds further: "Yes Bill. I heard you will come on Saturday. Make it one o'clock but be there. Don't let us wait. It is BLOODY awful to stand there in the cold. Will you be there at one o'clock?"

Again the danger signal is received. Bobbie then interjects:

> "My friends in the north were arrested by the Gestapo ... It is BLOODY awful being in this weather ... Probably one of the gadgets is in the hands of the Gestapo. Have you got that? Till Saturday and come on time. We are waiting for you and we don't like to stand outside waiting because it is very nasty ... It is BLOODY awful to stand there."

Bobbie provided his prearranged distress signal all through the contact and continued emphasizing the word BLOODY at every opportunity. After the first five minutes, Sawyer ordered the pilot to maintain vigilance for night-fighters as trouble was anticipated. No enemy aircraft appeared, and after completing contact, Knapp headed north climbing to 36,000 feet.

Upon return to Watton and after listening to the wire, OSS and Simpson confronted a dilemma. The Gestapo controlled Bobbie and the JE radio equipment was now compromised. Would future operations be jeopardized?

March 1945

The February-March moon period would end 4 March limiting time available for agent insertions. On the evening of 1/2 March, the HAMMER team and equipment were ready. Watton dispatched four reconnaissance flights to photograph the Berlin area and a weather mission reported conditions in the delivery area prior to A-26 takeoff. According to Miskho:

> "Previously, Commander Simpson approached Major John Walch and myself with a specific requirement of inserting two agents at a designated site deep into Germany. He provided the drop point and asked if we could assure delivering the agents there (within Berlin). After careful study and deliberation, we could not guarantee the request. The Commander then went to HQ Eighth Air Force and about two weeks later returned with another drop point, 7-10 miles outside the Berlin city limits. After careful consideration, we determined the pinpoint could be reached and promised the drop, weather permitting.
>
> "The mission would be flown in an A-26 at an altitude of 500 feet to Berlin and back. We had to contend with aerial balloons raised each night, often at different places and changing heights. Therefore, we checked the daily reports on their location and designed our flight plan accordingly. We studied Flak maps and determined a zigzag route necessary to avoid these sectors and known night-fighter locations.
>
> "Neither Major Walch or myself had flown a mission in an A-26. It was a very fast aircraft with limited range, therefore, having sufficient fuel was of vital concern and we confronted a very scary situation. We developed the final flight plan in a short period prior to takeoff, calculating fuel reserves to and from the target, allowing for evasive action should we encounter enemy fighters. It was a nip and tuck situation. In fact, we decided to taxi the A-26 out for takeoff and then fill the fueltanks again before leaving. In the back of our minds there was the possibility of a forced landing in a neutral country, and we prepared for this option."

1/Lt. Kingdon R. Knapp at Buckingham Palace to receive DFC from King George of England. *Photo National Archives of Canada*

Standing next to NS725 is navigator William Miskho on left, pilot Kingdon R. Knapp in the middle, and tall OSS JE operator William Sawyer on the right. *Photo W. Sawyer*

A rare photo at Watton of an all-black A-26 with lettering and serial number in red. During a training flight it went off the runway and became stuck in mud, and had to be towed out. Photo *W. Sawyer*

After loading the HAMMER team in the rear fuselage, the flight crew climbed the entrance ladder and moved/crawled through the interior to their positions. The A-26 cabin offered spacious room compared to the cramped Mosquito. Lt. Robert Walker started the engines, then taxied and swung the aircraft on to the runway for takeoff position and executed a preflight check. A-26 43-22524 shivered as both engine run-ups and magneto checks were performed, then groundcrew topped-off the wing tanks from fuel bowsers.

Walker waved to Simpson, then pushed the throttle levers forward. The aircraft rolled ahead in a deafening roar with blue flames belching from the exhaust stacks as it slowly accelerated down the runway. Easing back gently on the control column, lift off occurred at 2300 and all wheels retracted quickly. The engines labored to gain height as the altimeter turned slowly. The A-26 climbed for altitude and disappeared into the darkness.

Over the North Sea, the aircraft dove to gain speed, then flew on the deck to avoid German radar detection through the Flak-ringed, northeast Holland coast, and winged across a prearranged zigzag course over unfamiliar enemy territory. Walker continued flying on the deck at 100 feet to hedgehop across Germany, climbing when necessary to avoid tall obstacles.

From the air, northern Germany appeared like an endless crossword puzzle. Shafts of moon beams penetrated the clouds providing frequent periods of illumination across the landscape below, silhouetting towns, canals, roads, rail lines, and especially shimmering moonlight reflected from lakes and rivers. The entire flight path was planned to overfly such conspicuous and salient features to serve as checkpoints. Miskho's eyesight proved extraordinary keen for pilotage. Peering out the nose Perspex, he became absorbed with scanning ahead and below, searching for any distinguishing or recognizable feature to help determine their exact position. He called off checkpoints to Walch, the master navigator situated behind the pilot, initially navigating with Gee and then by an experimental design Loran plotting board, tracking readings from both UK and MTO chains. The crew placed solid reliance on his computation skills for an accurate course to the drop zone.

Meanwhile, both pilot and rear-gunner anxiously scoured the darkness, seeking a glint or indication of any nocturnal enemy; at the same time listening attentively to the voices of Miskho and Walch crackling in the earphones. While keeping a wary eye, church spires and chimney stacks flashed past as the countryside rolled by.

Approaching the Berlin area, the A-26 slowed its speed. The radio conversation sharpened as the pinpoint neared. Miskho:

> "Our plane took us fairly close to the original drop point, but when we arrived the whole area around it was completely flooded. We then feared having to abort, but continued on course for the second drop point a few miles away. We located the target, and on approach opened the bomb bay doors, and signaled the jump by my pushing the 'go' button. Both Joes and packages were released within a one-second interval – packages first, then Tony, then Phil. Visibility was three miles with 3/10ths clouds over the area. Having discharged our cargo, we slowly turned around and headed for home. After checking our fuel supply, it was determined we had barely enough for the return flight."

The night passed uneventfully after having managed to elude enemy detection. A thought crossed Walker's mind as he coddled the controls. They were briefed to land on the beach at Malmo, Sweden, in the event of an emergency. In preparation, Simpson had demolition charges prepared and attached to the electronic equipment. He provided Walker a telephone number in Malmo to memorize for assistance in returning to England. Would they have enough fuel for a safe return? With this thought in mind, Walker applied a lean mixture on both engines to conserve fuel and droned on continuously towards Watton at 225 knots.

The A-26 flew the shortest route back to England. Walker maintained the flight plan, crossed over the channel, and when the aircraft reached the English countryside, fuel became perilously low. Finally the Watton runway lights came into view on schedule. According to Miskho, the critical fuel situation required a straight-in approach to land. After what seemed an eternity, the A-26 executed a gentle touchdown at 0340 with less than 100-gallons fuel remaining in the tanks.

Mission accomplished. The 25 BG (Rcn) successfully inserted the HAMMER team near Berlin. That afternoon 654 Squadron flew a photo-reconnaissance mission for Simpson and photographed the drop area.

Overjoyed by the successful Berlin mission, OSS sent HQ Eighth Air Force a request for an additional nine A-26C aircraft having the clamshell canopy and bomber nose. The standard solid-nose model operated by Simpson's team lacked adequate forward visibility necessary for pilotage and would be returned. It was imperative two aircraft be assigned immediately and the remainder by 10 March, along with electronic equipment.

Pinetree replied the 492 BG at Harrington would receive all future A-26 allocations. The two A-26s at Watton would eventually transfer to Harrington, including pilot Lt. Robert Walker, his navigator, and ground maintenance personnel. One A-26 would train 492nd aircrews and the other employed for operations. The 492nd were familiar with the mission profile, and after a brief transitional training period would execute all future operations.

OSS discussed the new policy with Simpson who bitterly opposed both the transfer and dividing his JE operations. With no options available, he planned leaving two men at Watton with the Mosquitos, while other OSS personnel – communications men and dispatcher – accompanied the A-26s to Harrington.

On 2/3 March, Knapp/Jackson flew their first mission to TROY, with Layton as JE operator. They arrive over the pinpoint on time but all attempts to contact TROY prove futile. Previously only one of six missions established brief contact with the agent. And on the 7/8th, OSS launched the first Mosquito mission to ANZIO. After comprehensive directional dial tuning, no responding signal was heard and the crew returned to base.

Robert P. Walker on left with Kingdon R. Knapp. Both wear RCAF wings on tunic and Knapp has a DFC (RCAF) striped ribbon. *Photo. R. Walker.*

JE agent training continued the 9th. After one session, Emmel/Miskho and Ancrum (NS725) returned to Watton on the deck to buzz the base. Miskho reported a tree appeared from nowhere and "we went through it." Emmel initiated a single-engine emergency landing. Unable to lose sufficient speed the aircraft landed fast and went off the runway end without damage.

Sawyer recalls the following:

> "Plans were implemented to train George Fogarty as a JE operator. I was asked to take him up on a test flight (NS740) over the Watton airfield. The pilot was Knapp and I flew next to him in the navigator's seat. Fogarty was in the tail. We were at about 25,000 feet when Knapp asked me to reach behind him and turn on the emergency oil. When I straightened up, he had stopped one engine and feathered the prop, then informed me the other engine was low on oil, requiring him to sharply reduce power. He reported the emergency to the tower and was cleared for landing. Knapp asked Fogarty if he wished to parachute down or stay with the plane. In past experience, the tail had been known to break off in a bad landing. Fogarty made a rapid decision to stay.
>
> "The plane dropped rapidly with essentially no power. As we lined up on the runway, Knapp was lowering the landing-gear. As he throttled the one engine back further, the warning flags on the instrument panel indicated the gear was not locked down. Fortunately, it did finally lock down and Knapp pulled out of the steep glide and touched down in a beautiful landing, a most difficult task as we were essentially 'dead stick'. We rolled to the end of the runway and stopped. I looked around and Fogarty had already bailed out on to the ground and was on his hands and knees kissing Mother Earth, and I don't believe he was kidding around. We congratulated Knapp on a remarkable landing. The Mosquito was a difficult plane to land normally as it touched down too fast. It was later determined that, when the mechanics changed the cockpit heater, through which the engine oil circulated, they failed to tighten the hose clamps."

That evening a Mosquito flew to contact TROY. The aircraft circled the target for 20 minutes with no response. Unknown to OSS, a RAF bombing raid on Pforzheim destroyed TROY's lodgings and both JE sets. Trapped inside Germany with no communication devices, he continued gathering Intelligence and later surrendered to advancing American troops.

Simpson now accepted that Bobbie had succumbed to Gestapo control, a view shared by OSS. To avoid tipping off the enemy that OSS knew this fact, they planned to continue missions to him. On 10/11 March, Knapp/Jackson and Simpson flew to Holland arriving at midnight. Simpson made contact when the wire recorder malfunctioned. It took a minute to correct the problem, and Bobbie repeated "his friends in the north" had been captured, the new drop pinpoint was a "damn" good one, followed with, "They are trying to engage you." Simpson notified the pilot to expect trouble. According to Knapp:

> "Simpson reported he received a message the enemy was attempting to engage us. At his order we broke off, jettisoned the external droptanks and climbed to 36,000 feet … and headed towards England. At the English coast I requested Coltishall to see if any bandits were following. There were none."

Lt.Col. John H. Colby, 12 Army Group OSS Detachment, independently succeeded in convincing a Ninth Air Force unit to mount operations from Namur, Belgium, with the Douglas A-20 Havoc, (F-3 configuration) for short-range agent insertions behind enemy lines. The CO of 155 Night Photo Reconnaissance Squadron, 10 Photo Group, agreed to furnish four aircraft. The first A-20 assigned for conversion work had the rear escape hatch cover removed to allow for container storage and the agent a means for emergency exit. Strong points were installed for static line attachment.

An OSS member performed a daylight parachute test jump from 700 feet at a speed of 135 mph and found the opening shock from canopy deployment negligible. A container accompanied him to test for ground dispersion. Both exited through the hatch without problem and landed some 30 yards apart.

Two agents, a dispatcher, and one container fit easily in the aft fuselage. The agents sat on the floor facing the rear while the container hung over the hatchway. The dispatcher stood above the two men, allowing each to pass through his legs as they jumped out.

Pitch black conditions existed during a third test conducted the moonless night of 10 March. The A-20 took off, made a large detour then flew to the pinpoint. The open hatch prevented using light inside the aircraft, thus requiring perfect teamwork between agents and dispatcher. In addition, engine noise made conversation impossible. The navigator notified the dispatcher when five miles distant and called out each mile as they approached the pinpoint. At one mile, he began calling out numbers starting from ten, and at zero the dispatcher released the container. The first man jumped immediately behind the canister, followed by the second man. Touchdown proved uneventful.

Between 19 and 30 March, the A-20 flew ten insertion missions. Gee function limited operational radius to 300 miles during the moon period; dark non-moon intervals limited deliveries to a radius of 100 to 180 miles. Gee operated by line-of-sight transmission requiring the A-20 to fly higher on deep penetrations to receive the navigation beams. Due to limited navigation systems, it became imperative air crew locate clearly defined and easily recognizable landmarks near desired pinpoints to visually aid deliveries.

Plans were formulated to transfer A-20 operations to Euskirken, located 75 miles east of Namur. Until that time, OSS operations continue from Watton, Harrington, Lyon, and Namur.

Once US troops crossed the Rhine River it became imperative agent deliveries begin immediately. OSS discussed plans using both the A-20 and A-26 for insertion during the dark, instead of moonlight period of each month. OSS believed that on darkest nights, a good navigator-bombardier team could locate any pinpoint with fair accuracy. This assessment would later prove drastically wrong. The 492nd received orders to insert seven agent-teams during the 11-18 March period covering Hanau, Aschaffenburg, Wuerzburg, Augsburg, Ansback, and Munich.

Late evening 12/13 March, Knapp/Jackson and Ancrum left to contact HAMMER. The crew encountered Luftwaffe night-fighter activity in the Berlin area as 81 Mosquitos from RAF Bomber Command attacked the city. According to Knapp:

> "We approached the target at 25,000 feet and started to orbit. In a few minutes several fighter flares were dropped at our altitude, so I released the wing tanks and climbed to 35,000 feet. Flares of a different type appeared, but no other activity except two sweeping searchlights were seen. Ancrum made contact but communication was broken several times by either an electrical disturbance or enemy jamming."

JE operator Ancrum managed to establish agent contact:

> "Heinz replied to the first part of two questions, but I could not be certain that I understood replies to the latter half of each question. Heinz confirmed he understood instructions to listen to BBC for new frequencies and times. He indicated he preferred to stick to point one for contacts … Lt. Knapp climbed to 35,000 feet when I reported there was a lot of extraneous noise, making the contact difficult."

F-3 was the photo mission designation for the A-20 Havoc, 155 NPS, 10 PG. *Photo T. Ivie*

The Mosquito turned north for the Baltic area and landed at Watton. The wire underwent analysis at London OSS. They determined interference from the enemy jamming radio wavelengths to combat RAF raids on Berlin had ruined the successful contact. RAF night bombing operations exacerbated enemy Flak and night-fighter activity, exposing the JE Mosquito operating in the area to increased danger. A conference with RAF officials established guidelines to provide one designated OSS individual a five-day advance notice of contemplated Berlin raids. This allowed him, without divulging the RAF schedule, to alter OSS timetables to avoid conflicts. This also required advance notice in preparing BBC coded messages when contacting agents.[85]

Transfer to Harrington

Simpson was not privy to information regarding future Eighth Air Force plans. On 13 March, they summoned OSS to Pinetree for a meeting with Col. A.J. Pearce, Director of Planning. Pearce informed both Simpson and OSS officials that Eighth Air Force requested seven A-26s assigned to Harrington for special operations. He then took the position that all Carpetbagger operations be consolidated with one organization. Simpson questioned the maintenance problems created by assigning Mosquito aircraft to Harrington; OSS officials concurred. OSS emphasized the urgency to meet all March moon period operational commitments as more vital than any preoccupation with longterm organizational problems. Pearce appeared cognizant of the problem and promised to bring pressure on Harrington to fly operations regarded as restricted to A-26 aircraft with B-24s. Again he reiterated, the Watton OSS JE operations must transfer to Harrington.

Simpson sat stunned by this announcement which he found unfathomable. After the initial shock had worn off, he notified his men of the impending transfer. On this day OSS missions labeled Redstocking by the 25 BG (Rcn) ceased to exist. There is no documented evidence in 492 BG files that the Carpetbaggers continued the Redstocking label. The term is alien to official OSS files.

Eighth Air Force notified 25 BG (Rcn) of the OSS JE reassignment and CO Colonel Gray ordered all personnel and equipment to move immediately. The shift from Watton to Harrington began 14 March with Mosquito NS707 and NS725 flown in by Knapp and Emmel, and two trucks carrying electronic equipment arrived to untold confusion and spent two hours attempting to locate an officer to report for orders. No one knew anything about their reassignment, nor where to unload the material.

Simpson flew to Harrington and also learned no preparations had been made to receive his JE operations. Colonel Hudson H. Upham, CO 492 BG, informed him they lacked proper facilities, and Pinetree failed to provide full particulars of the transfer. Upham attempted to prepare billets, but unexpectedly a new unit, 416 Night Leaflet Squadron, also transferred in that same day. Until relocating his men, accommodating OSS personnel would temporarily be delayed. No quarters were available for anyone; men even slept in the mess hall, forcing Simpson's return to Watton for the night.

Colonel Hudson Upham, C/O 492 BG at Harrington. He became CO of the 306 Bomb Group from 16 April 1945 to May 1946. *Photo 306th BG Assoc.*

85 RG 226. Chief Secret Intelligence, OSS from OSS Operations, memo 14 March 1945.

Also this day, 492 BG dispatched several B-24s with support crews to begin operations at Dijon, France. The bulk of OSS missions previously flown from Lyon experienced problems. The base lacked administrative control over air operations, had no housing facilities for feeding or dressing agents, and the inferior field conditions resulted in two damaged B-24s. The 492nd agreed to maintain four aircraft at Dijon, and on good weather nights several missions could be flown, thereby preventing repetition of the Lyon experience.

In essence, the semi-autonomous OSS groups at Watton, Lyon and Namur would eventually be eliminated, and all OSS operations scheduled from Harrington. Watton was abandoned and England-based operations concentrated at Harrington; Lyon abandon and all France-based operations flown from Dijon. Namur operations would cease once reorganization at Harrington reached fruition. The USSTAF finally defined the 492 BG role and tactical mission.

Late morning 15 March, OSS Lt. Fogarty arrived at Harrington by station wagon, Lt. Knapp flew in Mosquito NS740, Lt. Emmel an A-26 with crew chiefs, and all eventually provided temporary quarters in Nissen huts in another part of the aerodrome. The 492nd dispatched a B-24 to Watton to collect Simpson and any additional equipment. The Liberator arrived and after visiting the 3 SAD side of the field returned to Harrington. Apparently no one issued the pilot instructions to pickup Simpson.

At the end of the day, Harrington temporally assigned the Mosquitos and A-26s to different squadrons. The following morning Harrington promised to schedule a B-24 to pickup Simpson. At 1500 hours Simpson called the 492nd to discover they forgot to dispatch the aircraft. Watton provided a Mosquito to transport him to Harrington.

On 18 March, Harrington scheduled their first JE Mosquito mission using a 25 BG (Rcn) crew. When preparing his navigational aids, Lt. John Jackson experienced difficulty obtaining weather data and learned no S-2 information was available. Nor were VHF frequencies, communication flimsies, colors-of-the-day, or emergency information provided. The mechanics from Watton searched all over the field to locate oxygen and gasoline trucks, and no alert crew were on duty.

Finally, the Mosquito left and when climbing, checked the JE radio with a Harrington ground unit; it functioned perfectly. Meanwhile, RAF Mosquito bombers attacked Berlin. Anticipating Luftwaffe activity similar to the previous mission, the pilot flew at 36,000 feet under a twilight sky for 45 minutes prior to reaching the target. They arrived in the Berlin area on time but failed to contact HAMMER even after comprehensive tuning. They turned for home.

Conflict at Harrington

From the beginning it became Colonel Upham's understanding that Simpson was merely at Harrington for a brief period in an advisory capacity. Upham intended placing his own radio personnel in the JE shop to learn the equipment. He also wanted all Intelligence, including handling the JE wire, dispatching missions, and picking routes based on signal plans, controlled by his Group S-2. This was not Simpson's understanding and if allowed, would compromise the secrecy of his JE operations. These plans piqued his displeasure, and he made London OSS aware of these deficiencies.

At the end of three days, complete chaos reigned at Harrington regarding Engineering. The six mechanics brought from Watton still came to Simpson for instructions. When the authority for aircraft maintenance and operations transferred from 856 to 858 Squadron, to Group Engineering officer, and finally to the Subdepot engineering officer, countermanding orders created further confusion.

Additional Mosquitos arrive at Harrington. Upham and Simpson agreed that Mickey Mosquitos NS584 and NS686 be used exclusively for training 492nd personnel, and only Watton pilots allowed to fly NS707, NS725 and NS740 for JE missions. But such was not the case. An untrained crew attempted flying NS707, forcing Simpson to halt taxing the aircraft while he called Upham. The colonel apologized but failed to comprehend why his men should not fly operational Mosquitos since they were good pilots, but agreed to prohibit his men practicing in these aircraft.

That same afternoon Simpson again discovered a new pilot taxing NS707 around the airfield. He again called Upham and requested he order his men out of the Mosquito and use only training planes for taxiing. The Merlin engines lacked necessary cooling and were not designed for any lengthy operation on the ground before sustaining damage. The pneumatic brake system air bottle held enough air for a circuit around a perimeter, requiring the aircraft either be run-up or flown to refill the air bottle. Upham resented the request and doubted Simpson's information regarding the Mosquito's shortcomings.

Two A-26s arrive 18 March increasing squadron strength to four aircraft – three operational plus one training ship. OSS Lt. Anthony Turano reported for duty as JE Operations officer to coordinate all Harrington OSS activities.

The CHISEL Debacle

In September 1944, OSS Lt.Cmdr. Steve Simpson established his Joan-Eleanor project at Watton. Besides furnishing aircrew, the 25 BG (Rcn) modified and maintained OSS A-26 and Mosquito aircraft for the JE program. They also provided Simpson instant access to current S-2 Intelligence data to schedule missions. Watton S-2 even prearranged both weather and reconnaissance flights to obtain information he required. The OSS project functioned completely independent and autonomous from HQ 25 BG (Rcn), yet obtained priority access to all base sections for complete technical support.

On 13 March 1945, Eighth Air Force transferred Simpson's JE project to Harrington under the 492 BG. The scenario completely changed with the AAF taking full charge of JE operations, much to Simpson's dismay. He virtually lost total control over his project. Besides fighting the war against Nazi Germany and HQ Eighth Air Force, Simpson became involved in yet another battle, this time with the 492nd. The following incident, five days after transfer to Harrington illustrates one of numerous developments that contributed to this ongoing feud and internal conflict.

During early March several attempts to deliver agent CHISEL to Germany were canceled. On the morning of 19 March the 25 BG (Rcn) received orders to send navigators Major John Walch and Lt. William Miskho to Harrington for an A-26 mission. However, Miskho was on leave and unavailable. Lt. John Jackson, the Mosquito navigator, would replace him as pilotage navigator in A-26 43-22524, which Lt. Oliver Emmel would pilot. Simpson learned from Walch that Colonel Upham ordered the CHISEL mission for that night, but by whose authorization? Simpson controlled and directed all previous JE missions and believed Upham infringed on his authority. After a heated debate, Upham defended his position that the AAF controlled the aircraft, and as Group commander, his organization will direct all OSS airlift operations.

Throughout the day orders were issued and canceled several times to fly the mission. Simpson brought to Upham's attention the aircraft required a 100-hour inspection. It currently was undergoing an operational check with considerable disassembly, and he doubted its current airworthiness. Upham finally decided the A-26 should remain in the hangar for repairs and further inspection.

At 1600, Major Walch informed Simpson that Upham inexplicably changed his decision again and ordered the mission to continue as scheduled. Simpson immediately went to headquarters to learn this was true. Upham indicated that during the late afternoon he received what amounted to a direct order from a general officer to proceed with the mission, so he issued orders to this effect. Upon being advised both engines were down, the radio altimeter needed calibrating, and both Loran and Gee required examination, Upham called the hangar and ordered the A-26 in commission and ready by 1930. The entire Watton crew – Emmel, Walch and Jackson – advised Upham it was impossible to ready a flight plan and organize the mission by that time, and besides, the weather reports were not encouraging.

Furious with their behavior, Upham indicated he would pilot the ship himself. In view of the colonel's brief flying time in the A-26 and the hasty flight preparations, Jackson doubted Upham's capability to pilot the mission and refused to participate. As he recalls:

> "The weather report for the flight route was atrocious. Flying a night mission below clouds at 500 feet without moonlight in adverse stormy conditions, it was a suicide mission! When he asked me why I refused to fly, I told Colonel Upham the mission did not stand any chance of success."

Angered by Jackson's response, Upham removed him from the crew roster and assigned Major Edward C. Tresemer, 492 BG senior Group navigator, in his place.

The agent to be delivered inside Germany was Karl Grimm of the CHISEL project. Besides gathering military Intelligence, OSS tasked him with establishing contact with German dissidents or underground and arrange reception committees for insertion of future agents. At 1525, the OSS section in London responsible for agents, received orders to dispatch CHISEL that evening. Previously, OSS attempted the mission during the dark period of 6 March and CHISEL reported to Watton that night. A scheduled weather reconnaissance Mosquito returned to Watton a half-hour prior to dispatch and reported poor weather in the drop area. Major Walch brought this to Simpson's attention and advised him to cancel the A-26 flight. Simpson consulted OSS officials who were in complete agreement to scrub the mission. OSS delayed the CHISEL insertion until 19 March when Harrington received notification scheduling the mission that evening. The agent left London at 2005, arriving 35 minutes later at the 492 BG Operations office.

This was the first A-26 mission from Harrington and no prior arrangements were formulated for an agent's flight preparation. After further deliberation, OSS chose the Intelligence Office of two small rooms having an open window-type passage in the wall between them; one room was empty and the adjacent office included a number of Operations personnel. Dressing immediately began and CHISEL, fully attired in indigent garments, donned a special flying suit. As preparations progressed, numerous 492nd officers not involved in the mission entered the room and remained present. Meanwhile, Operations personnel in the adjacent office continued to view the scene.

During adornment, both A-26 navigators Majors Walch and Tresemer and the pilot, Lt. Emmel arrived and proceeded to present the final pinpoint information. CHISEL confirmed his understanding and an OSS officer provided additional information concerning the drop area. Appropriate map sections involved were provided for the agent's use. Towards the end of this discussion Upham entered the room.

Earlier, both Walch and Tresemer studied the mission plan and weather reports and had misgivings. After discussion, the entire crew approached Upham requesting he reconsider his orders, in view of the available data. In their professional opinion the mission was not feasible. Upham disagreed and ordered them to proceed.

Once fully prepared, a limousine transported the agent and flight crew, along with OSS officials to the A-26 at 2150. Here, several vehicles parked in the light drizzling rain illuminated the immediate vicinity. The men walked to the aircraft but were requested to return and wait at the vehicle. Simpson and Fogarty continued checking equipment inside the aircraft, while dispatcher Plat. Sgt. Frederick J. Brunner USMC, arrived at the car and provided CHISEL with more preflight instructions. The men assisted with placing the parachute harness and pack on him, then snapping it tight.

At 2250, CHISEL climbed into the bomber and 15 minutes later both engines roared to life. The surrounding automobiles withdrew to the opposite sides of the runway. OSS officials discussed the prevailing situation with Simpson. In frankest terms, he declared adamantly he did not wish to share any responsibility for the dispatch of this mission. 1. Lt. Emmel had insufficient A-26 flight hours and flew it only twice at night. 2. Major Tresemer flew in the A-26 only once and became completely bewildered by ground pilotage at such speed. 3. None of the crew flew together as a team and were unfamiliar with the aircraft. 4. All had insufficient rest. 5. Simpson questioned the aircraft's airworthiness.

The prevailing base weather conditions reported 10/10ths clouds at 700 feet, visibility 2,200 yards with continuous moderate rain and winds west at 19-24 mph. The en route forecast predicted a ceiling of 400-600 feet in rain, lowering to 200-400 feet in stratus and drizzle over the North Sea, northern Holland, and Germany. The weather possibly changed inland at 50 miles to 4/10-6/10ths clouds at 1,000 feet. No target weather reports were available for the Munster area.

The aircraft lifted from the wet runway at 2317 hours and quickly disappeared into the low, dark rain clouds and never seen again. The expected target arrival time was 0245. Total possible time in the air according to fuel capacity was 6 3/4 hours. By 0615 when Harrington received no further word, they declared the aircraft overdue and possibly MIA.[86] Simpson ruefully recalls, "The mission proved woefully deficient in planning and inevitably deadly in its execution."

According to a Luftwaffe report, the A-26 crashed one kilometer from the farmhouse of Herr Olinde Kruse near the village of Schwege, a brief distance from Hunteburg. In the early morning of 20 March at 0100 local time, a loud noise awakened Frau Kruse. Quickly everything became quiet and she returned to sleep. Later that afternoon, Herr Kruse and a Ukrainian laborer employed on his farm discovered the strewn aircraft wreckage. They reported this to Bramsche authorities.

Evidence reveals the A-26 hit the ground at a shallow angle, then cartwheel several times before coming to rest in a small brook. Though the center fuselage remained fairly intact, the impact ripped off the wings, engines, nose, and tail section. Five badly mangled bodies lay scattered at the site: the agent and a crew member near the fuselage, a corpse 50 meters away in a meadow, while another floated in the water soaked moor. The impact hurled a fifth airman 150 meters distant from the main wreckage.

Sitting on left is then Capt. John "Knobby" Walch of Pittsburgh, Pennsylvania, KIA on 19 March 1945 CHISEL mission. Sitting to the right is then 1/Lt.. Lionel Proulx, ex-RCAF who took over Walch's position as Watton senior group navigation officer. Taken at Bovingdon in late 1943. *Photo L. Proulx*

86 MACR 19 March 1945.

Oliver Emmel of Kansas, KIA in 19 March 1945 CHIESEL mission.
Photo R.P. Walker

492 BG crew returned from first C-47 mission to occupied France, 9 July 1944. L-R. Col. Clifford J. Heflin, Maj. Edward C. Tresemer (KIA 19 March 1945), Capt. Stapel, Maj. Teer, and T/Sgt Krasevac. *Photo 801/492 BG Assoc.*

The aircrew wore military identity tags while the agent in civilian clothes carried papers and German passport identifying him as Pawl Nowak of Czech nationality. Nowak possessed letters of recommendation and documents from the Herman Goering-Works at Prague transferring him to a factory in Mulheim. The local police removed the agent's papers, stencils, small headphones, small microphone (Joan radio), and currency consisting of 7,000 Reich Marks and British One Pound notes. They transferred these items and the body to the secret police. The authorities then buried the four airmen in a military cemetery at Achmer and delivered their papers and personal effects to the Achmer Luftwaffe Air Field Kommodant.

According to 492nd Lt.Col. Robert Fish, whenever Tresemer functioned as pilotage navigator on low-level flights he always asked the pilot to fly lower to obtain better detail of the landscape below:

> "When we flew together at night, I ignored his requests to fly lower whenever we were within 300 feet of the ground. Oliver Emmel ... the pilot on that fatal flight was highly qualified, but he had not previously flown combat with Tresemer. He had no way of knowing Tresemer's idiosyncrasy about altitude. There is a distinct possibility that he allowed Tresemer to induce him to fly too low."[87]

Simpson always embraced a belief the ill-prepared A-26 suffered a major mechanical or electrical equipment malfunction resulting in the catastrophic crash that killed his men. He reported that Upham planned court-martial proceedings against Jackson for refusing to navigate the CHISEL mission. But when the A-26 went MIA, no charges were forthcoming.

Simpson greatly admired Major John Walch's creative genius developing novel solutions in resolving navigational problems and now lamented the loss of his intellectual rapport. The A-26 loss created an irreconcilable schism between Simpson's JE team and the 492 BG. There are two sides to each story. To be objective and impartial, this history of JE operations is based on Simpson's JE Log and letters to London OSS, detailing only one view of the events.

The first-quarter moon period began 20 March as OSS scheduled deploying over 30 agent-teams with wireless radios into Germany. Their goal: cover strategically important targets at Osnabruck, Hanover, Braunschweig, Magdeburg, Kassel, Halle, Leipzig, Dresden, Geeste, Erfurt, Aschaffenburg, Berlin, Nuremberg, and Regensburg. Dense cloud formations moving in both from the Atlantic and USSR covered western Europe, hindering OSS insertion operations.

Late night of 20/21 March, a Mosquito flew to rendezvous with HAMMER as RAF Mosquito bombers attacked Berlin. Heavy searchlight activity swept the area and all VHF channels received enemy jamming. After circling half an hour with no contact, the crew turned for home base.

l/Lt. Andre E. Pecquet arrived for duty 21 March as OSS JE operator. An American by birth, Pecquet lived in France, serving with a French Army antiaircraft unit. With the fall of France, he fled to England, then sent to USA served in the 11th Armored Division before transfer to OSS. He parachuted into France in 1944 and fought with the Maquis through the Vercors tragedy. After escaping to England he joined OSS operations at Harrington.

Meanwhile, at Namur, an A-20 successfully dispatched DAIQUIRI, while EGGNOG returned because mist covered the pinpoint.

To obviate using operational Mosquitos for JE training, Colonel Upham assigned two JE-equipped B-24s #531 and #568 for this task on the 22nd. Simpson notified London OSS that unless additional trained aircrews became available, further A-26 flights were not possible for the next two weeks. Upham expressed the view his men could fly any aircraft including the A-26 without transition training. Simpson scoffed at this assessment; the 19 March MIA mission using an inexperienced crew illustrated this critical problem. Two new A-26s 43-22513 and 43-22626 arrive at Harrington.

On 22/23 March, Knapp/Jackson and Ancrum took off near midnight. They climbed for altitude to attempt a contact with CHISEL, but 30 minutes into the flight Ancrum reported gas fumes in his compartment. Knapp took action:

> "I opened the bomb doors, switched to inner tanks, turned on the bomb bay fuel cell and attempted to use the gas rapidly. Lt. Ancrum soon reported liquid gas was flowing across the floor. I told him to put on his chute and jettison his hatch. We returned to change ships, but due to lack of ground cooperation we were barely in range at zero plus 30."

Knapp returned and landed at Harrington to find no night crew, nor fuel or oxygen truck available. The men transfer the JE gear to NS725 and took off again as Ancrum relates:

> "Knapp decided to continue the mission despite the fact we did not have more than 1,400 pounds of oxygen pressure, and there was not an operational load of fuel aboard. Our course was straight to the target area over a forbidden zone in the UK, over the enemy coast at 12,000 feet with both friendly and enemy flak encountered, but the target area was reached too late. An attempt to establish contact was made from 0125, but we were not in range until four minutes after the 30 minutes allowed ... we were within 50 miles of the target at 25 minutes after target time. There was insufficient fuel to return to base and a landing was made at Watton."

The night of the 23/24th a Mosquito left for a mission to ANZIO. After 20 minutes in the target area, radio interference assumed to be Jamming, came into strong evidence and no contact was established. The aircraft returned to Harrington where a radio communication problem occurred. The base control tower audited only the bomber radio channel even though equipped to monitor the fighter frequency. The pilot flew his Mosquito low over the runway to attract the tower's attention to switch on the fighter frequency for landing instructions. The lack of tower courtesy became a chronic and annoying problem among Mosquito crew. Unable to establish tower contact, Knapp landed anyway.

Also this night, Namur-based A-20s dropped EGGNOG, ALEXANDER, FRED, MAC, LIBERATION, and HIGHBALL in southern Germany.

Lt. William Miskho returned to Watton after completing his JE navigation duties. Meanwhile, numerous problems surfaced. Simpson recommended jacks, hydraulic kits, and Mosquito ground-handling equipment be obtained or he threatened to send the aircraft to Watton for repairs. Nor did he receive a list of new 492nd pilots and navigators who required training. To resolve these deficiencies, the OSS Air Operations officer arrived at Harrington, and Simpson provided him a tour of the facilities and discussed all the latest problems. Late that evening an A-20 from Namur dropped ABNOKI near Wesel, Germany.

Agent Executed

On 25 March, tragedy struck the OLD FASHION team of Sobiechowski and Bohdanowicz. They left Namur by A-20 for a pinpoint south of Hamburg. Unfortunately, the pilot issued the jump signal too early. Bohdanowicz jumped first, followed by Sobiechowski who spotted groups of houses below. The first agent descended directly on a hamlet illuminated by a full moon. Villagers were heard shouting as they apprehend and beat him. They later turned Bohdanowicz over to the Gestapo for eventual execution. OSS posthumously awarded him the Bronze Star and his family received a monetary stipend.

Meanwhile, Sobiechowski landed on the outskirts and immediately captured by two armed civilians. Forced to spread on the ground, they searched and stripped him of weapons. One stood guard as the other confiscated chocolate and cigarettes, but in the process dropped a pack on the ground. The armed civilian stooped to retrieve the item when Sobiechowski slugged him on the head and escaped. The agent sought refuge among bushes and reeds in the Lahn River and remained hidden until the moon disappeared. Under cover of darkness he fled westward toward the Allied lines and surrendered two days later to American forces.

87 Fish, Op. Cit., 243.

The A-26 Dispute

Tension mounted at Harrington over the A-26 loss. The relationship between Colonel Upham and his officers with Simpson became severely strained and verged on open conflict. It was Simpson's impression additional A-26 aircraft assigned to the JE program would undergo further conversion and operational development under his control. After all, his team evolved this concept, developed it through a difficult gestation period with exhaustive tests and proved its potential.

Simpson assigned Lt. Fogarty to assist 492nd radio and radar technicians resolve problems installing the electronic equipment. With A-26 43-22610 unserviceable due to both radio and Loran technicians bickering, they refused to follow Fogarty's orders. The next day A-26 43-22513 sat in the hangar with similar wiring mistakes. Fogarty gave up in disgust declaring neither ship would be ready at the current rate of progress.

Lt.Cmdr. Stephen Simpson standing next to OSS Mosquito. He is wearing a Navy jacket, AAF hat and pants at Harrington, April 1945. *Photo. W. Sawyer*

The following synopsis of A-26 problems is detailed by an irate Simpson in the JE Log:

"A-26 43-22524 was converted for dropping operations at Watton. The VHF, command radio, Gee, and intervalometer were modified and moved, Loran and radio altimeter, jump light and switch, new interphone installed, flooring installed in the bomb bay, and a wheel-up latch release installed. With the exception of the wheel-release installed by the Douglas field engineer, all other modifications were accomplished by Lt. Fogarty in about 100 hours maximum. Fogarty was asked to supervise the modification of the first A-26 at Harrington. Acting on the assumption that the Subdepot, communication, and radar sections had at least a modicum acquaintance with the procedure for performing mechanical work on airplanes, he submitted a list of changes to the Subdepot after having discussed the subject with Simpson, OSS Lt.Col. Bowman, Colonel Upham, and the Subdepot engineer. Work started on 43-22610 on 20 March. After two days of watching stupid blunders being made, Fogarty attempted to bring some order to the chaos wrought by the radar and communications sections. Since he had no authority, the only recourse was to complain to the Subdepot engineering officer. An attempt was made to coordinate the work of the men on the ship by suggesting an officer be appointed to take charge of radio and communication, and supervise the job.

"Five Captains ... came out to the airplane, gave out great cries that the work should have been done at a depot and departed leaving everything as confused as before. A-26 43-22610 is festooned with wire as a Christmas tree with tinsel, has been released for a test flight with everything allegedly in working order. A-26 43-22513 has been started on in the same way and is being worked on with the same total lack of cooperation between the assorted sections, and with the same crew of inexperienced incompetent mechanics, and no supervision by anyone with the necessary technical knowledge, except Lt. Fogarty, who has no authority over them. The only excuse for the mix-up as cried by the five Captains ... is that modifying airplanes should not be expected of them.

"It might be wise for someone in authority to point out to them that there are many men dying in Germany, whose army function up until a few weeks ago was to drive trucks and cook, but who are doing a good job as infantrymen because they are not handicapped by that frame of mind."

Simpson inquired at Operations if they scheduled a training flight for the afternoon and a Skywave mission that night. The Group Operations (Grp Ops) officer sharply informed him that Group directed both operations and training, and Simpson had no further authority in these matters. Simpson received another rebuke when he suggested Watton pilots and navigators be utilized as instructors to train Harrington air crews. The Grp Ops officer detailed Major Charles W. Schaeffer in charge of training, and if he required OSS assistance, he would so advise them.

The 492nd crews selected for Mosquito training were pilots Smith, Kuntz, LaDue and Webb; navigators Carson, Greene, Edwards and Kolacki. Simpson placed great reliance in Major Schaeffer's abilities as a pilot and officer and his efforts assisting the JE operations. According to Simpson, "Rest of outfit either doesn't know or care what is going on." Schaeffer concluded after his first Mosquito training flight with Simpson that his navigators required more preparation before flying combat missions.

Another mission to determine if CHISEL survived the MIA A-26 flight took off past midnight on 26 March. The Mosquito crew flew to the Munster area but failed to establish contact. En route home, a friendly night-fighter pursued them over Belgium, but Baggage ground control called them off.

OSS Operations officer Lt.Col. Bowman visited Harrington and submitted a review of his findings to London OSS:

"Those familiar with OSS problems lay most of the blame on the doorstep of the Group commander; not because he does not want to cooperate, but primarily due to the fact that he has not had the experience, nor has he had sufficient time to assimilate the capabilities of his available personnel. In search for the answer, he has changed his decision so often that officers of the 856 Squadron and other Group personnel are thoroughly confused and will not make decisions for fear of crossing the commanding officer."

Lt. Jackson reported more problems on the 27th:

"Absolutely no regard is shown in providing transportation for operational crews previous to a mission. On 23 March, we stole a Jeep at the mess hall for transport to the line. On the 26th, after waiting until 2300, we asked for transportation from the motor pool. Our takeoff was thus delayed 15 minutes, seriously cutting down our time safety margin. I also suggest that transportation be provided the alerted navigator the afternoon before a mission in order to arrange the mission, flight plan, clearance, etc. I have damn well spent my last afternoon walking all over the base for one thing or another while getting a mission prepared."

Both upper- and lower-gun turrets were removed from A-26 43-22626 and the second navigator, along with dispatcher, shifted to the old turret compartment. The 492nd formulated new agent dropping procedures. They removed the bomb bay platform and hung the supply canisters from bomb racks. The agent jumped from the rear compartment through the open bay doors immediately upon canister release.

During dinner Simpson accidentally discovered the Harrington tower radio frequency had been changed. He worked checking each Mosquito. Although the operational ship that night had a new crystal to communicate with Harrington tower, it was improperly tuned.

Despite numerous unsettled issues, JE operations continued the 27/28th. RAF Mosquito bombers attacked Berlin as Knapp/Jackson and Ancrum flew to the area and contact HAMMER. The Mosquito issued instructions for rendezvous on 1 April. Allied bombing destroyed the agents' residence and the team now sought new accommodations. HAMMER details coordinates for power plants on the Teltow canal and near Klingenberg supplying Berlin war production plants electricity; along with railroad and freight terminal traffic patterns. When difficulty is experienced maintaining agent contact, the aircraft signed off and turned for home.

Inclement weather shrouded England and confounded Knapp:

"Briefing instructions were to divert to Reims, France. When Ancrum informed me he received important Intelligence, I decided to return to the UK and hasten delivery of the wire. On leaving the French coast, I called Woodbridge, asked permission to land and received a favorable weather report. Soon after, Coltishall called and instructed me to land there. I proceeded there and was ready to enter the circuit when they told me to return to Woodbridge. By the time Woodbridge was reached the ceiling was only 200 feet with visibility 800 yards. Having lost time going to Coltishall, we were forced to come into Woodbridge through coastal Flak Barrage Batteries firing at V bombs, and Woodbridge was provided insufficient time to call off the firing. With excellent help from the navigator on Gee, I was able to land at 0630 with insufficient fuel to circuit the field – 20 gallons of fuel remained on landing. Ancrum successfully fulfilled his duties in spite of inaccurate flying caused by a hung droptank at altitude. Flying time was 6 hours and 25 minutes."

According to Ancrum:

"Lt. Knapp made a bold decision to return to England to prevent any delay in flow of important Intelligence. Fifteen minutes after our landing, I called the London office to request transportation. Woodbridge is two hours by automobile. After a lengthy delay, a car finally arrived at 1350 hours to pickup the wire."

Simpson admonished London OSS:

"The crew brought the wire back to England at great risk to themselves and the plane in order to get this information through quickly. The delay in securing transportation was criminal. This lack of cooperation does not encourage the crew to accept abnormal hazards beyond the line of duty."

OSS schedule a flight to Bobbie on the 28/29th, but late in the afternoon Harrington scrubbed the mission. Neither Simpson nor London OSS received notification until they accidentally called base Operations. Upon asking the reason for cancellation, the Grp Ops officer replied poor weather with no diversion points except Marseilles, which the Mosquito lacked range to reach. OSS consulted Watton who listed several diversion fields open, with Marseilles well within Mosquito range.

OSS contacted Harrington Operations and suggested using the diversion field list provided by Watton. The Grp Ops officer angrily replied, "it was their (492nd) business."However, he agreed to discuss the matter with Colonel Upham, and called back within ten minutes that their decision still stood – no flights. Simpson consulted both Knapp and Jackson who considered the mission safe and volunteered to fly. Since both had flown three nights in succession and lacked sleep, Simpson decided to allow them to rest.

After hesitation regarding weather, OSS scheduled a flight to TROY and ANZIO on 29/30 March. The Mosquito circled ANZIO's point at the specified target time for 10 minutes without contact. The aircraft then proceeded to TROY's location. Suddenly a brief signal was received, "Hello Vic! Hier spricht Willey. Kommen sie naeher Gagenau!" The JE operator notified the navigator and the Mosquito returned to the first contact point, but after repeated calls, failed to receive further transmissions from ANZIO. Fuel shortages caused by the diversion force return to Harrington. On approach, the tower provided no VHF for landing instructions.

Having no further need for the Watton crew chiefs, the 492 Engineering section requested they be returned. But Col. Leon Gray, CO 25 BG (Rcn), doubting Harrington men were qualified to handle Mosquito maintenance after only two weeks of instruction, ordered them to remain.

On 30/31 March, the 155 NPS operating from Namur finally transfer to Euskirken and delivered PINK LADY near Erfurt, the final A-20 (F-3) sortie. At Harrington two night missions were launched. The first carried the CUBA LIBRE team in A-26 43-22610. The supply container failed to release at the pinpoint, forcing the pilot to abort the mission.

The second to Holland involved the first 492 BG crew flying an operational JE Mosquito, with Simpson. Bobbie experienced difficulty receiving Simpson's signals and failed to properly respond to questions. Unknown to Simpson, Bobbie did this deliberately. The Mosquito spent 40 minutes over the pinpoint, when Simpson began to hear a steady suspicious signal on the JE radio frequency, possibly from radar. He immediately warned the pilot. The props churned to gain height as the Mosquito climbed steadily to 35,000 feet and headed home.

On the return flight Simpson wonder if the Gestapo compelled Bobbie to extend his contact time, thus providing the enemy sufficient interval to locate and intercept the Mosquito? On landing at Harrington, the brake pressure filter exploded creating a large hole in the aft bottom fuselage.

The night of 31March/l April, a 492 BG Liberator inserted JE agent CHAUFFEUR near Regensburg, Germany. Also, ANZIO became the target tonight. Knapp/Jackson and Ancrum reach Munster at midnight and attempt contacting him for 25 minutes. Ancrum felt ANZIO had the rendezvous time mixed up and ordered the Mosquito to fly away from the contact area, planning to return an hour later for a further contact attempt. Then the right engine began vibrating so violently, the pilot could not read the instrument panel. He feathered and shutdown the Merlin and called Baggage ground control to identify himself. Baggage warned that an enemy aircraft followed five miles behind the Mosquito. Soon they reported the enemy was now two miles off Knapp's left wing and asked him to attack. Knapp replied he was flying on single engine and unarmed, when both his radio and electrical equipment went dead. As the Mosquito continued through clouds, heavy ice layers coated the airframe making it difficult to maintain altitude on one engine. Slowly he descended through the clouds and safely landed at the first lit airfield, B-97 Rosiers-en-Saterre, near Amien, north of Paris.

April 1945

On 1 April, Simpson allowed the return all 25 BG (Rcn) A-26 crew chiefs to Watton. The 492nd took delivery of Mosquitos RG157 and TA614. NS740 finally arrived at Harrington after the emergency landing in France due to engine failure from an oil leak into a magneto. Simpson complained to London OSS:

> "These problems indicate poor maintenance. The Engineering section believe they can handle the Mosquito maintenance and want to return the Watton men (crew chiefs). They cannot learn in a week, and mechanical difficulties the last three missions have caused near disasters. Reason for oil leak on ANZIO flight was the bolts were only finger tight on cam case cover. Just discovered VHF frequencies were changed again without telling us. We are not kept informed on navigation aids. Lt. Jackson discovered two new Gee chains – Munster and Metz. They don't know about them here. No place for night alert crew to sleep. Operations (is) locked."

SHAEF informed OSS that information on Allied POW camps in Germany rated equal in importance to gathering Intelligence on the current enemy disposition. They request OSS issue priority to dispatching agents, or instructing those in the field to report conditions at specific camps.

On 1/2 April OSS schedule another mission to ANZIO. Knapp/Jackson and Ancrum leave for Germany, but one hour from rendezvous lost both Gee and Loran function. They reach the pinpoint but are unsuccessful in establishing agent contact. Knapp banked for Harrington:

> "We returned to base in a 200-foot ceiling and made very poor contact on VHF with the tower. I was unable to receive a QFE, and when I asked 'How are your oranges?', they must have thought I was kidding for they replied 'They haven't been picked yet.' In the circuit, the landing-gear warning lights failed to lit when the gear was fully extended. I called the tower to have them check our undercarriage but received no reply. I buzzed the field and caravan in an attempt to have them notice us on approach, to see if the gear was down. I selected down and had Lt. Jackson keep up pressure with the hand pump … until the ground locks were on. It was found the lights were shorted out and the landing-gear functioned properly."

Later that day, Knapp flew the Mosquito to further test and evaluate the undercarriage. OSS Lt. Anthony Turano witnessed the following spectacle:

> "He came in for a landing and he intuitively sensed something wrong as he touched down. Actually, one landing-gear was not fully extended. He shot his plane upward again, but the tower was not too cooperative. After failing in several attempts to obtain some help from the tower, he decided to hit the runway hard, and by so doing, the retracted wheel dropped down and he was able to land safely. A complaint followed, of course.
>
> "On his next flight, Knapp called the tower to check his landing-gear, and again he was given no attention, whatsoever. At this point, Lt. Knapp began to 'beat-up' the field. To be perfectly honest and quite frank, it proved to be the most impressive bit of flying I have ever witness! He proceeded to buzz the tower and every hangar on the field. He had mechanics and other field personnel running for cover. There were pyramidal tents scattered around the field, and after diving down on them, he would go into a steep climb, with the propwash pulling the tents (pegs and all) out of the ground. It was an awesome sight to behold. Every piece of fire apparatus was called out, ambulances, chaplain, etc. When it was over, Colonel Upham came directly to me and said he was going to court-martial Knapp.
>
> "At the time … I did not know the reason for Lt. Knapp's performance. When I did, I requested to see Upham. By that time, he too was fully aware of what precipitated the incident. It meant no matter. His own men would not have been allowed to get away with that type of protest. We had further meetings, but I made no headway.
>
> "I finally had a meeting in London with our OSS Executive Officer. I kept him informed of our position and hoped maybe some of our 'big brass' could convince Colonel Upham to drop the charges. OSS suggested having him for lunch at the American Embassy at Grosvenor Square which was nearby. Colonel Upham was eventually persuaded to accept the invitation."

Shortly after midnight on 2 April, A-26 43-22513 successfully inserted CUBA LIBRE in southern Germany. A second A-26 43-22610 also took off to deliver JE agent MALLET north of Berlin. The crew were pilot Capt. Willard H. Smith with l/Lt. Allen T. Wayne as pilotage navigator, l/Lt. William C. Ecklund as navigator and Sgt. Robert S. Chapman the rear-gunner/dispatcher, all 492nd members. As Wayne relates:

> "Our aircraft's name was 'Ace of Spades' and had a painting of an ace of spades on the fuselage. For this long flight, it was necessary we refuel on the extreme east coast of England before heading for Germany. My station for landing was the same as my duty station, in the nose. The aircraft touched down at 0040 on runway 21 at Rackheath, Norfolk, home of the 467 BG. As the plane decelerated, the nose of the aircraft slowly continued downward until the props struck the runway. The resulting sparks caused overflowing fuel to ignite in the left engine nacelle and burn fiercely. At this same time, the nose section slid against the concrete as we proceeded down the runway trailing smoke and sparks, until coming to a halt.
>
> "The entry door to the nose section was under my position and now against the runway. I turned toward the pilot's compartment, expecting to exit out the top hatch above him. He was experiencing difficulty getting the canopy hatch open. I immediately returned to the nose and retrieved an escape ax to chop open the Plexiglas that encompassed the upper portion of the nose section. Lo-and-behold, all of the Plexiglas had shattered, and with one step I was on terra firma. Needless to say, I wasted no time departing the immediate vicinity, yelling to the pilot and navigator my good fortune in finding a quick exit. But what really saved my butt was a thick piece of armor plate on the bottom of the nose. The navigator subsequently received the Soldiers Medal for an outstanding act of heroism in rescuing the spy from the burning aircraft. An alert fire unit prevented the aircraft from being totally destroyed. No one was injured, other than some singed hair."[88]

Shortly before midnight on the 2nd, a Mosquito went airborne and flew to the Berlin area. That same night the RAF dispatched 54 Mosquitos to bomb the city. The JE operator received enemy radio interference at the pinpoint, hindering all attempts to contact HAMMER. On return the aircraft diverted to Watton because of low ceiling at Harrington.

492nd BG crew members in front of Queen of Spades A-26. Rear standing: Unknown crew chief, Capt. Willard H. Smith and Lt. William C. Ecklund. Kneeling: Sgt. Thomson and Lt. Allen T. Wayne. *Photo A.T. Wayne*

88 801/492 BG Association newsletter, The "Carpetbagger", No. 40 (December 1992): 9.

An irascible Simpson dispatched a letter to London OSS claiming the 492nd were:

> " ... flying it (A-26) without making the changes which we recommended. In the same way, maintenance has gotten so bad on the Mosquitos that we are afraid to fly in them, and I am probably pretty close to correct in saying that we will not have a JE organization (for) more than two or three days. Because of meddling here at headquarters, and somebody trying to get us two squadrons of ships overnight, Eighth Air Force realizes they didn't know what they were talking about, and moved the whole damn thing (JE Ops) to Harrington Field ... They (492nd) have completely taken over the A-26 job and allow us no say whatsoever ... and they are trying to make us fly Mosquitos with green pilots who have been flying Libs. I have one Mosquito crew left from Watton. They have been flying every night and each night having an emergency landing due to poor maintenance. They (492nd) ... have no spare parts and no ground handling equipment. I came back with a hole in the fuselage of my compartment the other night. They pasted a piece of fabric over the hole and expected me to fly it again? ... In the meantime, I just learned that Fogarty has been ordered to fly a Mosquito mission with one of their green pilots and navigators, over on the other side sometime this morning ... and I am going to try and stop it."[89]

Early morning of 3 April proved an event-filled day for Lt. Fogarty at Harrington. The 492nd scheduled a training flight to Dijon at request of OSS London office. They assigned Fogarty to ride aft in Mosquito NS740 with a new flight crew, both of whom recently completed a brief instruction course at Watton. Concern centered on Fogarty being ordered to fly with an inexperienced crew who lacked Mosquito poor weather flying experience and single-engine emergency landing skills under hazardous conditions. Also, the tower radio proved unreliable for contact during landings and takeoffs. Simpson initiated efforts to prevent the flight, but to no avail. Forgarty recalls the following:

> "I climbed in the rear to find the oxygen pressure checked zero. On this particular aircraft the oxygen bottles are located forward and were turned on by the pilot – the valve was located on his left side. The pilot started the engines and taxied the aircraft without going through our preflight checklist. The Merlin engines heat rapidly on the ground, therefore, it was necessary to quickly taxi around to the runway, perform the engine check, then takeoff before they became too hot.
>
> "At the runway, the pilot began to test run one engine, and I'm hollering on the intercom, but there was no response. The interphone did not work, so I immediately hit the emergency signal light to contact the pilot. He still could not be reached. Apparently, the navigator's table covered the emergency light, and both were unaware of my attempt to signal.
>
> "Finally, he started revving up the second engine, and this was my last chance, as far as I was concerned. I had no intention of going to 30,000 feet without oxygen. I pulled the red handle, the side hatch door fell off, and I quickly squeezed through and fell to the runway. During my hasty exit, the D-ring snagged and the parachute deployed. A blast of propwash from the revving engine bloomed the canopy, and I was dragged along the runway pulled by the 'chute. The tower fired multiple red flares and the caravan signaled red lights to attract the pilot's attention, as fire trucks and an ambulance quickly sped to the scene."

After side door replacement, oxygen adjustments, and interphone check, the pilot took off again, but failed to retract the undercarriage. He circled, came back and made a heavy landing, frightening Fogarty in the aft section. Apparently, the pilot had been opening and closing the bomb bay doors, not knowing which lever controlled the landing-gear. This was his first solo flight in a Mosquito, although having minimum dual time (eight hours) flying a Mark III Mosquito at Watton. A policy change required 40 hours minimum flying time before allowing new pilots to fly with OSS personnel.

Later that day during takeoff, Lt.Col. Robert Boone ground looped Mosquito NS584 causing extensive damage. A "court and boards" officer called the electronics shop to inquire if anyone saw Knapp exit from a Mosquito the previous day. No one reported seeing anything.

On 4 April, Simpson and Fogarty visited deputy CO Lt.Col. Fish to discuss numerous problems: training JE operators with pilots who lack Mosquito proficiency; return of mechanics to Watton and Warton; and who is responsible for Mosquito conversion work? The conversation soon became an openly hostile shouting match, forcing Simpson to physically remove Fogarty from the room. When tempers subsided the three officers went to discuss the issues with Colonel Upham.

Upham agreed that Simpson's men should supervise the new Mosquito aircraft conversion; only Knapp and Jackson will fly JE missions until new pilots and navigators are qualified; assign two new pilots and Mosquito NS686 to Watton for training purposes; and all Watton mechanics be returned to their home station. Colonel Gray, CO 25 BG (Rcn), accepted the mens return except for one crew chief, who must remain with Knapp/Jackson as long as both continue the JE project and take orders only from them. According to Fogarty:

> "Knapp had many acquaintances and connections who could assist with the conversion work. He took Simpson and myself to de Havilland Aircraft and directed us to the proper authorities for assistance. He brought in both RAF mechanics and workers from de Havilland to assist with maintenance and Mosquito modifications."

492nd BG officers in March 1945, left is Lt.Col. Robert L. Boone, CO of 858 Squadron, and on right is Lt.Col. Rodman A. St. Clair CO 856 Squadron, standing under bulbous nose of former Mickey Mosquito with dark blue spinner, used as a trainer. *Photo R. Boone.*

89 RG 226. Steve Simpson, letter dated 3 April 1945 to London OSS, London.

Harrington men towed two new Mosquitos into a hangar where Simpson and Fogarty began modifications by supervising an RAF crew. The officer in charge of the British mechanics spoke with Colonel Upham and requested the 492nd ground personnel receive two more weeks of training.

Lt.Col. Fish notified Simpson that 492 BG headquarters canceled the impeding court-martial proceedings against Knapp. Simpson previously warned Upham that if legal action against Knapp continued, he would initiate countercharges of negligence, mismanagement, and incompetence against the colonel and his staff. This antagonism reflected the fragile and volatile situation at Harrington. Instead, Knapp received a reprimand with a 104th Article of War warrant entered into his permanent record. Meanwhile, when Knapp learned the 492nd continued to employ operational Mosquitos for training new pilots, he declared adamantly refusing to fly airplanes previously used for this purpose. He demanded all new pilots be restricted from operational aircraft until fully Mosquito qualified.

That night B-24s from Dijon inserted JE teams LUXE near Munich, and PICKAXE near Landshut in Bavaria. An attempt to deliver JE agents FARO and BUZZSAW experienced navigation problems, forcing the aircraft return to base; and the MALLET insertion mission aborted when the A-26 undercarriage failed to retract.

By 5 April, the Allied armies advanced deeper into Germany and now appealed for Intelligence on the Berlin and Leipzig areas. In view of this urgency, OSS requested Eighth Air Force permission to attempt dark period insertions. Pinetree agreed and furthermore, consented to risk B-24s flying north to Leipzig.

OSS initiated Project Joshua to promote the immediate JE application in all OSS theaters. The AAF supplied B-17, B-24 and B-29 aircraft for experiments to further refine JE technique for difficult operations covering vast distances in the Far East. B-17 44-8295 arrived in England the first week of April with JE equipment and communications personnel. It flew to Dijon to train three new JE teams while four more awaited preparation.

On 5/6 April, a Mosquito left for another attempt to contact ANZIO, with no response. The aircraft returned to Harrington with a faulty engine to find the Operations room locked, no place for the crew chief, and no alert crew on duty.

OSS scheduled a mission the night of 6/7th to LUXE and PICKAXE. The aircrew arrived at the aircraft but found no duty members available. Knapp called the tower who promised to summon someone. After waiting 45 minutes, the duty crew arrived claiming that Engineering informed them the mission had been scrubbed. The delay forced aborting PICKAXE and rendezvous only with LUXE. Finally, Knapp/Jackson and Ancrum (NS725) takeoff. En route, Knapp encountered problems:

> "During the climb to altitude, the right engine was 100 rpm short. At 14,000 feet the prop and boost both started to surge slightly. After climbing through 17,000 feet of cloud with this condition, the starboard prop began to surge violently from 2,200 to 3,100 rpm. I changed to inner tanks to no avail, then feathered the prop when it seemed it would runaway at any moment. A single-engine emergency landing at Manston was made at 0120.

Photo taken at Manston, England. Discernible serial number is NS72?, taken as NS725, with call letter L on tail. NS725 made Manston landings on 4 February, 7 and 16 April 1945. Harrington photos show NS725 had call letter F. *Photo D. Strange*

After landing, I went back to open the rear door for Lt. Ancrum. I told him to jettison it, which he did after much difficulty. The jettison system was placed in backwards which would have made an emergency jump extremely dangerous."

Flying Control notified OSS that Knapp landed at Manston with the starboard engine inoperative. OSS then informed the Engineering section, who intimated Knapp deliberately aborted on purpose. Operations sent a vehicle for the aircrew return and later dispatched a mechanic by B-24 to Manston with a new propeller governor.

Lack of serviceable aircraft compromised a mission late evening of 7 April. Knapp spoke with the Grp Ops officer and learned NS740 and NS707 were available:

> "I explained why 707 should not be used for operations; Mossie naturally weak in tail section, and the rear fuselage hole (from 31 March explosion) had not been repaired. I informed him of learning NS740 could be flown with a hydraulic leak if the engineering officer signed the exceptional release. He (Grp Ops officer) attempted to have me sign it but I refused. He then said, 'Of course, if you're afraid to fly the plane, etc. etc.' He informed me Colonel St. Clair could sign the exceptional release and would call me back."

Knapp checked NS740 and found it leaked a cupful of fluid with each wheel cycle operation. He discovered an oxygen pressure leak known to a crew chief for three days had not been repaired. Simpson requested the flight be scrubbed because the oxygen supply would last only two hours on a 5 1/2 hour flight. Meanwhile, a B-24 departed Harrington to deliver JE agents BUZZSAW outside Leipzig, and FARO near Plauen in Bavaria.

On 8 April, Knapp flight tested NS740 that afternoon and found both landing-gear failed to lower. Prior to touchdown Jackson hand pumped the undercarriage down while Knapp salvoed both underwing tanks for speed and balance. After repairs and another checkout flight, with more repairs, the ship was finally ready.

OSS scheduled missions to LUXE and PICKAXE and issued a late order adding HAMMER. The Mosquito staged through Luneville, France, but en route, Knapp/Jackson and Ancrum divert to Brussels-Melsbroek instead. After refueling they takeoff. Initial attempts to contact LUXE failed, so they called PICKAXE. Then LUXE interjected five minutes later, almost simultaneously when PICKAXE answered, and demanded communication immediately. Ancrum asked PICKAXE to wait.

A LUXE member suffered a ruptured hernia and begs the emergency delivery of a truss (support). The agent reports tank traps and barriers constructed along roads to Garmisch-Partenkirchen and the mountains; rail traffic pass through Weilheim all night towards Berchtesgaden; and two aircraft factory locations. He pleads that Raisting be spared bombardment for the entire population is anti-Nazi and trust the LUXE agents.

Ancrum instructs LUXE to monitor BBC radio messages for next rendezvous. He then attempted to contact PICKAXE but received no reply. The Mosquito then flew northward and found sections of Berlin a maelstrom glowing fiercely under RAF bombardment. Ancrum establishes communication with HAMMER who reports the previous pinpoint is now Wehrmacht territory. He is presently located 34-kilometers southwest of the original contact area. Slowly his signal becomes garbled and vague.

About this time the pilot accidentally released the emergency valve on his oxygen outlet and became unconscious. The Mosquito entered a power dive but the navigator managed to maintain control until the pilot regained consciousness at lower altitude. The Mosquito left Germany, landing at Brussels and Manston before reaching Harrington. The wire was shipped to London for further deliberation.

Knapp recommended flying NS740 to RAF Middleton for complete RCAF inspection to determine if Harrington crews had properly performed aircraft maintenance. Fogarty recalls:

> "Knapp arranged for an RAF Squadron Leader to lecture the Harrington air men on the Mosquito performance and flight operations. The S/L gave a great presentation, and then came the question and answer period. It was a typical American question period, where the inquiries served not to gain information, but to show how brazen and arrogant the questions were. Eventually, near the end of the discussion period, someone stood up, held his middle three fingers in the air, pointing to them one-after-the-other, and said, 'Squadron Leader, will you tell me why in Christ's name you designed the f*****g airplane with the gear, bomb bay doors and flap handles, side-by-side with no distinguishing marks?'
>
> "The Squadron Leader had about enough by this time and said, 'Yes Sir,' and held up his three fingers and pointed to them, one after the other, 'because in the RAF we do not allow anyone to fly unless they can tell the difference between the left, the middle, and the right.'
>
> "That ended the lecture."

As the Allied armies fought deeper into Germany, the daily tempo of missions increased. OSS received Intelligence reports the Nazis planned establishing a siege fortress or National Redoubt in Bavarian. They deemed it imperative more agents parachute into the region to further evaluate the Intelligence situation.

OSS officers and JE crew pose next to pranged Mosquito NS584 at Harrington. L-R: 1/Lts. George Fogarty, Calhoun Ancrum, Kingdon R. Knapp, John K. Jackson, and Andre Pacquet. *Photo A. Pacquet*

1/Lt. John K. Jackson in front of damaged Mosquito NS584 from 3 April 1945 accident. *Photo J. K. Jackson*

Late evening of 9/10th, a Mosquito left for FARO. The aircraft skirted known Flak areas and arrived over Plauen on time, but failed to establish contact. This night also heralded the first successful operation of an ingenious "drop seat" in the A-26. There had been hesitation and even refusals of some agents at the moment of dispatch. The new method consisted of a seat that swiveled downward. Operated by the dispatcher, it released the agents involuntarily in the correct jumping position insuring both they and the container landed in close proximity. Two A-26s thus modified proved thoroughly satisfactory; 43-22626 delivered PLANTERS PUNCH and 43-22513 ORANGE BLOSSOM, successfully.

FARO was again target for tonight, 10/11 April. Nearby Munich blazed in flames from a large-scale RAF raid. FARO could not remain long for fear of being observed. He spoke softly, confirmed instructions for next contact and would monitor the BBC for schedule confirmation. The Mosquito then attempted to contact LUXE, unsuccessfully. Meanwhile, A-26 sorties from Harrington proved successful; 43-22626 inserted JE agent MALLET near Berlin while 43-22513 deposited TOM COLLINS west of Laba, Germany. Training continued during the day. Pecquet qualified for Mosquito training, and new JE operator Lt. Jean Nater arrived for duty.

Tonight, 11/12 April, Schaeffer/Corson and Fogarty took off, but returned to replace a defective TR-6 Eleanor unit. They resume the mission and arrive ten minutes late over the Leipzig area to contact BUZZSAW. Fogarty radioed for 20 minutes with no reply. Fogarty recalls:

> "I noticed something shiny on the floor, pulled my mask off and sniffed what smelled like gasoline. In addition to the JE equipment, I had a VHF set also. When the channel switch was turned, all kinds of sparks occurred inside the set. So I was not really happy. I kept the mask on with 100 % oxygen, but the thought of those fumes and those sparks did not make me the least bit comfortable. We returned safely, and I earned my money that night."

Early that morning, Simpson accompanied an OSS official to HQ Eighth Air Force and met with Colonel Pierce of Air Operations. The men voiced concern of the current situation and problems encountered at Harrington that unless rectified would compromise the JE program. Pierce transferred all training control from 492 BG to Simpson, but he refused Simpson's plea to return the entire OSS JE operations to Watton: they must remain at Harrington. OSS also requested Colonel Upham be removed or transferred from 492 BG command. HQ Eighth Air Force relieved him for reassignment to 306 BG. Lt.Col. Robert Fish then became CO 492 BG. According to Simpson, "I finally got rid of Upham ... but the insurmountable problem was how to return the JE operations to Watton?"

An OSS report documents the following conclusion:

"Colonel Upham, a very fine officer ... with no experience in our type of activity and unwilling to make definite decisions, caused considerable confusion. He felt pickup crews could accomplish this work, a theory proven false by the 25 BG (Rcn) at Watton. No manner of persuasion could convince him that coordinated crews were necessary to accomplish OSS-type operations. Mechanical failures immediately cropped up and frequent cancellations and aborts resulted. In addition, the colonel returned the experienced crew chiefs on loan from Watton, as well as the mechanics from BADA. This was possibly due to the lack of knowledge by the colonel regarding Carpetbagger operations and with the type of flying necessary in JE operations."

Simpson ruminated over the hectic pace of operations:

"Making improvements was a continual headache. Planes and personnel were flying so many missions it was impossible to keep both in good shape. The strain, especially in high-altitude work was awful. One of my navigators, a normal full-grown man, broke down and cried, and he didn't even know why. Exhausted and overwhelmed by a grueling schedule, he had flown seven missions in eight days, all above 25,000 feet. I'm afraid I was not as tough or prepared as I should have been for this work. I hated to shake the crew awake after a mission to obtain a complete report from them. I even had to inventory the personal effects of bunkmates (Emmel and Walch) who never returned."

Tonight, 13/14 April, Knapp would visit LUXE to deliver a truss. He test flew a Mosquito to determine in-flight accuracy dropping a package from the aft JE compartment. The test proved successful. That night Knapp/Jackson and Ancrum climb for altitude with 150 magneto drop on both engines and rendezvous on time. According to Ancrum:

"When contact was established, LUXE was informed that we had a package for him, and he gave his position. I asked if it would be all right to return four and a half hours later to make delivery? I asked if he had knowledge about evacuation of troops from Austria through Munich, especially the *Sechste SS Panzerarmee*? (Sixth Armored Elite Guard Army). LUXE reported having no such information and indicated having very important cipher messages to pass on first. He continued providing a series of coded messages that were wire recorded. LUXE had several more series of ciphers to transmit but was forced to finish quickly because three kilometers from him, a car of the *Abwehr* (German Intelligence) had been seen the past several days. LUXE continued providing his messages in cipher codes."

The aircraft arranged to drop the medical equipment (truss) at 0430 and informed LUXE they would return in half an hour. The Mosquito left LUXE to rendezvous with PICKAXE. Ancrum heard PICKAXE but failed to establish satisfactory radio contact with him. His voice was recorded for recognition purposes. The Mosquito returned to LUXE, but he too failed to point his antenna properly to hear the aircraft. It was necessary to leave LUXE to refuel. Knapp landed the Mosquito at Dijon for petrol and took off at 0350 to contact LUXE:

"The target was identified visually at 0430. We flew low at 100 feet altitude but did not drop (package) as there were no signals. We returned home on the deck."

After examining weather reports, a JE Mosquito staged from Brussels the 14/15th, arriving near Leipzig as searchlights crisscrossed the horizon. The navigator reported increase air activity south and east of the target area. They radio BUZZSAW for 25 minutes, then flew to the second contact area and call FARO for 30 minutes, both with no response. The Mosquito diverted to the Woodbridge FIDO field because of low ceiling at base.

On 15/16 April, OSS schedule another Mosquito mission to visit MALLET and HAMMER near Berlin. Navigation instrument failure at the Denmark coast hindered making positive pinpoint identity. However, they attempt contact when in the estimated target area. After 15 minutes, an electrical power failure adversely disrupted the interphone, all cockpit lights, and heated flying suits. Because of an extreme low temperature (-38' F) in the rear JE compartment, the pilot lost altitude immediately and returned, landing at Manston.

Believed to be Mosquito NS725 with newly installed unpainted 100-gallon underwing tanks, at Harrington. *Photo A. Pacquet.*

Mosquito NS725 "F" with new underwing tanks taxing at Harrington.
Photo J. Nater

A ground delay staging through Brussels the 16/17th, force cancelling a JE mission to BUZZSAW and proceed directly to FARO. After 10 minutes on target, the JE operator's oxygen system failed, requiring a switch to his emergency supply. After a further 20 minutes, contact attempts proved futile and the pilot descended to 10,000 feet for return to base. Unknown to OSS, advancing U.S. forces previously repatriated agent FARO. Also this night a B-24 inserted JE agent FARMER near Straubing in southern Germany.

OSS scheduled a mission the 17/18th to contact PICKAXE. Fuel pump repairs delayed takeoff by ten minutes. Upon liftoff the Mosquito streaked towards Germany and successfully establish contact. The aircraft relayed instructions requesting PICKAXE to monitor all railway traffic for scheduled times, direction, car numbers, and cargo-type carried. The agent failed to direct his Joan antenna properly and the aircraft lost contact after several minutes.

The Mosquito then began a forty-minute session with LUXE. The agents await air delivery of the LUXE II team and supplies, giving pinpoint, approach directions, and signal light codes. Briefly, LUXE issues details of an Me 262 aircraft factory in an underground tunnel on Olympia highway to Munich; French Chief of State Petain and the Hungarian Chief of State Horthy, and other officials located in Hirschberg Castle under strong SS guard; and Dachau SS guards killing prisoners and burning corpses. LUXE pleads for a leaflet warning prior to bombing Weilheim to spare harm to citizens who support the agents. The aircraft left Germany with both communications safely recorded.

Again, inclement weather on 18/19 April required the Mosquito to stage from Brussels for Berlin. They arrive at both pinpoints on time but fail to contact either HAMMER or MALLET.

As Patton's Third Army arrived at Nuremberg, OSS schedule a Mosquito mission on the 19/20th to CHAUFFEUR and FARMER. Unsuccessful in contacting FARMER near Straubing, the aircraft altered course to CHAUFFEUR. The JE operator noticed the agent's signal wandering and instructed him to move clear of the woods, dramatically improving aircraft reception with the latest Intelligence recorded. After all briefing points were conveyed, the Mosquito turned for England.

On 18 April, a Ninth Army OSS detachment requests London OSS assistance with inserting agents into Germany by A-26 from an air strip near Kitzengern. They have agents RAYMOND and GODFREY II awaiting airlift. After issuing pinpoints, Harrington dispatched an A-26 that brought along two package-chutes and three bags containing dropping equipment.

The following 20/21st, OSS dispatchers presumed Eighth Air Force cleared the mission with the air base, then occupied by a night-fighter group, only to learn the unit knew nothing. Once informed, they provide all possible assistance including AAA information for one pinpoint, forcing last minute change in drop zones. Using the two package-chutes the dispatcher rigged both agents for the drop.

Meteorological information forecast excellent weather within the next hour. With agents loaded on board, A-26 43-22513 took off. At 0010 on 21 April a radio signal indicated RAYMOND dispatched, and the A-26 returned at 0120 after delivering GODFREY II. (492 BG records list only one A-26 mission, yet OSS files indicate two A-26s involved?)

JE Mosquito NS740 named *Greex* or *Greek*. Photo A. Pacquet

A-26 serial 43-22513 *King of Spades* parked at Harrington. *Photo A. Carnot*

OSS scheduled two Mosquito missions this evening. The first (NS740) staged through Brussels where engine trouble force the crew to cable Harrington for a replacement aircraft. The second Mosquito (NS707) soon arrived. But two unsuccessful takeoff attempts from right engine aneroid malfunction force aborting the mission to HAMMER and MALLET.

The second mission involved Knapp/Jackson and Ancrum to PICKAXE and LUXE. They contact PICKAXE at 2200 hours and Ancrum instructed him to keep his antenna directed towards the airplane, dramatically improving the signal. PICKAXE reports the 16 April AAF (446 BG) air bombardment closed rail traffic through Landshut, and military transports convey troops to Munich from Landshut using side roads. He provides coordinates of a building with a vehicle placarded, 'Communication Service of the German Luftwaffe.' At this period, both aircraft intercom and Loran failed. The pilot and navigator utilized the VHF sidetone for operation, but quality of JE transmission and reception decreased.

After sign-off, the Mosquito flew to locate LUXE. The agent details his latest surveillance including coordinates for antiaircraft railway gun-train and ground antiaircraft batteries; secret Reich's Air Ministry files from Berlin stored at Villach castle near Bad Aibling; increase in day and night vehicle traffic from Munich to the mountains; and promises coordinates of Himmler's HQ on Monday. The agent witnessed the AAF bomb Pasing Railway Station with track beds totally destroyed; Munich suffered severe bomb damage; fleeing refugees stream toward Garmisch and the mountains; and two air attacks on Weilheim achieve its objective.

The Mosquito flew to Brussels, Belgium, and after JE equipment removal, the JE operator climbed into the cockpit nose section armed with a camera. The aircraft took off at dawn and on entering Nazi-occupied Germany, Knapp flew below 500 feet. And upon reaching Ammer Lake, skimmed over the sparkling surface heading south for Weilheim, then east to Kaufbeuren, and returned towards the northwest:

Flying low over the rail yard at Weilheim, JE operator Calhoun Ancrum took pictures with a hand-held camera through the Mosquito nose. Small figures seen in the photo indicate they were taken by surprise by the low-flying aircraft. *Photo. J.K. Jackson*

> "We performed two low-level ... passes taking 100 photos of the areas Weilheim and adjacent marshaling yards, Augsburg, Donauworth, and Crailsheim. Also airfield installations and oil refinery, and miscellaneous factories all taken from altitude between five (5) and five hundred (500) feet. No enemy opposition, except for one member of Wehrmacht, who thumbed his nose at the aircraft."

OSS schedule two Mosquito missions for evening the 23/24th. The first flew to visit FARMER and CHAUFFEUR. After reaching the pinpoint, the JE operator tried contacting FARMER without success. He then reached CHAUFFEUR. The agent briefly reports Russian, Romanian and Bulgarian troops transfer to Augsburg, leaving the SS to pillage Regensburg; French and Hungarian Waffen-SS troops defend the city; analysis of defensive weapons; German General staff permanently quartered at Hotel du Pare on Maximilianstrasse, opposite the destroyed railway station; Italian POWs in Regensburg allowed return to Italy; British, American, and French POW columns evacuated from Regensburg pass through Abensberg area; guards execute stragglers and many corpses lay scattered along the road. The JE operator inquired about two parachutists arrested in Neustadt on 8 April (FARMER team). CHAUFFEUR reports SS troops found parachutes and supply canister, and he lacks knowledge of the agents disposition. A pool of gasoline on the JE compartment floor force an early sign-off and return to base.

The second mission targeted Berlin. After takeoff an air test revealed a dead Eleanor unit, so the pilot landed to replace the defective unit. The Mosquito then shot across to Germany arriving 30 minutes late. They called HAMMER for 15 minutes, then MALLET for 30 minutes, with no response. The aircraft returned to Harrington.

492nd BG officers assigned to the JE project. L-R, Standing: Lt. Samuel Webb, Lt. Ralph Smith, Lt. James E. Kuntz, all pilots; front row kneeling, navigators 1/Lt. Edward P. Kolacke and Marvin R. Edwards. *Photo M. Edwards.*

The Battle for Berlin continued with Koniev and Zhukov's troops linking up in the southern suburbs on 24 April. That night, the 492nd dispatched a B-24 from Dijon inserting JE agent teams BOYARD near Bregenz, and SULTANE near Munich. OSS also scheduled a Skywave training flight to the continent with pilot Lt. James E. Kuntz, navigator Lt. Richard L. Greene, and JE operator Lt. Jean M. Nater (NS707). During the flight they encountered an unidentified attacking aircraft that whizzed past trailing a yellow exhaust. After completing the mission the Mosquito turned for England.

The aircraft entered the coast at 0155 hours, 10,000 feet over Beachy Head for descent to 5,000 feet. Approaching near Odium the port engine began backfiring. Switching fueltanks from inner to outer failed to help, so Kuntz throttled back and trimmed for single-engine operation. Upon feathering the prop, the Merlin erupted into fire.

The pilot radioed a Mayday requesting homing directions. Receiving vector instructions, he increased starboard engine power and held at 3,000 feet altitude. Soon the cockpit filled with smoke, and he ordered an immediate bailout at 0205 hours. JE operator Nater left immediately, as he recalls:

> " ... I had torn off my oxygen mask and flying helmet. My chest pack was firmly hooked to my harness – I always wore it while flying ... So, I went out feet first. I soon became wedged in the hatch with my legs banging outside on the fuselage – the chute pack was stuck on the hatch frame. I attempted to disengage it with my hands, fearful my searching fingers will spill the chute inside the aircraft. Then something told me to relax. The next thing I knew, I was outside the hatch, falling free from the Mosquito as it continued flying away, trailing a sheath of flames from the port engine. I pulled the ripcord, then felt a long, strong tug on my shoulder – my chute had opened. I'm out, I'm alive and on my way down. It was very dark, and I didn't know where I was. No ground in sight, but I knew it must be very close. Suddenly, I hit the ground and fell backward. I lie there in the quiet darkness, panting, licking my lips, and swallowing hard."

Greene the navigator experiences problems. After releasing the floor hatch cover, he yelled that his leg straps were unfastened. Kuntz shouted at least one leg buckle must be fastened before bailout. Greene attempted connecting the left leg strap, said nothing further, and jumped feet first from 2,000 feet.

Kuntz moved from his seat to the open floor hatch and jumped feet first through the opening at 1,500 altitude. His parachute safely deployed followed by normal descent and touchdown in a plowed field, not far from a dwelling. The occupants provided telephone access and he notified the closest aerodrome. They dispatched a vehicle for transport to Odium Flying Control. He notified Harrington of the previous events. Next morning they initiate a search for Greene. A farmer discovered a parachute canopy and harness, with both leg straps unfastened, draped in a tree at Sheldon. Next day, searchers located Greene's body.

The ex-Mickey Mosquito landed northwest of Winchfield, 150 yards from a farmhouse with wreckage strewn about the area. Before leaving for Harrington, Nater inspected the wreckage. NS707 struck the ground flying level, plowed through a hedge and cattle fence, while tearing itself apart. The fuselage broke in half at the JE operator position but did not burn. According to Nater:

> "Later at Harrington, I was told by Kuntz that the port engine was retrieved by RAF engineers and taken apart in an effort to discover why the engine failed so dramatically. He reported they found a 'steel projectile' pierced the jacket of the supercharger, lodged there, effectively sealing the jacket. Later, when we were back over England, the projectile had worked loose, causing the carburetor to explode. If this was true, Kuntz reasoned the projectile could have only come from the unidentified aircraft which had been on our tail. I'm not an engineer, but it sounded plausible at the time. As far as I know, there was no official confirmation of this story."

Harrington crewmembers, left to right: navigator Lt. Carson, pilot Lt. Richard Ladue, later with OSS Wiesbaden; and JE operator Jean Nater. *Photo Jean Nater.*

The Battle for Berlin continued Wednesday, 25 April, as Generals Zhukov and Koniev's forces reach Potsdam. Also that day, Soviet and U.S. forces meet at Torgau on the Elbe River, 75-miles south of Berlin. An unforeseeable event occurred late that evening. An A-26 left Harrington with the GITANE team. Upon arrival at Frankfurt for refueling, one agent fainted and was rushed to hospital for an emergency appendectomy. OSS canceled the mission and returned the other agent to Harrington.

This evening, 25/26 April**,** Knapp/Jackson and Ancrum attempt contacting PICKAXE and LUXE. En route, fuel leaked into the JE operators compartment, but Ancrum agreed to continue the mission. At 2100, Knapp received a radio call from Chairleg (ground control) for all friendly aircraft to report in. He complied for the purpose of identifying himself, and they replied requesting a transmission for a fix. Ancrum finally made agent contact:

> "The matter of PICKAXE moving to a new location was undecided. He was unwilling to commit himself without giving the subject some consideration. I suggested an early attempt to contact him. He advance the objection he would have difficulty obtaining food, and he was unsure of controls along the way."

Briefly, PICKAXE reports all railway lines through Landshut are paralyzed; the city holds thousands of combat ready troops in replacement units without weapons; they travel nightly along rural roads towards Munich; 400 Polish POWs passed through en route to Dachau; a barracks three-kilometers east of Landshut on road to Schoenbrunn has a sign "*Message Center for 151st Feldausbildungsdivsion (560 Volksgrenadier-Divisio*n)."

After completing communication, the Mosquito made rendezvous with LUXE. The agent seemed tired after the successful reception of the LUXE II team. Ancrum inquired about Allied POW movement southward for Nazi use as hostages? LUXE responds that no POWs are mistreated in Weilheim because the population support the agents; now located near the frontlines, combat may interrupt future JE contacts; forests astride Landsberg-Munich highway are occupied by troops retreating nightly to Munich. He provides coordinates for Munich and Augsburg rail junctions active at night; Landsberg bridges mined with explosives; munitions plant and airplane parts factory; and Himmler's mobile headquarters consisting of two armored trains. After signing off, the Mosquito turned for Harrington.

On 27 April, Patton's Third Army crossed the Danube River in the Regensburg area. Late that Friday evening a Mosquito flew to southern Germany. All attempts to contact FARMER and CHAUFFEUR end in failure. Severe weather and heavy icing conditions en route prolonged the return flight and eventual landing at Tangmere. This is the final JE mission involving a 25 BG (Rcn) crew. Both Knapp and Jackson return to Watton leaving the JE operations an all 492 BG effort. Simpson said he got drunk: "I didn't care what happened … the war was nearly over anyway." Early that morning, JE personnel held a special meeting to benefit from Kuntz and Nater's accident experiences. OSS cancel a training flight over London for want of aircraft.

A JE Mosquito staged from Dijon 28/29th. A solid undercast covered the pinpoint to 1,000 feet. During radio contact PICKAXE failed to properly orient the Joan antenna for maximum signal. He confirms having received a BBC message to transfer to Rosenheim but declines because of strict German army controls to halt massive desertions. Then the transmission signal is lost.

The Mosquito establishes communication with LUXE. Increase enemy patrols in the immediate vicinity endanger his position; he is located near the frontlines surrounded by tank fire. Some 160,000 Russian, French and British POWs, including 12,000 USAAF, are located at Moosburg near Munich. LUXE rescued one American POW who requires immediate medical attention. Then radio contact is lost.

492 BG Citation

Extremely pleased with 492 BG achievements, London OSS sent a letter on 29 April praising both Harrington and Dijon operations for successfully dispatching agents and supplies to German regions previously denied by HQ Eighth Air Force. OSS promise a special commendation to both Colonel Fish and his staff for their excellent work.

In Berlin on 29/30 April, Adolph Hitler married Eva Braun as Soviet troops fought street to street. That night a Mosquito left for SULTANE and BOYARD, the first mission to the new agents. They reach SUTLANE near Munich, and BOYARD near Bregnaz, but neither respond. Unknown to OSS, both teams lost their JE radios. The Mosquito rear section began to fill with smoke, then Loran, Gee, and all communications cease functioning. The pilot aborted the mission and returned to base. At Harrington the tower failed to respond to radio calls, so he buzzed the field four times before they finally switched on the runway lights.

During the afternoon of 30 April/1May, Hitler and Braun commit suicide, and Admiral Karl Doenitz appointed his successor. The JE operations continue with Webb/Kolacki and Ancrum leaving for PICKAXE and LUXE. The port engine ran rough during the entire mission. Ancrum received no response from LUXE, but PICKAXE came in with a fluctuating and fading signal, indicating either he operated near trees and buildings, or had weak batteries. Unable to receive anything discernible, Ancrum radioed instructions for the next meeting. As the Mosquito returned to Harrington, the crew were unaware they completed the last successful JE mission.

May 1945

On 1/2 May, Navigator Lt. Marvin Edwards prepared for his first JE mission. Kuntz/Edwards and Pecquet (RG157) left for CHAUFFEUR when they lost both VHF and Gee. After return and equipment transfer to NS740, they takeoff again. En route both pilot and JE operator lost communications with the navigator, requiring another return to base. After interphone repair the men took off a third time, but starboard engine trouble force aborting the mission. On Wednesday, 2 May, Stalin announces the fall of Berlin. Skywave training flights continue. And on 6 May an A-26 inserted agent GAULOISE near Kiel, the final 492nd mission.

Lt.Cmdr. Simpson submitted recommendations to HQ Eighth Air Force issuing DFCs and commendations to ten 25 BG (Rcn) officers for significant contributions in development and successful prosecution of his JE program; and one posthumous to Major Edward Tresemer, 492 BG.

Post Script: Following the war, OSS JE project veterans went their separate way.

Colonel Hudson Hutton Upham, a West Point graduate, class of 1934, gained his pilots wings from March Field. He obtained a Masters Degree in the French language from Quebec University, then returned to West Point as a language, and later, air tactical instructor. He became commanding officer 492 BG in December 1944, and later reassigned to 306 BG (H) circa 16 April 1945. Postwar he commanded a European Air Transport Service unit with headquarters at Wiesbaden, Germany. He died age 33 on 1 November 1946, when his B-17G 43-39338 disappeared during an IFR night flight from Capodichino Air Base, Naples, Italy, to Bovingdon, England. On 25 July 1947 a French Army alpine patrol discovered wreckage on Aiguille des Glaciers at 12,187 feet (3,750 meters), 15 miles southwest of Mt. Blanc. Flying in a heavy overcast, the bomber smashed into a near vertical face, 450 feet below the summit. It exploded on impact, scattering wreckage on to a glacier. Heavy winter snows covered and buried the heavier parts in the glacier. All eight crewmembers are buried in a common grave at Arlington National Cemetery, Virginia.

Steve H. Simpson, Jr. and his OSS involvement first appeared in Joseph W. Persico's book, *Piercing the Reich*. Postwar, Simpson returned to the army reserve and later retired a colonel. He established his own electronics company in San Antonio, Texas. During formal affairs or business events, he occasionally met retired 492 BG senior officers living in the San Antonio area and cordially spoke briefly with them. Still bitter over his experiences at Harrington, he feared losing his temper should they engage him in conversation regarding the JE operations. To avoid this embarrassment, he declined all invitations to fraternize with their members. The highlight of retired life involved attending 25 BG (Rcn) Association reunions, especially October 1990 in San Antonio. Here he celebrated with Bill Miskho, Robert Walker, Roy Conyers and Richard Geary. Steve Simpson died 10 March 1997 at age 90 with life-long friend Robert Walker at his side.

Princess Xenia Romanoff, grandniece of the Tsar, and Lt. Calhoun Ancrum. The 26-year-old bride trained as a nurse at a children's hospital in England, being known to the staff as Nurse Romanoff. *Photo Daily Sketch*

JE veterans at 1990 San Antonio reunion. L-R. Robert P. Walker, William G. Miskho, Steve H. Simpson Jr., and Roy C. Conyers. *Photo N. Malayney*

Victor S. Doroski is listed Killed In Action on 8 February 1945 during a night photo Joker mission near Oberschonhage, Germany. He and navigator Captain Jackob Hochman died when a partially hung photo-flashbomb ignited in the Mosquito bomb bay, immolating both men.

DeWitt R. Goddard retired from RCA. During World War II, OSS classified his JE equipment as top secret. After the war he attempted to patent his invention but learned the company building the JE sets had previously patented the radio under their name. He died 26 May 1993 at age 89.

William L Sawyer remained in the OSS and then CIA involved with secret satellite espionage, retiring after 30 years service. He died 13 December 1997.

George H. Fogarty retired from the electronic engineering field and died 6 July 2004.

Brunnon C. Normand maintained the JE project radio equipment and postwar worked for the Georgia State Police. He died 12 October 2006 near Atlanta, Georgia.

Richard M. Geary became district school superintendent in LeGrand, California. He died 24 October 2004.

Norman Magee, like many veterans, returned to active duty during the Korean War. Later he joined the Air Force Reserve, and when flying an F-86H aircraft on a high-altitude exercise over the Gulf of Mexico, disappeared, never to be seen again.

William G. Miskho became a pharmacist and lived in Colorado. He died 16 February 1991 from leukemia.

Calhoun Ancrum Jr. married Princes Xenia Romanoff, of the Royal Imperial Russian Family on 17 June 1945 in London. Navigator John Jackson served as best man at the wedding. Postwar Ancrum returned to private life, divorced the princess, remarried, and later became an ordain minister. He died 21 February 1990.

Anthony Turano postwar became a professional photographer in private business.

Marvin Edwards is an 801/492 BG Association member and lives in Florida.

Jean Nater married a young English lady and lives in Devon, England.

Andre E. Pecquet postwar returned to his native France. He died 16 January 1997.

Victor J. Layton, an aeronautical engineer fluent in French and German, was recruited by OSS, and in early 1944 parachuted into northern France. After France's liberation he returned to England to recover from wounds, then joined the JE team. After leaving the project, he parachuted into Austria on OSS assignment. Postwar he remained on active U.S. Army duty serving in both Korean and South Vietnam wars. His last military assignment was Chief of Research & Development for Army Aircraft. He retired a colonel in 1967 and died 8 August 2010.

Roy C. Conyers returned to Rhode Island and in 1954 established Conyers Construction Company. In 1988 this writer placed him in contact with his former pilot, Robert Walker, and both men renewed their long lost friendship. On 13 November 1997, Roy went home from work. After lunch he stretched out on the sofa and passed away in his sleep.

Robert P. Walker made the USAF a career. He flew C-54 transport aircraft in the Berlin Airlift and Korean War. After graduating from test pilots school at Edwards AFB, he became a USAF acceptance pilot for the Douglas B-66 program and test flew aircraft off the production line. During the Vietnam War he flew RB-66C aircraft, and on 25 February 1966, a SAM missile struck his aircraft near Vinh, North Vietnam. He managed to fly the crippled jet to the Gulf of Tonkin where crewmembers ejected. A navy helicopter from the aircraft-carrier USS Ranger rescued five of six members from the sea. One officer is listed Killed In Action. Having served his country in three wars, he passed away 7 January 2010.

Colonel Hudson Upham with children on Red Cross tour at Harrington. *Photo 492 BG Assoc.*

John K. Jackson attended university postwar and became an aeronautical engineer. He worked for Thiokol and later Boeing Aircraft Company before retiring. He died 7 May 2001.

Kingdon Rodger Knapp married a lass from Scotland, Catherine Marjorie Pybus and had one daughter. Recalled to active USAF duty for the Korean War, he flew Douglas A-26 Invader aircraft with 8 Bombardment Squadron (L), 3 Bomb Wing, Kunsan (K-8), Korea. On 7 June 1951 he and two crewmembers took off in A-26 44-33277 on a night mission to interdict a highway between Pyongyang and Sariwon, North Korea. The A-26 disappeared and never returned. He is listed Killed In Action.

History of JE agents operating in enemy territory.

The previous chapter detailed the OSS JE operations chronology and events at both Watton and Harrington. The following supplements additional information to correlate specific JE missions with agents' activities inside enemy territory.

TYL

Born in Surabaya, Dutch East Indies in 1917, Anton Schrader graduated in 1938 as a civil engineer from the Technical University in Bandung. In 1939 he traveled to Holland to continue studies when German troops invaded the country. He became involved with the Dutch underground and later fled to England. After working for the Netherlands government-in-exile he requested agent status and came under OSS scrutiny. Now 27-years-old, they accepted him to the TYL project, establishing an underground network infiltrating OSS agents from Holland into Germany. OSS issued him the pseudonym Bobbie.

On the night of 10 November 1944, Bobbie parachuted at 2330 hours from a B-24, one-half mile north of Ulrum, Netherlands. Upon landing, he placed two Joan sets, regular radio receiver, batteries, pistol, ammunition, and clothing in a rucksack The canister with remaining supplies was hidden in nearby brush. Several hours later, shivering bitter cold force him to seek shelter at a nearby farmhouse. After warming himself by the fireplace and a quick change into dry clothing, he went to sleep quartered in a barn.

Early that morning Bobbie traveled to visit a friend named DeBoer. He divulged full particulars to his confidant who agreed to assist. Physically exhausted, Bobbie rested and slept before retrieving the luggage. Unknown to him, someone found and vanished with the supplies.

On 15 and 16 November, OSS flew two unsuccessful missions to contact him. Simpson warned Bobbie to transmit in an open field, away from trees or objects. Instead, fearing someone might observe, he communicated near a dwelling, thus attenuating signal transmission. Unfortunately, Bobbie accompanied DeBoer to visit Zwolle and this commitment prevented maintaining the second JE rendezvous.

The first successful contact occurred 22 November and Simpson formulated plans to replace the missing items. A B-24 arrived early 5 December in a raging gale forcing delivery cancellation. Later, technical radio problems or inclement weather thwarted numerous supply attempts.

Double agents menaced the Dutch underground. One of these, Bobby van de Zeemanshuis, traveled from one area to another portraying himself as an Allied agent. After galvanizing the trust of local opposition members to assist in covert activities, he informed the Gestapo, who arrested them. These tactics demoralized the underground movement.

A senior resistance official questioned Bobbie's credentials and intent. The promise of providing further agents and supply deliveries never materialized. Through a special courier service crossing the Allied-German frontlines, he dispatched a message to *Ordedienst*, the Dutch underground HQ in Eindhoven, hypothesizing Bobbie is the double agent Bobby van de Zeemanshuis. After deliberation, *Ordedienst* sent a letter back through the lines, approving his execution. The German *Sicherheitsdienst* (SD) or SS Security Police, intercepted the message and became determined to find and protect Bobby van de Zeemanshuis from assassination. The letter identified the double agent as living in Ulrum at the DeBoer residence. Unknown to both SD and Dutch underground, the real Bobby van de Zeemanshuis had been liquidated while using an alias identity.

On the morning of 20 February, the DeBoer family sat at the kitchen table for breakfast while upstairs Bobbie slept in a bedroom attic. At 0900 an automobile arrived and four men brandishing Sten guns and revolvers stepped out and surrounded the residence. On signal, they simultaneously burst through the doors. The leader accosted DeBoer and rudely explained the nature of the visit. He revealed a letter from *Ordedienst* to a deputy in Zoutkamp warning that a Gestapo spy successfully infiltrated the Dutch underground and to execute Bobby van de Zeemanshuis the suspected agent. The four men claimed they represented the Dutch underground and demanded DeBoer surrender Bobby van de Zeemanshuis for execution. By coincidence, Anton Schrader was also known as "Bobbie" and the intruders mistook him for van de Zeemanshuis.

Startled by this revelation DeBoer attempted to convince the intruders otherwise, that Anton Schrader is not a Gestapo agent. To add credence to this claim, he displayed the JE radio equipment and detailed Bobbie's clandestine contacts with OSS aircraft. Having extracted this information, the intruders dropped their disguise and revealed their true identity as the German SD. They demanded to know Bobbies' whereabouts.

The shouting and commotion downstairs awakened Bobbie. Overcome with anxiety, he immediately burned all incriminating papers in his possession. Now confronted with a horrible situation, he gave thought to escape by shooting his way out. Fearing the Germans would retaliate and murder the DeBoer family, he decided not to resist arrest.

The SD searched the premises and soon located him. They asked if he was Bobby van de Zeemanshuis. Unaware of what DeBoer disclosed, he stubbornly denied being this person. Unsatisfied, they repeatedly punched him in the face and jabbed his body with their weapons. When they threatened to kill the DeBoer members, Bobbie moaned he would confess providing no harm came to the family.

As they hauled him downstairs, he immediately spotted the JE radio equipment spread on the table. It seemed further denials were obviously impossible. After DeBoer and his eldest son were taken away to Groningen, he learned from SD boasting of what actually transpired during the interrogation. They allowed Bobbie to read part of the letter referring to Bobby van de Zeemanshuis who worked with 12 Americans in the Zoutkamp area distributing arms. This information fitted in accurately with information previously revealed to him by underground members.

The SD arrested and transported Bobbie to Groningen, the most northern provincial capital in the Netherlands that served as regional SD HQ. Now faced with an explicitly intellectual cat-and-mouse game with SD, he attempted to solve the conundrum before him. He requested to view all telegrams referred to in the letter and was shown several SD intercepted from Eindhoven involving Bobby van de Zeemanshuis. Bobbie knew the Dutch underground had already executed someone similar. He decided to accept the biggest challenge at risk taking and play the role of Bobby van de Zeemanshuis. Bluffing, he professed with great indignation, the traitorous manner in which the organization in Eindhoven treated him. He proposed working for SD against his people under certain conditions: release of DeBoer and his son without further prosecution; he receive treatment as an allied officer in prison; and punishment for the men who struck him in the face.

Having convinced them of his sincerity, SD also put forward conditions; demanding to know everything about his organization and assistants in Holland; he not reveal to OSS his now aiding the SD; and he would be executed if found lying.

After interrogation, SD transported him to a prison. Next day, 11 February during interrogation, he gave details of the BBC code, rules for concealing coordinates, and one of the two security checks for operating the Joan radio – the spelling of his name. He did not explain the lights to signal the aircraft, nor the Joan "swearing'"security check. He informed SD of a scheduled contact set for 12 February in Zwolle to provide information for a new drop zone. Confirmation could be obtained by listening to BBC radio messages sent out that day.

Late 12 February, he arrived at SD HQ in Zwolle. The SD again warned him not to play a double game and promised to meet all previous demands if he proved honest and sincere. Everyone then climbed into two staff cars and drove off to contact the OSS aircraft.

Bobbie successfully established contact with the Mosquito. Two English-speaking SD men stood nearby to overhear the conversation, though unable to listen to the Mosquito reply. To protect his friends, he gave new coordinates 10 kilometers south of the previous drop area; was unprepared to receive more material (more friends); and wanted only packages, repeating "packages only" several times during the conversation. Bobbie refrained from using the duress signal because he deemed expediting the packages as imperative in establishing his credibility and sincerity as a "turned"agent. After completing the contact, the SD detained him in prison at Zwolle without further interrogation.

On 16 February, SD staff summoned him to the interrogation room. Before him lay a container with supplies spread on a table. The previous night delivery proved successful. The B-24 pilot flew a wrong direction diagonally across the signal lights. Bobbie stood astonished the delivery succeeded so well because he failed to provide the correct signal letters for SD to flash. In appreciation, he received a piece of chocolate and carton of cigarettes as reward. As a result of the success, SD released both DeBoer members. The post-mission conference lasted three hours ending after midnight whereupon he returned to his cell, now in good grace with the Germans.

On 17 February, SD notified Bobbie that a recently received BBC message scheduled another contact that night. They provided written instructions in German of specific details to communicate, comprising accurate military Intelligence of a general nature. They expressed hope this ploy would incite OSS to request additional information, thus allow SD to determine the area of Holland the Americans expressed interest for details.

He accompanied SD officers to the previous contact point. The Mosquito arrived on schedule. Bobbie confirmed the successful supply delivery, then proceeded to verbally communicate the prepared SD message. He withheld the duress warning signal because he felt the SD were not completely convinced of his sincerity. He requested another B-24 supply drop of personal items, including a wind jacket, rubber boots, and so on. Two days later, the SD provided Bobbie with an inspection tour of military installations as visual proof they indeed conveyed accurate information to OSS.

Another contact 20/21 February arranged a supply delivery the following evening. At 2100 the SD transported Bobbie to a new location for the expected drop. Twenty SS men fully armed with submachine guns surrounded the drop point. Bobbie stood between two SD officers on the edge of the field while four men in civilian clothes held the signal lights. The rendezvous plane never arrived. BBC announced another attempt the following night at the same time and the whole procedure repeated. Again no aircraft appeared.

The BBC broadcast a third contact for the 24th. SD summoned Bobbie to their office for a conference. They had become suspicious because the deliveries never materialized, even in weather considered good. SD wanted to know: 1. Why the drops were not made. 2. If the OSS suspected Bobbie had been arrested. 3. He was to assure them he was perfectly all right. The SD prepared an extra headphone set that plugged into the Joan unit to monitor communications. Bobbie decided on 28 February to use the duress signal.

The Mosquito arrived as scheduled. OSS explained they postponed previous attempts because of unfavorable weather. It was at this point Bobbie gave the duress signal, agreeing that it was "BLOODY bad weather." As luck would have it, the spare earphone malfunctioned and the previous conversation not monitored. It became common practice for an SD man standing next to Bobbie to use one earphone from time to time and audit the conversation. Fearful that reference to "friends" would reveal future agent insertions, he replied that his friends in the north had been arrested.

SD provided Bobbie a new pinpoint for next delivery. He calculated and enciphered the coordinates, listing the correct minutes but erroneous degrees, placing the drop area off the Dutch coast in the North Sea. This glaring error would serve as a distress signal – his mission had been compromised. SD failed to verify the coordinates and allowed him to continue with his communication. The rendezvous was a success.

A long period of inclement weather followed. Everyday for the next two weeks, Bobbie underwent intensive interrogation with questions covering the gamut from inquiries about his personal life and training as an agent with American and British Intelligence organizations, to the Dutch underground and personalities in the Netherlands government in London. Only Bobbie and the interrogation officer, who sat at a typewriter taking down details, were present during the sessions. At first, the questioning was purely impersonal, but after a few days, both became quite friendly, and with an easy and reassuring manner, the interrogator told many stories of his Intelligence exploits.

On 10 March, OSS schedule another JE contact. A casual remark during briefing alerted Bobbie of an attempt to track the aircraft by radar. SD wished to discover if other agents were using JE. He must warn the Mosquito. The SD provided more military Intelligence to read and requested another supply drop.

As planned the Mosquito arrived. Recognizing Simpson's voice, Bobbie swore at the beginning of contact. He received Simpson's response, "I understand. What can I do for you?" to which he replied, "Try to have another damn drop at the place I mentioned to you last time. But be careful, they are trying to engage you." Hopefully, the last sentence would warn Simpson that radar plotted the Mosquito's course. Bobbie managed to open an electrical circuit in the battery case without SD being aware, thus abruptly breaking off contact. He then repeated the prearranged information knowing Simpson would not receive it. Following the successful contact came the usual SD debriefing.

SD became suspicious when BBC messages arrived early, rather than late evening. Bobbie now confronted a dire situation requiring his powers of persuasion. He underwent a third series of interrogations and received two political theory books to read. The sessions evaluated and determined if he was pro-Nazi and his trustworthiness as a turned-agent. Now pitted against his enemy in a test of skill and strategy, he mustered forth all his intellectual talents and rather cautiously proceeded to provide a convincing display only a turned-agent could render.

The next JE contact was crucial for Bobbie because SD doubted his sincerity. On 30/31 March, with a sigh of relief, he establishes communication with Simpson in the orbiting Mosquito. The contact demonstrated to SD satisfaction that Bobbie was both reliable and dependable. After 20 seconds he interrupted the Joan radio function and broke contact. He continued the mock transmission reading the prepared material while SD looked on oblivious that nothing was sent. The contact saved his life.

SD transported him to facilities at Groningen, Leeuwarden, Alkmaar, and finally arriving at Scheveningen on 15 April. SD issued notification of his release from custody within a few days. When asked his future intentions, Bobbie requested permission to cross the frontlines and collect Intelligence as an SD double agent. Agreeing to his request, SD formulated plans for a cover story with necessary identification papers and passes. Two days later they provide temporary accommodations in The Hague for official release. In the interim, checking that no one followed him, Bobbie cautiously proceeded to a friend's home and reunited with friends and former Dutch government officials.

SD issued Bobbie identification papers and a special pass to cross restricted battle areas. Arriving in Utrecht, SD drove him to Veenedaal to cross the frontlines. On 3 May, an SS officer accompanied him through German minefields and pointed the way toward the Allied lines. He gave the SS officer the special pass for return to SD HQ.

Waving a white handkerchief, he surrendered to a Canadian soldier and taken to 1st Canadian Army HQ, then transferred to British Army Intelligence at Ede. The British placed him under arrest charged with collaborating with the enemy. While investigating his story, German forces in Netherlands surrendered 5 May. OSS officials arrive 11 May and obtain his safe release. Postwar he graduated from Yale University.

TROY

Agent Rene Carle parachuted in to the region of Gechingen, Germany, the night of 21 January 1945 from a Lyon-based 492 BG B-24 aircraft. Equipped with two Joan sets, he received instructions to settle in Pforzheim where he developed and established many excellent Intelligence resources. Attempted contacts by Mosquito on 29 January, 1, 3 and 7 February proved unsuccessful. He finally established a 45-minute JE contact 13/14 February and communicated general Intelligence information including rail traffic through Pforzheim and coordinates of an aircraft factory at Calw. An RAF bombing attack on Pforzheim the night of 23/24 February created a blazing inferno with enormous loss of life. During bombardment, Rene's two Joan sets were destroyed along with personnel belongings and a substantial sum of money. Although isolated from OSS, he continued gathering Intelligence surveillance. When American troops finally arrived in Pforzheim, he provided valuable information on enemy troop positions in the region.

ANZIO

A Lyon-based 492 BG Liberator successfully dispatched ANZIO east of Esslingen, near Stuttgart, Germany, 28 Feb/1 March 1945. Agents Bob and Emmy landed safely and packed essential provisions and a Joan set in a rucksack before burying the canister with remaining supplies and reserve Joan unit. Early that morning en route to Buch, Emmy confessed to Bob that the Gestapo issued arrest warrants against her and she refused to enter the town. The unexpected news force the team to separate with plans to rendezvous later. As Bob proceeded to Buch, several civil security members confronted and escorted him to the local police office. Previously he concealed the codes, crystals, and Joan set wrapped in a paper package addressed to a German firm. En route, he distracted the guards and threw the package over a brick wall of a nearby house. Any incriminating documents he disposed in nearby bushes. At the police station Bob provided a convincing account with his cover story. After a brief interrogation the authorities released him.

Emmy and Bob later met and obtained temporary accommodations with her aunt in Gaggenau. Late that evening, Bob returned to the landing zone to retrieve a buried suitcase, then went to Buch and recovered the package with Joan set, and began searching for the disposed documents. He found all except the Decode Sheet. Children playing nearby found the paper and notified the mayor. This unexpected revelation compelled the team to flee by train to Stuttgart arriving the 5th. They contacted a mistress of the city's mayor but the Gestapo monitored her every move, forcing the team to leave the following day. They eventually arrived in Gent where a friend provided temporary living accommodations.

Mistakes marred Mosquito communications. A 7/8 March contact proved unsuccessful, and on the 9th, Bob heard "Hallo Paul, ici Jacques" instead of "Hallo Willey, heir spricht Vic." When attempting to correct the error, the aircraft failed to reply. Next day, no contact was made though Bob heard an aircraft drone high above. OSS files list no scheduled JE flights to ANZIO for either 9th or 10th. He may have accidentally listened to the Mosquito communicate with another agent in the area.

After Bob retrieved the reserve Joan set buried in the woods, the team proceeded to Lorelei, east of Stambul. The 23 March contact failed, so the team returned to Gent to monitor BBC broadcasts for further instructions. Meanwhile they recruited numerous civilians to assist in espionage activities.

JE communications continue to plague the mission. Bob made contact 29/30 March and heard, "Hallo Paul, ici Jacques" and replied in German, "Hello Vic! Hier spricht Willey. Kommen sie naeher Gagenau," but received no reply. Signal attempts on 30 March proved fruitless because OSS did not schedule a JE mission. And when the Mosquito arrived the nights of 31 March, 1 and 9 April, they receive no agent response. It appears OSS failed to coordinate Mosquito schedule and rendezvous times with ANZIO.

When BBC radio failed to provide rendezvous messages, Bob infiltrated the US Seventh Army front and reported to OSS. Emmy traveled to Ulm and joined another OSS associate patiently listen to BBC for further instructions. Finally, Emmy and the agent crossed the frontlines northwest of Stuttgart on 12/13 April. They surrendered to French forces and later handed over to OSS at Heidelberg.

HAMMER

OSS developed the Faust plan to insert thirty agents working in teams tasked with contacting dissident underground groups. Originally this required introducing agents peripherally across enemy lines bordering France and Switzerland to friendly reception committees assisting with transportation to larger German cities. Unfortunately the rapid advancing Allies continually moved the frontlines towards Germany and overran the reception committees. To overcome this handicap OSS decided to insert agents blindly deep inside Germany by aircraft. The Watton OSS detachment brought these plans to fruition with A-26 aircraft development in this role.

The first priority involved agent insertion in Berlin under the HAMMER project. The proposal involved contacting underground groups in Berlin, secure and transmit Intelligence via JE communications,

L-R. 1/Lt. Ralph Smith and Lt. Marvin R. Edwards next to Mosquito Patty at Harrington. Is this aircraft all-black? *Photo M. Edwards*

L-R. Lt. Edward P. Kolacke, navigator, and 1/Lt. Jerry Kuntz, pilot, in front of an all-black Mosquito? *Photo M. Edwards.*

and arrange the reception of agents. The two men chosen had extensive experience in anti-Nazi underground work before fleeing to England in 1938.

On 2 March 1945, the HAMMER team of Phil Land and Toni Reu successfully parachuted at 0215 from an A-26, 50-kilometers northwest of Berlin near Alt Friesack. Collecting the canisters they divided the supplies into separate food, weapons, and radio equipment caches, buried in three different locations. The men set off to the nearest railroad station concealing only one Joan set on their person.

After trekking three hours, they arrive near dawn at Grieden, a small station on the main Berlin line. Railway officials briefly examined their papers. Phil posed as a German toolmaker deferred from military service for war essential work; bombing attacks destroyed his factory in Koingsberg and the labor exchange directed him to Berlin. Toni posed as a Czech welder sent by the Breslau labor exchange to Berlin. They found it advantageous if Toni, as a Czech, pretended not to understand German, thus allowing Phil to speak on his behalf. They continued this ruse whenever asked for identity papers.

Late that evening, the men stepped on board an overcrowded passenger train. The blacked-out windows kept the compartments in total darkness as the conductor moved through the crammed coaches with a lamp, selling tickets. For security purposes the men avoided purchasing passage, thus eliminating evidence of where they came or were going; and remained on board until the stop past their intended destination before disembarking.

The train arrived in Berlin near midnight 2 March. A once thriving city now comprised a vast landscape littered with skeletons of buildings. The men trudged through darken streets surrounded by a landscape of rubble and ruin. Phil knocked on the door of his parent's residence who enthusiastically welcomed him. He acquainted them and a married sister with his secret mission. They provided the men living quarters in a nearby empty dwelling in their care. The agents remained here four weeks until Toni changed residence to another section of Berlin.

The team spent the first week acquainting themselves with social conditions including how to live and behave like Berliners. The men avoided registering with the police to prevent labor conscription; forced employment allowed little opportunity for Intelligence work. Stringent military controls conscripted males of all ages into the *Volkssturm* to defend the city. Berlin officials alerted citizens observing men between ages 16 and 60 years wandering with no apparent purpose, to report them to the nearest police station immediately. This mandated creating a cover story to allow the team free movement about Berlin. Therefore, they always carried rucksacks packed with dirty clothes to appear as having recently arrived in the city.

On 8 March the team returned to the Alt Friesack to retrieve half the equipment, weapons, and food supplies. A thorough survey of the surrounding area proved the pinpoint excellent for unobserved delivery of both agents and canisters. Unfortunately, an incorrect train schedule forced the men to hike 50 kilometers the second night. They arrived in Berlin late evening 10 March, tired and exhausted as RAF Mosquito bombers attacked the city.

The first prearranged Mosquito contact took place 12/13 March. During discourse the Joan set malfunctioned preventing further communication. A thorough examination revealed a damaged antenna and repairs rectified the problem. The agents utilized a German gasmask container to conceal the Joan radio.

OSS provided HAMMER a list of anti-Nazi sympathizers to contact, but unfortunately most perished during the war. Toni recruited his sister who provided information on one sympathizer who became an active confederate, investigating pinpoints for future JE rendezvous. Phil recruited his former teacher who agreed to furnish special reports on conditions and troops in Wensdorf; data on a Wehrmacht camp southwest of Berlin; enlist a physician for medical emergencies; and secure living quarters for Phil or other agents. They successfully achieved all objectives. Several weeks later, Phil's son-in-law arrived from the eastern front on leave. He too volunteered to assist.

OSS failed to anticipate daily Allied air bombardment hindering HAMMER operations. The only functioning transportation available, the streetcar system, broke down after each attack. On several occasions this prevented the men traveling through the city to reach the designated location for JE contact. Unwittingly, OSS scheduled contacts on weekends when limited train service made transportation difficult to secure. A coded BBC radio message scheduled contact for 18 March at a secure pinpoint two miles south from the original contact site. Early dawn light prevailed at the 0400 rendezvous time. Phil received the Eleanor calling tone on his Joan unit but failed to establish verbal communication with the aircraft.

The immense destruction by Eighth Air Force bomber formations on 18 March attacking rail centers, armament plants, and an industrial facility, set Berlin ablaze. The raid also destroyed the team's fixed address, utilized as a "letter box" by collaborators to drop-off and pickup reports. An RAF Mosquito night attack the 20th, disrupted the urban tram system and stranded the team in northern Berlin until midnight, thwarting Mosquito rendezvous efforts. Both day and night bombing attacks prevented the men reaching the rendezvous area 26 March for JE contact.

By the end of March the agents succeeded in locating two new rendezvous points, a new letterbox, and recruit two more sympathizers. Finally they successfully contact the JE Mosquito 27/28 March communicating Intelligence information for 25 minutes. The Mosquito set 31 March as next contact.

Conflicting BBC messages with rendezvous instructions for two different days at different points confounded the agents. No trains operated past one location on the appointed day. The agents incorrectly assumed OSS scheduled two contacts and chose to meet the original 31 March schedule. They waited past midnight but no aircraft arrived. It soon became apparent the BBC message for 2 April contact was correct. As dawn approached, the team sought concealment to rest. In the gaining light, pitched tents slowly began to appear around them. Reconnoitering further, more tents with troops came into view. Unknowingly, they spent the night in a Wehrmacht bivouac area. After skirting several sentries they discreetly exited the pinpoint, but not before recovering urgently needed food and supplies from buried caches. The men transferred the dirty clothing to Toni's rucksack while packing Phil's with supply items.

They trekked through the woods towards a railroad station 20 kilometers distant. A half-hour later, upon reaching open terrain they encountered German troops constructing defensive positions. Continuing unmolested another half-hour, a young lieutenant on a passing motorcycle demanded their identification papers. Phil again acted as spokesman. The lieutenant returned their papers, pondered a moment and demanded to examine the rucksacks. Toni began emptying the odorous dirty laundry, waving it in the officer's face. Phil gripped a hidden revolver with one hand while showing his Nazi Party membership card with the other. Quickly satisfied, the lieutenant waved the team on their way.

En route, frequent roadside document inspections continued until they approached a crowded crossroads where armed SS troops examined every vehicle and bicycle. Overcome with despair, the men confronted their greatest challenge. They composed themselves and cautiously proceeded forward, slowly walking past the distracted SS men preoccupied with examining papers from a crowd of civilians. The agents safely escaped notice and passed through the checkpoint. With a sigh of relief they continued to a railroad station, and after hiking 80 kilometers in 36 hours without sleep, entrained for Berlin. That evening a BBC message scheduled a supply drop at 0300 the following morning, 2 April. Current circumstance prevented HAMMER meeting the rendezvous.

Advancing Soviet armies slowly closed around Berlin resulting in new German defensive positions encircling the region. This required locating new JE rendezvous points closer to the city. The men found a field near Wintermark plus another at Eichwalde, south of Berlin. Unable to comply with a BBC message to rendezvous at a compromised area, the team decided to signal the aircraft from Wintermark. Still within JE radio range, they planned vectoring the aircraft to their new position. On 8/9 April they contacted the Mosquito. Both distance and weather conditions interfered and the aircraft slowly lost radio transmission with the agent.

To overcome travel restrictions and hazards reaching the rendezvous area, the men planned contacting the Mosquito from within Berlin. This involved lengthening the antenna to extend through a rooftop of a high building to communicate with the circling aircraft. Though they acquired the necessary equipment the plan never reached fruition. Locating a structurally intact and secure tall building proved elusive.

An RAF air raid on Berlin delayed Phil's departure on 15/16 April. Lack of public transportation forced hiking to the rendezvous point, arriving 30 minutes late. Toni arrived earlier, heard the circling aircraft above, but Phil had the radio. Meanwhile the Mosquito experienced instrument and electrical failure in the cockpit with loss of intercom and aborted the mission. During next contact 18/19 April near a forested area, HAMMER heard only the Eleanor calling tone. The agents disobeyed a JE cardinal rule – communicate only from open fields – avoid nearby structures and wooded areas. The Mosquito left without making contact.

Hammer attempted a final contact late night 21 April. Toni lived with friends in northwest Berlin that now endured heavy Soviet artillery fire. After Phil arrived both pedaled their bicycles to the pinpoint. Pandemonium reigned in the district as evicted civilians vacated dwellings and Wehrmacht troops dug in. The men remained secluded in a forest near the field to await the Mosquito's arrival. Unknown to them, mechanical failure caused the aircraft to abort the mission.

When no aircraft appeared the men left for Berlin, but a mile away encountered loud crashing sounds in the pitch darkness from adjoining forest areas. They hid among trees as unidentified tanks churned and ploughed through the surrounding woods and underbrush for fifteen minutes heading in a westerly direction. Soon afterward when heavy fighting erupted they realized the armor force comprised a Soviet spearhead advancing towards Berlin. They safely pedaled to Berlin as Soviet forces overran the pinpoint preventing further Mosquito contact.

CHISEL mission agent Karl Grimm, KIA on 19 March 1945 mission. *Photo NARA*

On Sunday, 21 April, a coded BBC radio message directed one agent to crossover to American lines while the other remain in Berlin for U.S. troop arrival. This proved dangerous: any male not participating in defending Berlin was subject to indiscriminate execution. Next day, German army defensive positions west of the city thwarted Phil's attempt to reach American lines. Meanwhile the JE Mosquito arrived as scheduled without making contact. Had OSS provided the team a wireless radio to communicate with England, the contact rate may have increased dramatically.

On 25 April, Toni and Phil surrender to a Soviet command post in Johnnistel, a suburb south of Berlin. They reveal their identity as American OSS agents on an Intelligence mission and request the U.S. Army be notified. The Soviets strip them of watches and pens, then transport the men by vehicle to a headquarters south of Berlin for interrogation. Incarceration followed until transfer to Dresden. On Saturday 16 June, they were driven to a bridge over the Mulde River near Grimma and transferred to American guards.

CHAUFFEUR

The first JE agents equipped with both JE and W/T radio prepared for dispatch. The Belgium team of Michel Dehandschutter and Andre Renaix left Dijon at 2200 hours, 31 March 1945, on board a B-24 for insertion near Regensburg. Following container release, both men jumped immediately and landed safely near Eining, five kilometers from the original pinpoint. They hauled both canisters to the nearest forested area, remove all supplies and hid the containers in thick underbrush. Now approaching sunrise, they remained in the woods until noon, then proceeded to reconnoiter the area.

The following days they roamed the surrounding region including Neustadt and Abensberg, seeking sleeping accommodations and to establish a base for operations. Fleeing refugees overcrowded the towns and villages eliminating prospects of locating living quarters, forcing them to sleep in the forest at night. The agents befriended several Belgium POW laborers for future assistance. Andre attempted contacting London by W/T but the battery became weak. A French POW in Straubing had it recharged. On 11 April, Andre began dispatching Intelligence to London by W/T when the battery again failed.

From 4 to 14 April they remained at a forest retreat but maintained daily contact with several POWs in Straubing, who communicated with a POW in Abensberg, through an intermediary driver. In this manner the agents continued to receive regional Intelligence information.

Michel traveled to Abensberg and met several Belgium POWs employed at a local dairy. He confided his secret mission and desperate need for electricity to dispatch coded W/T messages. The workers agreed to assist. CHAUFFEUR transferred their equipment to the dairy cellar.

The team visited Regensburg to recruit sympathizers. Michel met two young French women who labored in a brothel. Local police ordered the harlots to report deserters or politically unreliable individuals and visited the women regularly to obtain this information. Many Nazis and military personnel frequented the brothel. Michel confessed the true nature of his mission and asked if "they would work for France." Both eagerly volunteered to assist. He remained several days, enjoying their hospitality. Secretly hiding in a cupboard, he listened attentively during the amorous sessions with military and local officials, noting information of Intelligence value. Though both women acted in goodwill, he provided a small monetary stipend and cigarettes for barter.

One day a police official ordered the harlots to obtain information on Michel. The situation now became too dangerous for him to remain. He quickly departed promising to return in a week. The women proved a valuable source for reliable Intelligence on military installations, troop movements, and both Nazi and anti-Nazi citizens in Regensburg.

Andre received a BBC message on 19 April announcing a JE rendezvous. That night he successfully contacted the Mosquito and detailed Intelligence the next hour. The following night he operated the W/T radio from the attic with antenna extended from the roof. Immediately he established communication with England and began transmitting and receiving coded message groups (QSL) the next two days.

On 23 April, near midnight, Andre established JE radio contact with the Mosquito. A few days later, a BBC message directed the men to evacuate to Landshut, but advancing American forces blocked roads leading south. This prevented Mosquito contact the 27th. US Third Army entered Abensberg next day and CHAUFFEUR surrender to G-2 as OSS agents.

PICKAXE

On 4 April 1945, a B-24 left Dijon for Landshut, located between Munich and Regensburg in Bavaria. Prior to dispatch both men and Liberator crew studied a regional map for route orientation and discussion. PICKAXE agents Bill and Eric jumped blind and landed safely. The agents spent all night groping in darkness seeking each other and the supply container. Finally at dawn Eric located and hauled the canister to a heavily forested area. He remained secluded that day while pondering how to locate Bill. Remembering the pre-mission map briefing, he set off the following day to the JE rendezvous pinpoint. En route, he converge with agent Bill proceeding in the same direction. The agents discussed future plans and returned to the original landing site. After burying the canister they proceeded to Landshut.

Once a city of 25,000, Landshut doubled in size, overcrowded with fleeing refugees, foreign workers, and soldiers. Local authorities controlled all housing, and available accommodations were virtually non-existent. Regulations required foreign workers to register weekly at the police or labor exchange in order to receive ration cards. The authorities either assigned work on local farms or transferred laborers to other cities; both options restricted Intelligence gathering activities. The agents chose to avoid registering and instead, establish a residence in a nearby forest as a base for operations. Bent saplings and twigs, covered with raincoats formed a living and sleeping enclosure. The woods surrounding Landshut supported refugees and local citizens who slept in the forest nightly in fear of RAF bombing attacks. Thus the agent's presence inconspicuously blended in among the populace.

Technical problems force OSS to scrub the 6/7 April rendezvous. PICKAXE established contact 8/9 April; one agent spoke while the other maintained surveillance. The Mosquito interrupted communication and asked him to hold, while contacting LUXE. PICKAXE heard the Mosquito operator quite clearly conversing to an agent in French and listened to the conversation. On 13/14 April, PICKAXE failed to correctly align the Joan antenna towards the aircraft. They finally achieve successful communication relaying Intelligence information.

L-R. 1/Lt. Samuel Webb pilot and 1/Lt Edward P. Kolacke, navigator. *Photo M. Edwards*

Landshut harbored a major military medical center for recuperation of wounded troops. After recovery the men transferred to a training and replacement battalion. Trucks transported several hundred combat-ready troops nightly towards Munich. Landshut contained a major railroad junction. An Allied air attack 16 April totally destroyed the tracks shutting down all rail movements northward. On 17/18 April, the Mosquito requested further train traffic Intelligence but poor agent antenna technique disrupted communications.

OSS trainers at Harrington failed to permit PICKAXE practice to initiate aircraft contact. Only after the instructor established communication with the training aircraft did he allow the agents to converse. This lack of practical application hampered all future operational PICKAXE communications.

Near midnight 20/21 April, Hitler Youth members were seen entering the nearby woods prior to JE rendezvous. The agents trod several hundred meters distant before establishing JE contact. On 25/26th the Mosquito petitioned the agents to move their operations to Rosenheim but lack of rail access and stringent military controls prevent transfer. Landshut resigned itself to defeat and the *Volkssturm* lacked motivation to man defensive fortifications; they resented Gestapo officials labeling them cowards.

On 30 April/1 May, the agents found themselves stranded between American tank fire and opposing German artillery batteries dueling near Tiefenbach. That night poor antenna technique resulted in a garbled JE contact.

When American tanks rolled into Landshut, the agents surrendered and reported to US Army G-2. PICKAXE returned with LUXE to OSS HQ, St. Germaine, Paris, France. The agents bitterly complained of not being allowed sufficient time to exchange Intelligence before the aircraft broke contact.

LUXE

The LUXE team of Leon and Fred parachuted from a B-24 on 5 April 1945 at 0235 near the village of Raisting in southern Germany. They landed in a heavily wooded area and Fred's canopy caught in a high tree. After cutting the shroud lines, he tumbled to the forest floor inflicting numerous injuries including an abdominal rupture and sprained ankle. Leon assisted him to a nearby meadow hay shed. As daybreak approached Leon dragged both containers into nearby woods. Overwhelmed by exhaustion he relaxed and immediately fell asleep.

Early that morning Fred awoke in the hay shed. In the far distance an ox team appeared, driven by a 17-year-old Ukrainian farmhand who cultivated acreage for an elderly couple in Raisting. Using a stick as a crutch, Fred hobbled towards the lad and introduced himself. With the parachute canopy conspicuously draped from a nearby treetop, he had no option but confess he parachuted from an American aircraft. The youth became overjoyed at hearing the word "American" and declared his employer an anti-Nazi.

A woman approached from far-off across the meadows. Fred awaited her arrival as the young lad began working his oxen in the field. After greetings, Fred engaged the Frau in conversation to determine her political persuasion and views on the war. Ruefully she responded the war should end tomorrow and reviled the Nazis responsible. When she asked why Fred was in her field, he replied having parachuted from an American airplane to end the war. Pointing to the parachute she urged its immediate removal before local police discovered it. Both she and the farmhand slowly and steadily pulled the canopy off the branches and buried it.

The Frau and young farm worker returned to Raisting. After observing Fred all day, Leon appeared late evening and both shared the latest events. Mechanical failure at Harrington aborted a scheduled Mosquito rendezvous on 6/7 April.

The agents initiated plans to establish a viable espionage network. Three days later the young farmhand guided Fred to Raisting to the Frau's residence. Fred sought her assistance in recruiting sympathizers. She left briefly and returned with her brother, Hans Huttner, the world's champion shot-putter in the 1936 Olympics. A mechanic by trade, he served as local *Volkssturm* leader. Fred explained his mission and warned that should harm come to LUXE, the Americans would bomb Raisting in reprisal. Huttner suggested a meeting next evening at the hay shed. The youth led Fred back to the meadow where the agents discussed what transpired.

Late night of 8/9 April, LUXE established JE contact and communicated their activities. Leon ended the transmission pleading delivery of a truss for Fred's rupture. Next day Huttner arrived and announced enlisting two relatives. Leon accompanied him to Raisting and met with the village blacksmith and a tavern owner. Leon stressed the importance of disrupting railroad traffic, thus hindering cattle export from the area. Raisting would remain safe from American bombing if they cooperated. Passwords were issued and each promised to enlist additional men.

LUXE establish headquarters in a church at Unterstillern. That night the volunteers transport the agents' supplies in a hay wagon through Raisting to the new location. Meanwhile, LUXE carried the Joan radios cross-country through the woods. During the move they miss a Mosquito contact, 10/11 April.

A local family provided haven to a German Army deserter named Sommer. He joined the collaborators and recommended two prominent Landsberg men. Two days later, Sommer drove Fred to Landsberg to meet the gentlemen. They promised to recruit the town garrison commandant and reported the mayor issued supplies to Heinrich Himmler's sister living in Landsberg. Fred requested more information on the SS leader's sibling.

The agents displayed innovative skills contriving a special cloak for JE communications. To hasten concealment during contacts, LUXE built a special cover from a rubber sleeping bag. The Joan antenna passed through an opening on top, around which a blanket covered the hole to muffle any conversation. Under this shroud, with flashlight hanging from his neck, LUXE could read off written communications without being seen. A Mosquito made an attempt to deliver a truss on 13/14 April but LUXE decided to forgo the rendezvous for security reasons. The truss would accompany next agent delivery. During the day Leon visited Weilheim to investigate an aircraft factory producing replacement fighter and bomber components. LUXE communicated this Intelligence on 17/18 April. The agents request that warning leaflets be dropped prior to any Weilheim bombing attacks to spare civilians from injury.

Fred volunteered to accompany the blacksmith to Munich for monthly work supplies. Next morning 19 April, he and Fred left Raisting by truck. At Pasing near Munich, they pause to observe a USAAF bombardment. Continuing to Munich, they load the supplies then perform a grand tour inspecting everything of a military nature. Late that evening they arrive in Raisting and Fred returned to the church. That same afternoon, Leon visited Weilheim as air raid sirens sounded and P-38s approached at low-level. The bombs destroyed several factory buildings. Forty minutes later, a second wave struck a railroad depot.

Next day, Fred arrived in Landsberg to meet both mayor and garrison commandant. LUXE demanded the Lech River bridges not be destroyed; garrison commandant surrender to U.S. troops arriving in Landsberg; deliver all Nazi party members to the Americans; and finally, all food supplies remain in town. In return, Fred guaranteed sparing Landsberg and garrison troops from American attack. Both officials agreed to all terms.

The volunteers announced the mayor expected SS leader Heinrich Himmler to visit his sister in Landsberg. Fred requested immediate notification of the arrival and ordered Himmler placed under surveillance. Next night, 20/21 April the agents established JE contact. LUXE details his eyewitness account of the Pasing railway station bombing and requests JE contact in two days for information on Himmler's headquarters. The men receive notification that LUXE II members would arrive soon.

Unable to cancel delivery, the collaborators prepare a reception four nights in succession. On the final night 25/26 April a Liberator discharged two agents and five containers over a forested area. The reception committee fanned out and worked laboriously in trees and thickly grown underbrush, eventually finding two canisters, then agents Jack and Erik of LUXE II. Near daybreak they locate the last of three containers by which time all nine members faced physical exhaustion. Besides assisting night drops, they all held regular daytime jobs. The sympathizers transported LUXE II to Etterschlag where a local farmer provided shelter and assistance.

News arrived that American POWs en route from Nuremberg to Moosberg passed through Raisting. That night LUXE visited Raisting and observed seven prisoners under guard. At dawn six scraggly POWs shuffled past while one remained behind, an AAF officer. Now ill and malnourished, the agents transported him to the church for safety. LUXE communicated a report of this POW to the Mosquito on 28/29 April.

On Sunday, 29 April, an American armored column advanced along the road from Landsberg to Weilheim. Fred rode ahead on a bicycle waving a white cloth. In poor English he asked for G-2. He proceeded to detail tactical Intelligence on a Weilheim region map, pointing to the castle where SS held Hungarian Regent Horthy under guard; location of Himmler's armored train; regional AAA defenses; church location with the ailing POW. Two hours later, G-2 issued orders to retrieve the POW.

After further interrogation, Fred accompanied two Intelligence officers to Landsberg to arrest Himmler's sister. Next day, he joined G-2 to visit a concentration camp near Landsberg. They found 400 smoldering bodies; survivors swathed in filth stated the camp commandant ordered the installation burned down. Later, Fred discovered the commandant disguised in civilian clothes hiding among German prisoners. G-2 transferred him back to the concentration camp. Here an American colonel summoned all Landsberg males to view the macabre scene. After a speech in German, the colonel forced the camp commandant to lay among the putrid corpses and ordered the Landsberg men to spit on him. The colonel placed the commandant in custody of the few surviving camp victims until the following morning. Next day, two Intelligence officers transport Fred via Raisting to Unterstillern to recover Leon.

Leon's Story

On Sunday afternoon, an American armored column pushed through to the church and retrieved the ill POW. They refused to transport Leon to G-2 and instead, confiscated his radio, weapons, and ammunition, leaving him behind defenseless. A German SS colonel in civilian clothing concealed in a nearby farmhouse viewed the proceedings. The following night a group of disguised and armed SS men approached the church. Lacking weapons for protection, Leon fled to Raisting. When he returned next day, the SS men had disappeared.

LUXE accompanied Intelligence officials to search Himmler's armored railroad train and heavily armed cars. While searching through the command coach, the G-2 officers discarded various papers into cartons. Fred and Leon later reexamined one box and found a bundle of single-use carbon paper. When trans-illuminated with a flashlight, a sheet read quite clear, "Personal Instructions of *Reichsführer-SS* Himmler, Directions for the Organization of the Werewolf." They brought this to the officers' attention and the documents immediately dispatched to HQ; the sheets contained vital Intelligence information. Next day, LUXE rode by Jeep to Augsburg and joined other OSS agents including LUXE II. OSS attempted though an Act of Congress to obtain U.S. citizenship for both LUXE members.

BUZZSAW

OSS selected agent Fred Wagner (student name) for Leipzig, Germany. A Dijon-based B-24 dispatched BUZZSAW on 7 April 1945. An OSS file only details his incarceration in a POW camp.

FARO

On 7 April 1945, a Dijon-based B-24 dispatched Karl Pfeiffer, known as FARO, to Plauen, Germany. The aircraft accidentally delivered him 14 kilometers southwest of the pinpoint on a moonless night. The agent landed safely in a meadow and spent till daybreak trying to locate the supply container. One Joan unit and regular radio with the supplies were buried nearby; the second Joan radio packed into a rucksack. He proceeded to Plauen by road as B-17 formations bombed the city. The attack totally destroyed the marshaling yards as uncontrolled fires raged throughout forcing a general evacuation. He joined Plauen refugees fleeing along the autobahn, while gathering Intelligence observations and sleeping on farms at night.

The first JE rendezvous 9/10 April proved unsuccessful. The second, 10/11 April at 0100 hours, he established communication and provided his observations.

FARO attempted to retrieve the regular radio to monitor BBC messages for JE rendezvous times. Unfortunately, encamping German soldiers thwarted accessing the buried cache. That night he luckily managed to retrieve the second Joan unit, but without the regular radio he could not listen for BBC messages directed to him. Unforeseen circumstances stranded him behind enemy lines. Even so, he continued roaming the region ever attentive to military activities. Unknowingly, he missed the 14/15 April JE contact.

That morning while strolling the autobahn, five German soldiers emerged from nearby woods and questioned him. He identified himself as a lost refugee and requested directions. Karl's left hand gripped a loaded pistol concealed in an overcoat pocket, prepared to shoot all five should they attempt to search him. Instead, they gave directions, recommending he avoid traveling west because of advancing American troops. After walking out of sight, he hid in a thick stand of trees.

An hour later shots rang out. As he emerged from cover to investigate, American troops in a Jeep shouted "Halt." In broken English FARO identified himself as an OSS agent on a secret mission. After providing the location of the five German soldiers, whom the Americans quickly captured, the GIs escorted him behind the lines to company headquarters. FARO explained to G-2 his covert activities and relayed his Intelligence observations. MPs escorted Karl to rear Division HQ where OSS officials greeted him.

MALLET

The MALLET project with 31-year-old Dolf Bach (Adolph Buchner) would be inserted on the Berlin outskirts. A previous delivery 3/4 April with A-26 #513 failed due to a solid cloud overcast covering the area. On 10 April 1945, A-26 #626 inserted the agent. OSS never achieved JE radio contact with MALLET. His file details confinement in a prisoner-of-war camp.

FARMER

The night of 16/17 April 1945, a 492 BG B-24 dispatched the FARMER team of DeWaals and Smeets near Straubing, northeast of Munich, Germany. The aircraft flew too high resulting in wide dispersion of both men and cannister. Smeets touched down near Hunderdorf in a marshy meadow near a dwelling. He began searching for DeWaals but located the supply container instead. After dragging it into the woods, he removed provisions and two valises, then packed the parachute inside and hid the canister in heavy brush. One valise with JE equipment was placed near a dribbling brook and camouflaged by covering it with

Andre Pacquet standing in front of Mosquito with crewman's' leg hanging from JE station in rear of aircraft. *Photo A. Pacquet*

OSS JE Mosquito undergoing refueling at Harrington. *Photo A. Pacquet*

mud; the other containing provisions, he buried in another forest area. Now tired, Smeets slept in the woods and next day continued the search for his companion without success.

Meanwhile, the wind carried DeWaals' parachute quite a distance, before landing in steep terrain and injuring his ankle. Excruciating pain limited mobility. He concealed the parachute among nearby trees and remained secluded till morning. OSS issued the team a contact address for assistance. The painful, swollen ankle required immediate medical attention. At daybreak DeWaals hobbled along the road to Windberg to seek sanctuary at an Abbey. Arriving late afternoon, he disclosed himself to a prelate, the OSS contact. The cleric wrapped the swollen ankle then introduced him to three Dutch workers. All members present agreed to assist OSS Intelligence activities.

Two days later Smeets knocked on the Abbey door and learned DeWaals had arrived earlier. The prelate reported the *Volkssturm* received SS warnings of enemy parachutists. The situation became too dangerous to conceal both agents; DeWaals may remain but Smeets must go.

After leaving the Priory, Smeets conducted a cursory walking survey of the region. Returning to the forest to retrieve the radio units, he discovered the hidden cache missing. An observant farmer noted fresh impressions in a clover field where Smeets had landed. Following the furrow trail from dragging the supply container, the German soon discovered the buried valise with Joan radios. The police were notified and SS troops now scoured the area. The FARMER team found themselves trapped behind enemy lines without radios for Mosquito contact: 19/20, 23/24, and 27/28 April.

Smeets obtained employment on a farm near Hunderdorf. Three days later DeWaals joined him, and both recruited local and town officials wishing to surrender. The team continued gathering Intelligence on military fortifications and troop encampments. When U.S. Army forces arrive, FARMER provide G-2 with maps detailing Straubing defenses, antitank lines, mined bridges, communication units, SS and Nazis officials. An OSS field team later retrieved both agents.

BOYARD

Agents, Antoine Claude (Louis Leopold) and Roger Stockley (Roger Bohrer) left Dijon on 24 April 1945 at 2200 hours GMT by B-24 and parachuted southeast of Bregenz. Claude came down near a dwelling while Stockley landed on top a barn roof. After hauling the supply container to a hay barn, they packed their rucksacks with money, rations, radios, and clothing, slung the packs on their back, then carried away and buried the canister. Returning to the barn for the remaining supplies, they found several civilians searching through the material. Startled, the agents immediately fled in a westerly direction. Consulting a map, it soon became apparent the aircraft inserted them in a different area than planned.

Walking through a dark valley, they followed a railway line bordering a mountain stream called Aachs. Claude's rucksack contained the radio sets while Stockley's held clothing and rations. Several hours later when approaching a small railroad station five kilometers from Bregenz, they unexpectedly encountered an SS patrol. The soldiers immediately arrest and escort the men to a nearby house to examine their identification papers. After collecting the rucksacks and passports, the SS locked both agents in a small bedroom under guard, awaiting arrival of a duty officer.

Gripped by fear, Claude banged on the door to gain the sentry's attention. When he entered, Claude bolted out the doorway. When the sentry attempted to seize him, Stockley grabbed the guard from behind and slit his throat with a knife. Bleeding profusely, the wounded soldier collapsed to the floor. As Stockley dashed out, he grabbed a rucksack on the corridor floor and followed Claude through the rear door. Both men fled into the forest, ducking and dodging through trees, fleeing further in the darkness. Fully exhausted, they stopped to rest and remained secluded all day and the following night. Unfortunately, during escape Stockley grabbed the rucksack containing only clothing. They were trapped behind enemy lines without radios for JE communication. The night of 29/30 April, a Mosquito contact failed because the agents lacked Joan radios.

Andre Pecquet sitting on top of Mosquito nacelle. *Photo A. Pecquet.*

The following day they learn that Germany had opened the Swiss frontier to French, Dutch, and Belgium citizens. A French worker provided them with an evacuation order. Sequestered among throngs of Frenchmen, the agents cross in to Switzerland and board a special train to Zürich. Next day they journeyed to Annecy, France, and present themselves to the French Securite Militarie, who transfer the team to OSS on 7 May 1945.

SULTANE

The team of Vogel and Deville parachuted from a B-24 at 0340 on 25 April 1945 near the Augsburg-Munich railway line in Bavaria. Vogel landed next to a railroad track with three parked trains guarded by troops; Deville touched down on the Augsburg-Munich highway. After gathering and burying the parachute equipment in a forest area, the agents began searching for the two widely dispersed canisters. As dawn approached, the men hauled both to the nearby forest.

Opening the canister they began removing radios and essential supplies when unexpectedly a young girl appeared among the trees. Dense saplings partially obscured her view. She then turned and departed. Deville slowly followed behind and observed her converse with three German soldiers. Securing their weapons the agents maintain constant lookout on the surroundings and buried the supplies in caches. Near noon, a dog began barking, forcing them to silently move deeper into the forest. Next, a voice urged the dog forward, so SULTANE drew their weapons. The animal approached, stopped, and after an outburst of barking, turned and rushed away. They waited an hour before slowly retreating through trees and undergrowth away from the road, and later slept till morning.

The following day the men went to retrieve both JE and W/T radios from the caches. They heard a dog bark and quickly fled the scene sprinting through the woods. Too dangerous to return for the supplies, they hiked further to a convent. Introducing themselves as impressed French laborers, the men beg the mother superior for shelter and work. Instead, the nuns provide a small meal then urged both men to leave. They spent that night in a forest.

Next morning while passing near Gunzelhofen they met several French POWs. Overcome with hunger, SULTANE prevailed upon the POWs for assistance, identifying themselves as American secret agents and that the enemy discovered their buried equipment. The POWs willingly volunteered to provide shelter and food. Fitted in POW uniforms, the agents hid under straw in a granary above the prisoners' quarters. That night a heavy rainstorm provided cover as the POWs retrieved some of the buried equipment.

Early Saturday morning 28 April, a large SS contingent of 14- to 16-year-old conscripts armed with machine guns entered the village. Standing at attention the youths pledged to die for the Fuehrer, then began searching houses for young men hidden by family members. On Sunday a U.S. Army tank column rumbled into Gunzelhofen as SULTANE emerged from hiding to surrender. The lack of Joan radio units prevented Mosquito contact on 29/30 April. Next morning SULTANE rendered assistance with arresting suspect Gestapo and Dachau SS guards disguised as civilians. Army Intelligence accompany them to unearth both W/T and JE radios, and provide transportation to OSS HQ, Paris.

Robert P. Walker on board USS Ranger, 25 February 1966 after rescue from Gulf of Tonkin. *Photo R.P. Walker.*

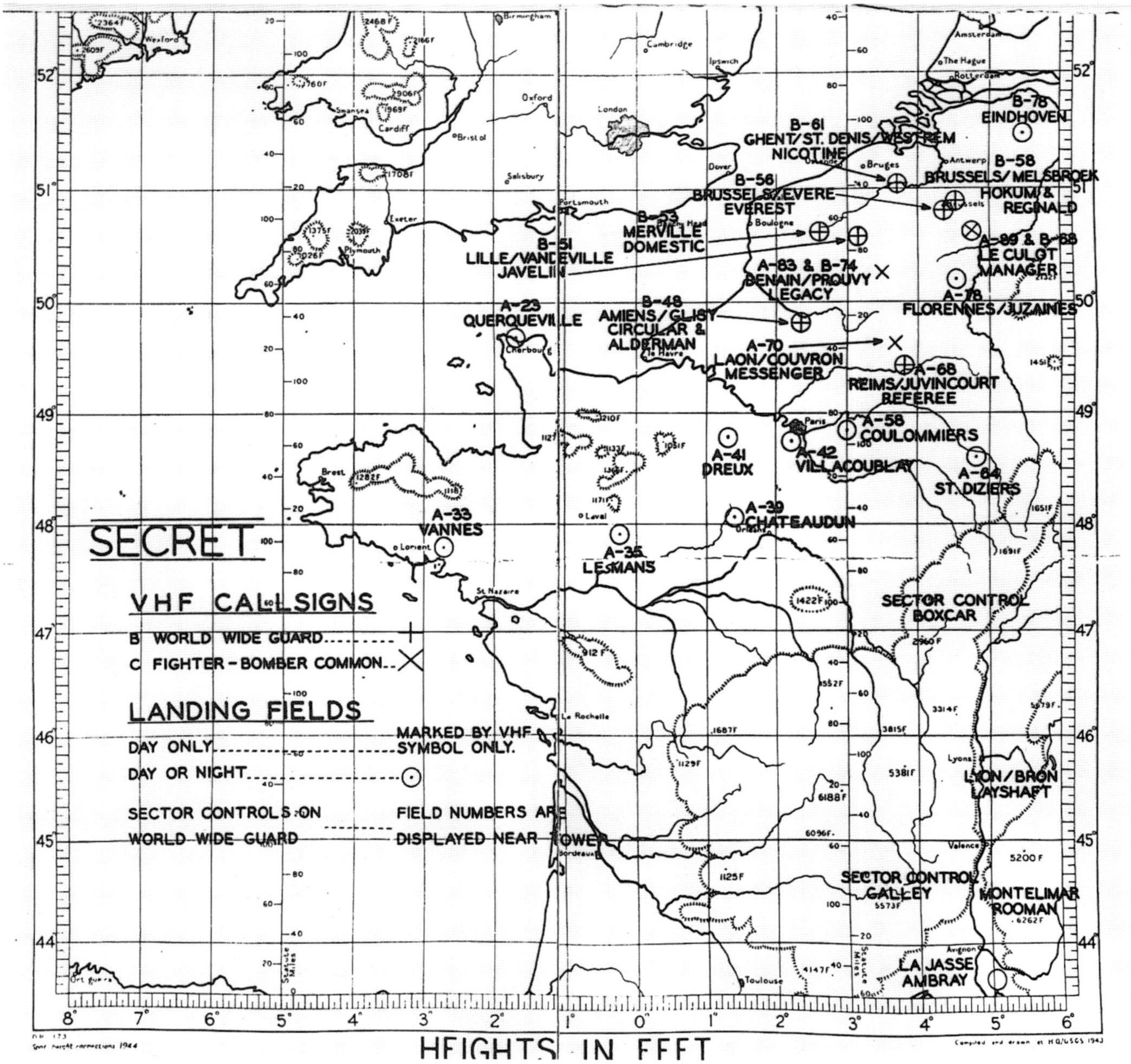

Navigator's reference map to landing fields in France and low-countries. *Photo P. Lovett*

Officers of the 653rd Bomb Squadron (Light Recon.). L-R. Front row: J. E. (Gene) Goodbread, Charles R. Bottorf, J. Keith Werthman, Harold L James, James D. Spear (KIFA), Robert L. Lee, Willis D. Locke, Rodney C. Peeke, Robert D. Schoenhair. Middle row: James M. Davis, Earl B. Rodriguez, George C. Hand Jr. (KIA), Thomas S. Blood, Ralph E. Murray, Clarence W. Jodar (KIA), Kenneth D. Dunaway, Ralph E. Fisher (KIA), Ford D. Porter Jr., William E. Scharbach, Edward B. Scruggs. Back row: Warren R. Barber, Milford B. Hopkins (KIA), Shaffer E. Horne, Hubert R. Farrens, Robert C. Grimes (KIA), Henry F. Davis, James W. Keiger (KIA), Unknown, John W. Green, Unknown. *Photo AAF-Borges.*

Officers of 654 Squadron (Special)(H) taken 13 June 1945, Watton, England. L-R. First row: Charles Finley, William Miskho, Richard O. Foster, Robert P. Howle Jr., Cornelius A. Engle, Roger W. Gilbert, Robert M. Balser, James C. Via. Second row: Lucien Peters, Robert B. Hastie, Kenneth McGriffin, John G. Morrison, Albert D. (Jack) Rasmussen, Ben O. Shaver, Raymond G. Spoerl, George M. VanderLeest. Third row: Robert Morrow, Frank C. Catalano, Clarence C. Korthase, Kenneth H. Kenyon, James C. Evans, Jack H. Wendling, John W. Grassel, John Coyne (Cone), Robert G. Spaight. Fourth row: Joseph Ward, Richard W. Tyhurst, John K. Jackson, ? Luse, Len A. Erickson, Lyndon F. Lakeman, Roger W. Krakow, John E. Murphy. Fifth row: Sgt. W. Stuck, Sgt. E. Turner, Cpl. Charles P. Barbero, Sgt. M. Smith, Cpl. R. Myers, Sgt. J. Walford, Cpl. L. Doctor. Absent: Peter D. Dustman, Charles A. Veldon, George Henion, Warren Borges, William E. Grey, Claude C. Moore (WIA), T/Sgt. J. Greene, S/Sgt. Temple B. Lowe. *Photo AAF-Kenyon*

25th Bomb Group (Rcn. Sp.) Headquarters Staff and Squadron commanders. L-R. Front row, ground level: Major John W. Walch (Navigation)(KIA), Maj. Ernest Patterson (CO 652nd, Lt. Col. John Larkin (CO 654th), Maj. Alvin Podwojski (Asst. CO 25th BG), Col. Leon Gray (CO 25th BG), Maj. Albert S. Straff (Air Executive), Capt. John Van Dolson, Jr. (Adjutant), Maj. Roy Ellis-Brown (CO 653rd), Capt. Lionel A. Proulx (Navigation-Asst. Group), Maj. Warren Ringgold (Operations); Second Row, ground level; Lt. Albert W. Holcomb, Capt William Whitehead (Personnel), Capt. Finis D. McClanahan (Intelligence), Maj. Rodney Peeke (Group Operations), Lt. G. Sesler, Maj. Delacroix Davis (Engineering), Capt. Harlow Dean (Special Services Off., Personnel Relation); Third and fourth rows on steps: Lt. Loyd L. Snyder, Lt. Howard E. Kaplan, Maj. Philip T. Mulligan (Flt. Surgeon), Maj. George G. Martin (Group Weather Off.), Maj. Ralph Espositio (Supply), Maj. Henry Walcott (Station Claims Off.), Capt. David W. Doane, Capt. Carl G. Durbin (Chaplain), W.O. Robert German (Inspector General), Lt. Robert Herzog (Supply). *Photo AAF-Sesler.*

Glossary

AAA Anti-aircraft Artillery
AB Air Base
AB Army Air Base
AAC Army Air Corps
AAF Army Air Force
A/C Aircraft
ACAS (I) Assistant Chief of the Air Staff (Intelligence)
AC in C Air Commander-in-Chief
ACIU Allied Central Intelligence Unit
AD Air Division
ADG Air Depot Group
AF Air Force
AEAF Allied Expeditionary Air Force
AEF Allied Expeditionary Force
AI Air Intelligence Branch (Air Ministry)
AM Air Ministry
AMC Aerial Mapping Committee
AOC Air Officer Commanding
AOC-in-C Air Officer Commanding-in-Chief
AM Air Ministry
ATC Air Transport Command
Avn Aviation
A-2 Intelligence, AAF

BADA Base Air Depot Area
BAC British Air Commission
BC Bomber Command
BD Bomb Division
BG Bomb Group
Bluestocking Weather reconnaissance mission.
BS Bomb Squadron
BW Bomber Wing
BW-1 Bluie West One, Greenland

Capt. Captain
CAS Chief of Air Staff
CBI China-Burma-India (theater of operations)
CCRC Combat Crew Replacement Center
CCU Combat Camera Unit
CCOS Combined Chiefs of Staff
CIA Central Intelligence Agency
CIU Central Interpretation Unit (Medmenham)
CMS Combat Mapping Squadron
CO Commanding Officer
Col. Colonel
COS Chiefs of Staff
Cpl. Corporal
CS Courier Squadron
CW Combat Wing

Det. Detachment
DH de Havilland Aircraft Company
DHC de Havilland Aircraft Company of Canada
DI Directorate of Intelligence
Do Dornier
DoI Director of Intelligence
DR Dead Reckoning
DZ Drop Zone

Eleanor Air radio unit
ERS Emergency Rescue Service
ETA Estimated Time of Arrival
ETO European Theater of Operations

F-3 PR configured A-20 Douglas Havoc
FAF French Air Force
FG Fighter Group
FFI Forces Francaises de Interieur (Gaullit Free French)
FIDO Fog Investigation Dispersal Operations
F/L Flight Lieutenant
Fld. Field
FMPU First Motion Picture Unit
F/O Flying Officer
Fort B-17 Flying Fortress
FS Fighter Squadron
FG Fighter Group
Fw Focke-Wulf

G-2 Intelligence, Army
GAF German Air Force
GCI Ground Control Interception
GCA Ground Control Approach
Gee British navigation system
Gee Whiz Gee calibration flight
Gen. General
B/G Brigadier General
M/G Major General
L/G Lieutenant General
Grp Group
G/C Group Captain

He Heinkel
HE High Explosive (bomb)
H&HS Headquarters and Headquarter Squadron
HQ Headquarters
HW Heavy Weather Squadron
H2S British radar equipment
H2X Radar equipment, Mickey

IFF Identification Friend or Foe

JCS Joint Chiefs of Staff
JE Joan-Eleanor mission
JIC Joint Intelligence Committee
Joan Ground radio unit
Joker Night PR mission using PFBs
JPRC Joint Photo Reconnaissance Committee
Ju Junkers

KG Kampfgeschwader
KIA Killed In Action
KIFA Killed In Flying Accident
KPs Kitchen staff

Lib B-24 Liberator aircraft
Lt. Lieutenant
Lt. Col. Lieutenant Colonel
Lt. Cmdr. Lieutenant Commander
Loran Long Range Navigation equipment
LOP Lines of Position using Loran
LW Light Weather Squadron

MAAF Mediterranean Allied Air Forces
MAPRU Mediterranean African Photo Reconnaissance Unit
Me Messerschmitt
Met/Obs Meteorological Observer
MI-5 Military Intelligence-5 (British Service for Home Security)
MI-6 Military Intelligence-6 (British Intelligence Service)
MIA Missing In Action
Mickey H2X radar PR mission, H2X radar.
Mk Mark
Mossie Mosquito aircraft
MOS Military Occupation Specialty
MP Military Police
mph Miles per hour
M/Sgt. Master Sergeant
MTO Mediterranean Theatre of Operations

NAPRW Northwest Africa Photo Reconnaissance Wing
NAS Naval Air Station
NCO Non-commissioned officer
Non-coms Non-commissioned officers

Ops Operations
OSS Office of Strategic Services
OTU Operational Training Unit

(P) Provisional
PFB Photo Flash Bomb
Pinetree HQ Eighth AF, High Wycombe, Buckingham
PG Photo Group
PIS Photographic Interpretation Section
PPI Plan Position Indicator, radar scope
P/O Pilot Officer
POW Prisoner of War
PR Photo reconnaissance
PRG Photo Reconnaissance Group
PROTU Photo Reconnaissance Operational Training Unit
Prov Provisional
PRU Photographic Reconnaissance Unit
PRU Daylight photo reconnaissance mission
PS Photographic Squadrons
PSP Pierced Steel Plate runway panels
PW Photo Wing

QDM Request for magnetic heading to steer.
QFE Request for airfield pressure to set altimeter.
QFF Base atmospheric press converted to MSL.

RAF Royal Air Force
RCAF Royal Canadian Air Force
RAAF Royal Australia Air Force
Rcn Reconnaissance
RDF Radio Direction Finder
Recce Reconnaissance
Redtail Command observation mission
Redstocking Label for OSS mission at Watton
RNZAF Royal New Zealand Air Force
RON Remain Over Night
RW Reconnaissance Wing

(Sp) Special
S-2 Section concerned with Intelligence
S-3 Section concerned with plans and operations
SAAF South Africa Air Force
SAD Strategic Air Depot
Scout Advance weather mission
SD Sicherheitsdienst, SS security service searching for political enemies
Sgt. Sergeant
S/Sgt. Staff Sergeant
SHAEF Supreme Headquarters Allied Expeditionary Force
SOE Special Operations Executive (British)
SOP Standard Operating Procedure
SL Searchlight
S/L Squadron Leader
M/Sgt. Master Sergeant
SP Special Squadron
Sq. Squadron
SW South West
SWPA South West Pacific Area

T/Sgt. Technical Sergeant
T/W Twin Engine

UK United Kingdom
US United States
USA United States of America
USAAC United States Army Air Corps
USAAF United States Army Air Force
USN United States Navy
USSAFE United States Strategic Air Forces in Europe
USSTAF United States Strategic and Tactical Air Forces
USSR Union of Soviet Socialists Republics
Unt Off. Unteroffizier

VHF Very High Frequencey

W/C Wing Commander
WIA Wounded In Action
Widewing Headquarters sof General Carl Spaatz.
W/O Warrant Officer
WR Weather Reconnaissance
W/T Wireless Telegraphy radio using Morse code..
WX Weather

Y Radio Intelligence service

1/Lt 1st Lieutenant
2/Lt 2nd Lieutenant

653rd Squadron Mission by Type, Number, Aborts, and Operational Hours Flown

Month	Mission Type	Number	Abortive	Hours Flown
May 1944	Bluestocking	7	2	12:00
June	Bluestocking	48	1	144:00
	Scout	35	1	100:00
	Frantic	4	0	18:25
July	Bluestocking	82	1	283:50
	Scout	44	4	138:55
	Frantic	2	0	24:45
August	Bluestocking	71	3	249:55
	Scout	68	0	243.50
	Gee Whiz	1	0	2:50
September	Bluestocking	71	3	249:55
	Scout	27	1	94:10
	Frantic	4	0	60:21
	Pee Dee	1	0	5:50
October	Bluestocking	100	3	348:10
	Skywave	2	0	26:05
November	Bluestocking	126	3	475:00
	Redtail	4	0	16:00
	Route Survey	3	0	9:20
December	Bluestocking	91	3	333:45
	Redtail	6	0	25:45
	Grapea	13	0	90:35
January 1945	Bluestocking	113	3	433:15
	Graypea	13	1	106:15
	Redtail	2	0	8:45
	Skywave	1	0	5:42
February	Bluestocking	112	0	476.30
	Graypea	10	0	117.20
	Redtail	4	0	18:20
	Skywave	4	0	16:15
March	Bluestocking	137	2	511:45
	Redtail	37	1	157:35
	Skywave	7	0	35:30
April	Bluestocking	113	1	461:55
	Skywave	15	2	69:35
	Redtail	11	1	47:45
May	Bluestocking	19	0	71:10
	Skywave	1	0	4:35

654th Bomb Squadron Line Inventory
Watton, England - 1944-1945

Serial Number	*Call Letter*	*VHF*	*GEE*	*Chaff*	*Camera*	*Crew Chief*
A Flight, Flight Chief M/Sgt. Simonds						
NS 590	S-B	X	X		X	Montgomery
NS 551	K	X	X		X	Austin
NS 552	G	X	X		X	Miller
NS 553	P	X	X		X	Rider
NS 554	V	X	X		X	Allen
NS 559	I	X	X		X	McKee
NS 581	Q	X	X		X	Mitchell
B Flight, Flight Chief M/Sgt. Krause						
MM 385	C	X	X		X	Hanson
NS 569	N	X	X		X	Garcia
NS 595	A	X	X			Goodwine
NS 619	U	X	X		X	Koskela
NS 651	F					Paternoster
NS 711		X				Stallins
NS 725	L	X	X		Navy	Tompson
NS 740	M	X	X		Navy	Shewmake
C Flight, Flight Chief T/Sgt. Jordan						
NS 707	E	X	X		Navy	
RF 979	D	X	X	X		Darilek
RF 983	O	X	X	X		Landers
RF 986	W	X	X	X		Berge
RF 988	J	X	X	X		Lachuisa
RF 992	R	X	X	X		Whorf
RF 996	Y	X	X	X		Locke
RF 999	T	X	X	X		Whinnery
D Flight, Flight Chief M/Sgt. Hall						
RG 113	S	X	X	X		Anderson
NS 650	H	X	X	X		Gebicke
NS 743	Z	X	X	X		McGeehan
NS 785	B	X	X	X		Goff
NS 793	X	X	X	X		Cummings
LR 534		X				Ritchie
B-26-195	AX	X	X		X	Stone
B-26-205	AY	X	X		X	Nasief
B-26-226	AT	X	X		X	Stone
A-26-500		X	X		Navy	Kominsky
A-26-524		X	X		Navy	Goodwine

Roster of OSS Agents

Team	Dispatch Date	Target	Delivery		Communication	
			Air	Land	Type	Contact
Downend	1 September 1944	Bochum	Hudson		Courier	Yes
Ruppert	2 November	Berlin		Yes	Courier	Yes
Ragweed	4 November	Ruhr		Yes	Courier	Yes
Tyl	10 November	Meppel	B-24		JE	Yes
Eclipse	7 December	Dusseldorf		Yes	Tourist	
Marietta	22 January	Amsterdam		Yes	W/T	Yes
Alex	16 January 1945	Gustorf		Yes	W/T	Yes
Pitt	21 January	Karlsruhe	B-24		W/T	Yes
Troy	21 January	Pforzheim	B-24		JE	Yes
Hofer	23 January	Austria		Yes	Tourist	
Rubens	30 January	Stuttgart	B-24		W/T	Yes
Mimi	30 January	Stuttgart	B-24		Tourist	
Student	9 February	Euskirchen		Yes	Tourist	
Lulu	21 February	Schwarzenbach	B-24		Tourist	
Cocco	21 February	Gravbenstetten	B-24		Tourist	
Joshua	23 February	Mainz	B-24		Tourist	Yes
Brahms	24 February	Hamburg		Yes	Courier	Yes
Schiller	24 February	Berlin		Yes	Tourist	
Mickie	24 February	Pforzheim	B-24		Tourist	Yes
Gogol	26 February	Friburg		Yes	Tourist	
Anzio	26 February	Stuttgart	B-24		JE	Yes
Hammer	1 March	Berlin	A-26		JE	Yes
Crocus	3 March	Brenner Pass		Yes	Tourist	
Florida	10 March	Braunschweig		Yes	W/T	Yes
Godfrey	16 March	Dinsslaken		Yes	Walkie Talkie	Yes
Katga	16 March	Oberhausen		Yes	W/T	Yes
Sidecar	18 March	Ansbach	B-24		W/T	No
Martini	18 March	Augsburg	B-24		W/T	No
Painter	19 March	Munich	B-24		W/T	No
Chisel	19 March	Essen	A-26		JE	No
Conte	19 March	Darmstadt	B-24		Tourist	Yes
Sally	20 March	Mannheim	B-24		W/T	Yes
Daiquiri	21 March	Aschaffenburg	A-20		W/T	No
Serpent	22 March	Rotenburg	B-24		Tourist	
Tiger	22 March	Stuttgart	B-24		W/T	
Doctor	23 March	Kufstein	B-24		W/T	Yes
Alexander	23 March	Osnabruck	A-20		W/T	No
Eggnog	23 March	Hanau	A-20		W/T	No
Highball	23 March	Kassell	A-20		W/T	No
Mac	23 March	Munster	A-20			

Team	Dispatch Date	Target	Delivery		Communication	
			Air	Land	Type	Contact
Fred	23 March	Hamm	A-20			
Dubuque	23 March	Frankfurt	B-24		W/T	
Liberation	23 March	Wiesbaden	A-20			
Abnoki	24 March	Wesel	A-20		W/T	
Old Fashioned	25 March	Giessen	A-20		W/T	No
Pink Lady	30 March	Erfurt	A-20		W/T	Yes
Fat (Pope)	30 March	Hechingen		Yes	W/T	
Zombie	31 March	Regensburg	B-24		W/T	
Chauffeur	31 March	Regensburg	B-24		W/T & JE	Yes
Hamilton	31 March	Regensburg	B-24		Tourist	Yes
Hamilton	31 March	Regensburg	B-24		Tourist	Yes
Raleigh	31 March	Babenhausen	B-24		Tourist	Yes
Vacuum	31 March	Munich	B-24		Tourist	Yes
Morris	1 April	Baberg	B-24		Tourist	Yes
Cleaner	1 April	Regensburg	B-24		Tourist	Yes
Kangaroo	1 April	Donaueschingen	B-24		W/T	
Camel	1 April	Schwenningen	B-24		W/T	
Cuba Libre	2 April	Gottingen	A-26		W/T	No
Hound	2 April	S/Regensburg	B-24		Tourist	Yes
Pickaxe	4 April	Landshut	B-24		JE	Yes
Luxe	4 April	Weilheim	B-24		JE	Yes
Hare	4 April	Munsingen	B-24		Tourist	Yes
Dolly	4 April	Pforzheim	B-24		Tourist	Yes
Comet	5 April	Bremen		Yes	Tourist	
Manhattan	7 April	Chemnitz	B-24		W/T	
Buzzsaw	7 April	Leipzig	B-24		JE	
Faro	7 April	Plauen	B-24		JE	
Balto	7 April	Ingolstadt	B-24		W/T	
King	7 April	Ravensburg	B-24		Tourist	Yes
Gallagher	7 April	South Munich	B-24		Tourist	Yes
Rabbitt	8 April	Loffingen	B-24		Tourist	
Planters Punch	9 April	Stendal	A-26		W/T	
Orange Blossom	9 April	Magdeburg	A-26		W/T	
Mallet	10 April	Wittenberg	A-26		W/T	
Singapore Sling	11 April	Passau	B-24		W/T	
Hot Punch	11 April	Passau	B-24		W/T	
Farmer	16 April	Straubing	B-24		JE	
Queen	16 April	South Munich	B-24		Tourist	Yes
Jack	16 April	South Munich	B-24		Tourist	Yes
Dubarry	16 April	Hechingen	B-24		Tourist	Yes
Bowman	16 April	Munich	B-24		W/T	Yes
Cachatto	18 April	Pocking	B-24		Tourist	
Zebra	18 April	Augsburg	B-24		Tourist	
Ace	18 April	Endorf	B-24		Tourist	Yes
Elephant	19 April	Muhleim	B-24		Tourist	
Arrow	19 April	Salzburg	B-24		Tourist	Yes
Isabel	20 April	SW Munich	B-24		Tourist	Yes
Waldorf	20 April	S. Cheim	B-24		Tourist	Yes
Sherwood	20 April	N Lake Constance	B-24		Tourist	Yes
Godfrey II	21 April	Bruck	A-26			
Raymond	21 April	Jutterbog	A-26			
Boyard	24 April	Bregenz	B-24		W/T	
Sultane	24 April	Munich	B-24		W/T & JE	No
Georgia	24 April	St. Johann	B-24		W/T	Yes
Virginia	24 April	Strass	B-24		W/T	Yes
Baer	24 April	Salzburg	B-24		Tourist	Yes
Leone	24 April	Weindorf	B-24		Tourist	Yes
Sea Horse	24 April	Fussen	B-24		Tourist	Yes
Hyena	25 April	SW Munich	B-24		Tourist	
Tobacco	26 April	Bergen	B-24		Tourist	Yes
Gauloise	6 May	Kiel	A-26		W/T	

Agent Contact Synopsis

TYL

15 November 1944	No Contact	7 January 1945	No Contact
16 November 1944	No Contact	10 January1945	No Contact
22 November 1944	Contact	22 January 1945	Contact
27 November 1944	Contact	8 February 1945	No Contact
1 December 1944	Contact	13 February 1945	Contact
10 December 1944	Contact	17 February 1945	Contact
12 December 1944	Contact	20 February 1945	Contact
15 December 1944	Contact	28 February 1945	Contact
16 December 1944	Contact	10 March 1945	Contact
3 January 1945	No Contact	21 March 1945	Contact
5 January1945	Contact		

TROY

30 January 1945	No Contact	22 February 1945	No Contact
2 February 1945	No Contact	27 February 1945	No Contact
4 February 1945	No Contact	2 March 1945	No Contact
7 February 1945	No Contact	9 March 1945	No Contact
13 February 1945	Contact	30 March 1945	No Contact

ANZIO

7 March 1945	No Contact	1 April 1945	No Contact
23 March 1945	No Contact	6 April 1945	No Contact
30 March 1945	Contact		

HAMMER

13 March 1945	Contact	8 April 1945	Contact
18 March 1945	No Contact	15 April 1945	Plane Aborted
21March 1945	No Contact	18 April1945	No Contact
28 March 1945	Contact	21 April 1945	Plane Aborted
2 April 1945	No Contact	23 April 1945	No Contact

CHISEL

19 March 1945	Plane MIA	26 March 1945	No Contact
22 March 1945	No Contact		

CHAUFFEUR

19 April 1945	Contact	27 April 1945	No Contact
23 April 1945	Contact	1 May 1945	Plane Aborted

LUXE

8 April 1945	Contact	20 April 1945	Contact
10 April 1945	No Contact	25April 1945	Contact
13 April 1945	Contact	28 April 1945	Contact
17 April 1945	Contact	30 April 1945	No Contact

PICKAXE

8 April 1945	Contact	25 April 1945	Contact
13 April 1945	Contact	28 April 1945	Contact
17 April 1945	Contact	30 April 1945	Contact
20 April 1945	Contact		

BUZZSAW

11 April 1945	No Contact
14 April 1945	No Contact
16 April 1945	Cancelled

FARO

9 April 1945	No Contact
10 April 1945	Contact
14 April 1945	No Contact
16 April 1945	No Contact

SULTANE

29 April 1945	No Contact

MALLET

15 April 1945	Plane Aborted
18 April 1945	No Contact
21 April 1945	Plane Aborted
23 April 1945	No Contact

FARMER

19 April 1945	No Contact
23 April 1945	No Contact
27 April 1945	No Contact

BOYARD

29 April 1945	No Contact

Aircraft Status Chronology for 802nd RG/25th BG (Rcn) (S)

Information compiled from 802 RG and 25 BG records, MACR, and photos showing serial numbers. This listing is not all inclusive and is a best case scenario. Chaff aircraft assignment has discrepancies The Mosquito serial number records and those listing B-17 serial numbers is wanting. The 652 Squadron mission reports list only the last three numbers of the aircraft serial. Fortress information obtained from archives of B-17 researcher George Wyatt.

Known P-38F serial numbers, 50th FS, Iceland, via Ragnar Ragnarson.
42-12572
42-12576
42-12577
42-12578
42-12581
42-12582
42-12583
42-12585
42-12586
42-12587
42-12589
42-12592
42-12594
42-12595
42-12596
42-12597
42-12599
42-12602
42-12603
42-12605
42-12607

Known B-17 aircraft operated by the Combat Weather Flight, Bovingdon

B-17F	4229811	Crash 28 September 1943
B-17G	237744	Crash 9 December 1943
B-17G	237869	Crash 25 December 1943
B-17G	230131	MIA 2 February 1944
B-17G	4239871	Crash 1 April 1944, Warton for repairs, reassigned 8th AF

2 March 1944. Mosquitos assigned unmodified for transitional training.
MM345 AAF Sta.. 131 Nuthampstead
MM346 131 Nuthampstead
MM342 113 Cheddington
14 March 1944. Additional aircraft for transitional training at AAF Sta. 113 Nuthampstead.
MM368, M370, MM371, MM384, MM385, MM364, MM391
4 April 1944. Training aircraft available and status: Operational
MM345 Sta. 113 Transition training
MM346 113 “
MM364 113 “
MM370 113 “
LF371 113 “ (MM371)
LF384 113 “ (MM384)

Out of Service
MM310, Sta. 582 Warton. Bowed landing gear, defective, being repaired.
MM340, Sta. 582 Warton. Recent crackup by test pilot, for RAF repair.
MM382, Sta. 113 Cheddington. Crashed on take off, being returned to RAF
MM344, Sta. 113 Cheddington. Crashed, ground looped at Cheddington, return to RAF
MM337, Sta. 131 Nuthampstead. Crashed on landing, return to RAF
LR516, Sta. 131 Nuthampstead. Mark III dual trainer crashed in training accident, return to RAF

4 April 1944. Status of Mosquito aircraft
Prototype:
MM338 Weather Prototype
MM308 H2X - Station 102 Alconbury.
MM311 H2X - Station 102 Alconbury.
MM342 Photo Prototype - Station 590 Burtonwood.
MM368 Weather Prototype - Station 590 Burtonwood.
For Modification:
MM387 - Weather Mod - Sta. 590 Burtonwood.
MM388 - Weather Mod - Sta. 590 Burtonwood.
MM391 - Weather Mod - Sta. 590 Burtonwood.
MM393 - Weather Mod - Sta. 590 Burtonwood.

802nd RG aircraft status as of 8 May 1944:

B17F Courier Sq (Heavy)
42-3074
42-3095
42-3135

42-3259
42-3292
42-3346
42-3465
42-30214
42-30380

L-5 Courier Squadron (Heavy)
42-98996
42-98997
42-98998

C-64 Courier Squadron (Light)
43-5361
43-5368
43-5396
43-5397

Mosquito Squadron aircraft

MM311
MM338
MM345
MM346
MM364
MM368
MM370
MM371
MM384
MM387
MM388
MM393
NS505
NS509
NS510
NS512
NS513
NS514
NS515
NS516
NS518
NS519
NS535
LR530

The following aircraft were assigned to Squadrons, June 1944. This statistical summary of aircraft assignment differs from that stated in the group history.

Weather (Light)	
Mosquito	23
UC-64	1
Weather (Heavy) Squadron	
B-17	14
B-24	1
A-20	1
L-5	1
UC-64	2
Special Squadron	
Mosquito	20
UC-6	1
Oxford	1
L-5	2
375 Service Squadron	
B-17	2
Total aircraft 802nd RCN Grp	69

Known Weather Squadron (Heavy) B-17F
42-102431
42-102521
42-102676
42-102970 *Hope and Grope* (J).
42-23436
42-30101
42-30131
42-30777
42-31130
42-31170 Our Baby.
42-31474 Stinky Weather
42-31674
42-31736 Crash 6 September 1944.
42-31766 AAF Sta. 505, Battle damage, reassigned.
42-3279
42-3301
42-37744
42-39771 or 42-39772, last digit unreadable as 1 or 2.
42-39777 *Stumble Butt*, salvaged 26 June 1944.
42-39959

B-24J mission reports list only last three digits of serial number. Known histories from John R. Beitling Archives

233 44-40233 Departed USA on 5-9-44 as original equipment for the 493BG/860BS, named, transferred to 25BG/652BS, and ZI on 7-9-45.

320 44-40320 Departed USA on 5-14-44 as original equipment for 493BG/860BS, named, transferred to 34BG/19BS, then to 25BG/652BS, ZI with 466 BG/785BS on 6-17-45.

383 44-40383 Departed USA on 5-7-44 as original equipment for 493BG/860BS, believed un-named, transferred to 34BG/18BS, then to 25BG/652BS, ZI on 6-21-45.

484 44-40484 Departed USA on 5-19-44 as original equipment for 493BG/863BS, believed un-named, assigned to 25BG/652BS, and Interned in Portugal on 12-2-44.

718 44-0718 Departed USA on 6-19-44, assigned to the 25 BG/652BS, named , and ZI on 6-29-45.

762 44-40762 Departed USA on 7-10-44, and assigned to 448BG/715BS, named , transferred to 25BG/652BS, and ZI on 6-21-45.

763 44-40763 Departed USA on 6-20-44, believed un-named and assigned to 93BG/330BS, transferred to 25BG/652BS, and ZI on 7-17-45.

789 44-40789 Departed USA on 6-20-44, assigned to 25BG/652BS, and ZI on 6-20-45.

879 44-40879 Departed USA on 6-24-44, and assigned to 25BG/652, named , carried squadron letter "H" on crash landing on 9-22-44. Believed to have served with 489BG/844BS prior to transfer to 25BG/652BS.

880 44-40880 Departure date unknown, assigned to the 25BG/652BS. Salvaged 5-29-45.

882 44-40882 Departure date unknown, named, and assigned to 25BG/652BS. Believed to ZI on 8-2-45

Further Chronology
B-24H 42-7472 42-7492**,** conversion training Watton**,** reassigned 8th AFCC, 4 July 44
July 1944 B-26 G Marauder**s** Major John Hoover delivered B-26 43-34193. Others are 43-34226, 43-34195, 43-34205.

21 July 1944 B-26 aircraft to be painted Carpetbagger-type camouflage.

Mosquitos 26 July 1944
H2X, 8 authorized, 4 at Watton now and 1 promised by 29 July and next by 8 August 1944.
Weather, 20 authorized and 25 now at Watton.
Night Photo, 12 authorized with 2 at Langford Lodge being prototyped.
On hand 24 not modified (10 of these for night photo, 14 replacement).
Total 55 aircraft
F-8's from US, 25 Canadian MkIV modified in US and 23 will or are on the way to UK (2 cracked up in staging area), 10 at Watton, other 13 coming. 8th RW reject acceptance and request those in staging areas be reassigned to another unit/theatre.
Monthly allotment of 7 aircraft for attrition reduced to 4 per month to relieve storage and maintenance on future schedule deliver.
26 July 1944, NS554 night photo Mosquito delivered to Watton.
16 August 1944, MM371 prototype to dispense chaff canceled. Aircraft will be transferred to Sta. 112, Director of Technical Services for prototyping.
18 August 1944, Mosquito PR XVI Tropical NS638, NS677 and NS686 allotted to Alconbury for H2X conversion.

28 October 1944 Proposed Chaff Mosquitos for 2 SAD, Abbotts Ripton.
MM344
MM345
MM368
MM371
MM386
MM393
NS508
NS509
NS512
NS515
NS516
NS519
B-17G Fortress beginning November 1944.
2102970 (J)
23301
23436
230131
237744
43-38555 Accident 16 December 1944
43-38573
338833 (O)
338867
43-38709 Accident 27 February 1945
44-6575 Accident, 1 Feb. 45
46577
46586
48520 (G)
44-8601 MIA 17 March 45

The 652nd Squadron mission reports list only last three digits of B-17 serial numbers:
240, 315, 477, 502, 507, 510, 530, 536, 652, 655, 656, 708, 0763, 836, 846, 1736.

9 November 1944 Mosquito status.
Written Off

MM308
MM377
MM344
MM388
MM370
MM385
NS518
NS533
NS535
NS537
NS538
NS555
NS557
NS558
NS574 (return, acquired in error)
NS625
Aircraft still charged to 8th AF
MM310

MM338

MM340

MM342
MM344

MM345

MM346

MM364
MM368
MM384
MM386
MM388
MM391
MM393*
NS509
NS510
NS513
NS514
NS515
NS516
NS519
NS520
NS551
NS552
NS553
NS556
NS581
NS559
NS568
NS569
NS570
NS582
NS583
NS584
NS585
NS590
NS591
NS592
NS593***
NS594

NS595
NS596
NS619
NS626
NS630
NS632
NS634
NS594
NS635
NS636
NS638
NS650
NS651
NS676
NS677
NS686
NS707
Mark III Mosquitos written off.
RF516
RF534**
Mark III Mosquitos charged to 8th AF.
LR530
LR553
Apdl 653*
Apdl 654**
MIA 18/9/44***
Note RF534 status is inconsistent with assignment.

Proposed night Photo Aircraft of 654 Bomb Squadron.
NS551
NS552
NS553
NS554
NS556 (Cr 22/1/45)
NS557 (Cr 5/8/44)
NS559
NS568 (Cr 6/11/44)
NS581
NS582 (Cr 28/11/44)
NS583 (MIA 8/2/45)

4 March 1945, 9 Mosquito aircraft completed with Chaff dispensers at 2nd SAD with 7 more in works.
14 March 1945, 492 BG request for 2 Mosquito Mk III to transition pilots denied.
18 March 1945, Transfer from 25 BG to 492 BG Mosquito NS707, NS725, NS740.
19 March 1945, Transfer from 25 BG to 492 BG Mosquito NS584 and NS686.
24 March 1945, 7 Mosquito aircraft to be assigned to 492 BG to maintain stock levels.
1 April 1945, A-26 aircraft: 43-22500, 43-22513, 43-22524, 43-22610, 43-22626.
1 April 1945, Transfer from 25 BG to 492 BG Mosquito RG157 and TA614.
Mosquito aircraft transferred to 492 BG, Harrington: NS584, NS686, NS707, NS725, NS740, RG517 and TA614.

Disposition of Mosquito Aircraft at Watton 25 May 1945 in excellent shape.
MM340
MM342
MM346
MM385
MM388
NS551
NS553
NS554
NS559
NS581
NS651
NS676
NS686
NS709
NS740
NS725
NS743
NS775
NS785
NS793
NS794
NS796
NS802
NS804
NS805
NS811
NS812
RF979
RF983
RF985
RF986
RF988
RF992
RF999
RG148
RG154
RG155
RG156
RG157

The following Mosquito XVI are Cat. B with 43 Air Group.
MM364
NS619

Following Mosquito XVIs are at Watton, reconditioned by 71 Maintenance Unit.
NS552
NS590
NS774
RG145
MM368

Following Mosquito XVIs at Marshall Flying School for major overhaul.
NS632
NS591
NS592
NS634
NS636

Following Mosquito XVIs at Martin Hearn Ltd. for reasons indicated.
NS744 Cat. B
NS585 major overhaul
NS626 major overhaul.

20 May 1945 76 Mosquito aircraft for return to RAF along with three Mk III trainers and two Oxfords.
9 June 1945, All Mosquitos grounded for return to RAF.
Air Speed Oxford aircraft assigned to 802 RG and 25 BG.
R6034, AB720, AB768, DF417, DF528, P8917, T1066, V3587, V3780, V3628
4 March 1944, Two Oxford and one Piper Cub assigned to 8th Photo Rcn Wing (P).
Oxfords at Sta 239 Maghaberry are AB720 and V3780 along with L4B #43-631 at Sta 236 Toome to be sent to Sta. 112 Bovingdon.
24 March 1944, T1066 and R6034 assigned to Sta. 113 Cheddington for transitional training.
27 March 1944, Two additional Oxfords obtained by Sta. 113 Cheddington: DF528 and DF417 and an additional Oxford P-8917 assigned to 8th RW(P), Sta. 634.
29 March 1944, 8th RW request repair action on AB720, V3780, V3628, and T1066 required parts.
22 January 1945, Watton requested Mark III Mosquitos to train new crews, but will accept Oxford aircraft. It was 8 AF policy to return and not acquire additional British training and communication aircraft.
19 June 1945, Oxford aircraft ferried from 325 PR Wing to 46 MU are V3587 and AB768.
21 June 1945, Oxford II R6034 flown to Portsmouth Aviation for reduction to spares and produce. Both R6034 and V3628 have 1200 hours each.

F-8 Mosquito aircraft serial numbers assigned to the USAAF.

KB. No.	US Serial
KB312	43-34924
KB313	43-34925
KB315	43-34926
KB316	43-34927
KB317	43-34928
KB321	43-34929
KB171	43-34930
KB306	43-34931
KB130	43-34932
KB131	43-34933
KB132	43-34934
KB133	43-34935
KB138	43-34936
KB139	43-34937
KB140	43-34938
KB141	43-34939
KB146	43-34940
KB147	43-34941
KB148	43-34942
KB149	43-34943
KB150	43-34944
KB151	43-34945
KB152	43-34946
KB153	43-34947
KB154	43-34948
KB155	43-34949
KB156	43-34950
KB157	43-34951
KB158	43-34952
KB159	43-34953
KB180	43-34954
KB181	43-34955
KB182	43-34956
KB183	43-34957
KB184	43-34958
KB185	43-34959
KB186	43-34960
KB187	43-34961
KB188	43-34962
KB189	43-34963

F-8 Mosquitos at Watton released to RAF on 18 September 1944.
43-34934
43-34938
43-34940
43-34942
43-34952
43-34954
43-34956
43-34959
43-34962
43-34963
Retrieved 12 October 1944
43-37950 (del. in error)
43-37954,
43-37959

25 BG (Rcn) OSS JE Mosquito and A-26 Mission, Along with 492 BG JE Mosquito Missions.

Agent	Date	Contact	Crew	Serial number	Unit
Tyl	15 November 1944	No Contact	Doroski/Miksho/Simpson	NS676	25 BG
Tyl	16 November	No Contact	Doroski/Miskho/Sawyer	NS676	25 BG
Tyl	22 November	Contact	Doroski/Miskho/Simpson	NS707	25 BG
Tyl	27-28 November	Contact	Walker/Conyers/Sawyer	NS676	25 BG
Tyl	1 December	Contact	Walker/Conyers/Simpson	NS676	25 BG
Tyl	10 December	Contact	Walker/Conyers/Sawyer	NS676	25 BG
Tyl	12-13 December	Contact	Walker/Conyers/Simpson	NS676	25 BG
Tyl	15 December	No Contact	Walker/Miskho-delivery	NS559	25 BG
Tyl	15 December	Contact	Magee/Conyers/Sawyer	NS676	25 BG
Tyl	16 December	No Contact	Walker/Miskho-delivery	NS559	25 BG
Tyl	16 December	Contact	Magee/Conyers/Simpson	NS676	25 BG
Tyl	3 January 1945	No Contact	Walker/Conyers/Layton/Simpson	A-26 43-22500	25 BG
Tyl	5-6 January	Contact	Walker/Conyers/Sawyer	NS707	25 BG
Tyl	7 January	No Contact	Walker/Conyers/Layton/Simpson	A-26 43-22500	25 BG
Tyl	7 January	Abort	Magee/Miskho/Sawyer	NS707	25 BG
Tyl	9-10 January	Contact	Walker/Conyers/Simpson	NS707	25 BG
Tl	22-23 January	Contact	Walker/Miskho/Sawyer	NS707	25 BG
Troy	29-30 January	No Contact	Walker/Miskho/Layton	NS707	25 BG
Troy	1-2 February	No Contact	Walker/Miskho/Layton	NS707	25 BG
Troy	3-4 February	Abort	Walker/Miskho/Layton	NS707	25 BG
Troy	7 February	No Contact	Walker/Miskho/Layton	NS740	25 BG
Tyl	8-9 February	No Contact	Walker/Miskho/Sawyer	NS740	25 BG
Tyl	12-13 February	Contact	Walker/Miskho/Simpson	NS707	25 BG
Troy	13-14 February	Contact	Walker/Miskho/Layton	NS725	25 BG
Tyl	17-18 February	Contact	Gray/Morrow/Simpson	NS740	25 BG
Tyl	20-21 February	Contact	Magee/Miskho/Simpson	NS725/	25 BG
Troy	22-23 February	No Contact	Walker/Miskho/Layton	NS740	25 BG
Troy	26-27February	No Contact	Knapp/Miskho/Layton	NS740	25 BG
Tyl	28 February/1 March	Contact	Knapp/Jackson/Sawyer	NS725	25 BG
Hammer	1-2 March	Delivery	Walker/Miskho/Walch/Derr	A-26 43-22524	25 BG
Troy	2-3 March	No Contact	Knapp/Jackson/Layton	NS740	25 BG
Anzio	7-8 March	No Contact	Knapp/Jackson/Ancrum	NS7225	25 BG
Troy	9-10 March	No Contact	Knapp/Jackson/Ancrum	NS725	25 BG
Tyl	10-11 March	Contact	Knapp/Jackson/Simpson	NS725	25 BG
Hammer	12-13 March	Contact	Knapp/Jackson/Ancrum	NS725	25 BG
Hammer	18 March	No Contact	Knapp/Jackson/Ancrum	NS725	25 BG
Chisel	19 March	MIA	Emmel/Walch/Tresemer/Brunner	A-26 43-22524	25/492 BG
Hammer	20-21 March	No Contact	Knapp/Jackson/Ancrum	NS725	25 BG
Chisel	22-23 March	No Contact	Knapp/Jackson/Ancrum	NS725	25BG
Anzio	23-24 March	No Contact	Knapp/Jackson/Ancrum	NS707	25 BG
Chisel	26 March	No Contact	Knapp/Jackson/Ancrum	NS740	25 BG
Hammer	27-28 March	Contact	Knapp/Jackson/Ancrum	NS740	25 BG

Agent	Date	Contact	Crew	Serial number	Unit
Anzio	29-30 March	Contact	Schaeffer/Kolacki/Simpson	NS707	492 BG
Troy	29-30 March	Contact	Knapp/Jackson/Ancrum	NS707	25 BG
Tyl	30-31 March	Contact	Knapp/Jackson/Ancrum	NS707	25 BG
Anzio	31 March/1 April	No Contact	Knapp/Jackson/Ancrum	NS740	25 BG
Anzio	1-2 April	No Contact	Knapp/Jackson/Ancrum	NS725	25 BG
Hammer	2-3 April	No Contact	Knapp/Jackson/Ancrum	NS725	25 BG
Anzio	5-6 April	No Contact	Knapp/Jackson/Ancrum	NS725	25 BG
Luxe	8-9 April	Contact	Knapp/Jackson/Ancrum	NS740	25 BG
Pickaxe	8-9 April	Contact	Knapp/Jackson/Ancrum	NS740	25 BG
Hammer	8-9 April	Contact	Knapp/Jackson/Ancrum	NS740	25 BG
Faro	9-10 April	No Contact	Schaeffer/Corson/Fogarty	NS707	492 BG
Faro	10-11 April	Contact	Knapp/Jackson/Ancrum	NS725	25 BG
Luxe	10-11 April	No Contact	Knapp/Jackson/Ancrum	NS725	25 BG
Buzzsaw	11-12 April	No Contact	Schaeffer/Corson/Fogarty	NS707	492 BG
Luxe	13-14 April	Contact	Knapp/Jackson/Ancrum	NS740	25 BG
Pickaxe	13-14 April	Contact	Knapp/Jackson/Ancrum	NS740	25 BG
Buzzsaw	14-15 April	No Contact	Schaeffer/Corson/Fogarty	NS740	492 BG
Faro	14-15 April	No Contact	Schaeffer/Corson/Fogarty	NS740	492 BG
Mallet	15-16 April	No Contact	Knapp/Jackson/Ancrum	NS725	25 BG
Hammer	15-16 April	No Contact	Knapp/Jackson/Ancrum	NS725	25 BG
Buzzsaw	16-17 April	Cancelled	Schaeffer/Corson/Fogarty	NS740	492 BG
Faro	16-17 April	No Contact	Schaeffer/Corson/Fogarty	NS740	492 BG
Pickaxe	17-18 April	Contact	Knapp/Jackson/Ancrum	NS740	25 BG
Luxe	17-18 April	Contact	Knapp/Jackson/Ancrum	NS740	25 BG
Mallet	18-19 April	No Contact	Schaeffer/Corson/Fogarty	NS740	492 BG
Hammer	18-19 April	No Contact	Schaeffer/Corson/Fogarty	NS740	492 BG
Farmer	19-20 April	No Contact	Knapp/Jackson/Ancrum	NS725	25 BG
Chauffeur	19-20 April	Contact	Knapp/Jackson/Ancrum	NS725	25 BG
Hammer	20-21 April	Abort	Schaeffer/Corson/Fogarty	NS740 &	492 BG
Mallet	20-21 April	Abort	Schaeffer/Corson/Fogarty	NS707	25 BG
Pickaxe	20-21 April	Contact	Knapp/Jackson/Ancrum	NS725	25 BG
Luxe	20-21 April	Contact	Knapp/Jackson/Ancrum	NS725	25 BG
Chauffeur	23-24 April	Contact	Knapp/Jackson/Pecquet	RG157	25 BG
Farmer	23-24 April	No Contact	Knapp/Jackson/Pecquet	RG157	25 BG
Mallet	23-24 April	No Contact	Schaeffer/Corson/Fogarty	NS707	492 BG
Hammer	23-24 April	No Contact	Schaeffer/Corson/Fogarty	NS707	492 BG
Pickaxe	25-26 April	Contact	Knapp/Jackson/Ancrum	NS740	25 BG
Luxe	25-26 April	Contact	Knapp/Jackson/Ancrum	NS740	25 BG
Chauffeur	27-28 April	No Contact	Knapp/Jackson/Pecquet	NS740	25 BG
Farmer	27-28 April	No Contact	Knapp/Jackson/Pecquet	NS740	25 BG
Pickaxe	28-29 April	Contact	Schaeffer/Corson/Ancrum	RG157	492 BG
Luxe	28-29 April	Contact	Schaeffer/Corson/Ancrum	RG157	492 BG
Sultane	29-30 April	No Contact	Webb/Kolacki/Nater	RG157	492 BG
Boyard	29-30 April	No Contact	Webb/Kolacki/Nater	RG157	492 BG
Luxe	30 April/1 May	No Contact	Webb/Kolacki/Ancrum	NS740	492 BG
Pickaxe	30 April/1 May	Contact	Webb/Kolacki/Ancrum	NS740	492 BG
Chauffeur	1-2 May	Abort	Kuntz/Edwards/Pecquet	RG157/NS740	492 BG

Personnel Losses during Operations and Training

Date	Name	Rank	Duty	Unit	Aircraft	Notes
17 July 1942	Dawson, Albert	2/Lt.	Pilot	50 FS	P-38F 42-12588	KIFA Collision with fuel trailer on takeoff, Bradley Fld., Connecticut.
1 August 1942	Goodrich, Herbert	2/Lt.	Pilot	50 FS	P-38 42-12589	MIA Disappeared during flight from Goose Bay, Canada to BW1, Greenland
25 April 1943	Eichman, Frederick E.	2/Lt.	Pilot	50 FS	P-3842-12575	KIA Crashed during a search mission, Iceland.
27 August 1943	Wood, William	S/Sgt.	Cameraman	8 CCU		Evaded Shot down St. Omer, France, evaded to Spain, returned 17 December 1943.
9 September 1943	Edwards, Robert H.	2/Lt.	Cameraman	8 CCU		MIA On his second mission when bomber shot down over Boulogne, France.
14 October 1943	Arnold, Alfred A.	2/Lt.	Pilot	50 FS	P-38F 42-12586	KIA Crashed into mountain during search mission, Iceland.
18 October 1943	Stengle, Harry P.	2/Lt.	Pilot	50 FS	P-38F 42-15276	KIA Killed bailing out from burning P-38 in Iceland
13 November 1943	Slate, Ervin G.	T/Sgt	Cameraman	8 CCU	B-24	MIA Shot down in B-24 during mission to Bremen.

Date	Name	Rank	Duty	Unit	Aircraft	Notes
9 December 1943	Rissinger, LaVerne	2/Lt.	Pilot	1 CCRC	B-17 4237744	KIA
	Holcomb, William H.	2/Lt.	Copilot			KIA
	Ahlers, Otto H.	2/Lt.	Navigator			KIA
	Costello, Peter M.	T/Sgt.	Engineer			KIA
	Cohen, Bernard	S/Sgt.	Radio Operator			KIA
	Ekes, Matthew	Cpl.	Gunner			KIA
	Buchanan, John P.	M/Sgt.	Engineer			KIA
	Harding, Ralph W.	Sgt.	Gunner			KIA
	Osborne, Jack H.	S/L	RAF	Met/Obs.		KIA
	Leigh-Clarke, John	F/L	RAF	Met/Obs.		KIA
						1 CCRC crew, take off crash from Bovingdon, UK.
25 December 1943	Neary, Richard T.	2/Lt.	Navigator	1 CCRC	B-17 42-37869	KIA
	Renner, Sherwood R.	T/Sgt.	Engineer			KIA
	Craig, Samuel L.	T/Sgt.	Radio Operator			KIA
	Panetti, Mario A.	S/Sgt.	Gunner			KIA
	Blanchard, Albany J.	S/Sgt.	Gunner			
						Crashed at Oakhampton returning from Atlantic WR Mission
11 January 1945	Barker Dave A.	S/Sgt	Cameraman	8 CCU	B-17 42-30248	KIA
						Shot down in 410 BS, 94 BG bomber, Braunschweig, Germany
2 February 1944	Grafton, Robert H.;	2/Lt.	Pilot	1 CCRC	B-17F 42-30131	KIA
	Williams, Ralph E.	2/Lt.	Copilot			KIA
	Durgin, Casper M.	2/Lt.	Navigator			KIA
	Feller, Louis	T/Sgt	Engineer			KIA
	Nicolette, Angel M.	S/Sgt.	Radio Operator			KIA
	Lawr, Tillman O.	S/Sgt.	Gunner			KIA
	Green Charles	Sgt.	Gunner			KIA
	Pye, Edward J. B.		RAF	Met/Obs		KIA
						Disappeared over Atlantic Ocean during weather mission.
5 May 1944	Thompson, Richard B.	Captain	Pilot	8 Wx Rcn Sq. (L)(P)	Oxford T1066	KIFA
	Hicks, David P.	1/Lt.	Pilot			Injured
						Crashed in East Dereham near Middleton Tower.
31 May 1944	Jenkins, Herbert C.	Sgt.	Cameraman	8 CCU		KIA
						Killed by flak while filming from a heavy bomber.
8 June 1944	Gernan, Walter D.	Captain,	Pilot	8 Rcn Sq. (S)(P)	Mosquito NS555	KIA
	Lynch, Ebbet C.	Sgt.	Cameraman	8 CCU		KIA
						Crashed while returning from D-Day invasion coast.

Date	Name	Rank	Duty	Unit	Aircraft	Notes
21 June 1944	Clounch, Richard	Captain	Pilot	8 Rcn	Mosquito NS558	KFIA
	O'Connor, C.	1/Lt.	Navigator	8 Rcn Sq.(S)(P)		KFIA
						Struck barrage balloon cable during landing at Grave's End.
7 July 1944	Mann, John J.	1/Lt.	Pilot	8 Wx Rcn (L)(P)	Mosquito NS537	KIA
	Davis, William L.	2/Lt.	Navigator	8 Wx Rcn (L)(P)		KIA
						Bailed out over North Sea during Scout mission.
5 August 1944	Arrowood, Robert F.	Sgt.	Cameraman	8 CCU	B-24 42-95190	POW Shot down in 707 BS, 446 BG bomber, captured at Wehlpage, Germany.
7 August 1944	Thompson, Walter W.	1/Lt.	Pilot	8 Rcn Sq (S)(P)	Mosquito MM311	KIA
	Edgard, Carl C.	2/Lt.	Navigator	8 Rcn Sq (S)(P)		KIA Shot down by Luftwaffe night-fighter during a Mickey mission.
12 August 1944	Nichols, Ronald	1/Lt.	Pilot	653 BS	Mosquito NS533	KIA
	Harris, Elbert F.	2/Lt.	Navigator	653 BS		Evaded Shot down by P-51 over Toulouse, France, returning from Frantic mission.
13 August 1944	Sanner, Dean	1/Lt.	Pilot	654 BS	Mosquito MM370	POW
	Kurjack, August	S/Sgt.	Cameraman	8 CCU		KIA Mosquito destroyed by exploding Batty bomb, filming attack on E-boat pens.
19 August 1944	Musgrsove, Raymond L.	1/Lt.	Pilot	654 BS	Mosquito NS625	KIA
	Fordham, Harold F.	1/Lt.	Navigator	6564 BS		POW Second Mickey Mosquito lost on operations over Germany.
6 September 1944	Stephenson, Clarence R.	1/Lt.	Pilot	652 BS	B-17G 42-31736	KIA
	Swianicki, Eugene W.	1/Lt.	Navigator			KIA
	Ward, Joseph	2/Lt.	Copilot			KIA
	Licurse, Leonard L.	S/Sgt	Radio/Gunner			KIA
	Westerholm, Earl N.	T/Sgt	Gunner			KIA Crashed on take off for weather mission to Azores

Date	Name	Rank	Duty	Unit	Aircraft	Notes
8 September 1944	Whitmer, Russell C.	Flying Officer	Pilot	654 BS	Mosquito NS584	KIFA
	Armstrong, Raymond C.	M/Sgt.	Radar Technician	654 BS		KIFA
						Aircraft exploded mid-air near Kennet during test flight
18 September 1944	Tunnell, Robert A.	1/Lt.	Pilot	654 BS	Mosquito NS593	KIA
	Cunney, John G.	S/Sgt	Cameraman	8 CCU		KIA
						Shot down by ground fire near Platunne, Germany.
22 September 1944	Barnett, Harry	Sgt.	Radio Operator	652 BS	B-24J 44-40879	KIA
	Knowles, Richard A.	Sgt.	Gunner			KIA
	Chiricio, Matthew	Cpl.	Gunner			KIA
	Macklin, John N.	2/Lt.		Me/Obs	18 Weather Detachment	KIA
	Stoodley, Richard W.	M/Sgt		Met/Obs	18 Weather Detachment	KIA
						Killed in landing accident at Watton returning from Atlantic WR.
24 September 1944	Vinyard, Clayborne O.	1/Lt.	Pilot	653 BS	Mosquito NS570	POW
	O'Mara, John J.	1/Lt.	Navigator	654 BS		POW
						Shot down over Germany by Luftwaffe night-fighter during weather mission.
11 October 1944	Hazlett, Stanley V.	2/Lt.	Pilot	654 BS	P-38-J-10L0 42-67903	KIFA
						Aircraft crashed five mile east of Watton during aerial training.
25 October 1944	Brooks, George M.	1/Lt.	Pilot	654 BS	Mosquito NS582	KIA
	Taylor, Richard C.	2/Lt.	Navigator	654 BS		Returned
						Pilot struck tail on bailout and fell to his death, Ypres, Belgium.
1 November 1944	Chipman, Vance J.	2/Lt.	Pilot	654 BS	Mosquito NS677	POW
	Cannon, William G.	1/Lt.	Navigator	654 BS		POW
						Third Mickey aircraft lost, shot down by flak over Schweinfurt, Germany
6 November 1944	Grimes, Robert C.	1/Lt.	Pilot	653 BS	Mosquito MM386	KIA
	Jodar, Clarence W.	1/Lt.	Navigator	653 BS		KIA
						Crashed on landing after return from weather mission over Germany
6 November 1944	Kaellner, Otto E.	1/Lt	Pilot	654 BS	Mosquito NS568	KIA
	Cerruti, Edwin, R.	1/Lt.	Navigator	654 BS		Injured
						Crashed on landing returning from Joker mission over Germany.

Date	Name	Rank	Duty	Unit	Aircraft	Notes
18 November 1944	Marshal, Benjamin, C.	1/Lt.	Pilot	652 BS	B-24J 44-40484	Interned
	Gisriel, James M.	2/Lt.	Copilot			Interned
	Gulick, James N.	2/Lt.	Navigator			Interned
	Dooley, Jack P.	S/Sgt	Radio/Gunner			Interned
	Mancuso, Americo J.	S/Sgt.	Engineer/Gunner			Interned
	Taylor, Maxwell R.	Sgt.	Gunner			Interned
	Smith John B.	Sgt.	Gunner			Interned
	Johnson, Adolph F.	1/Lt.		Met/Obs	18 Weather Detachment	Interned
	Zagol, Stanley	T/Sgt		Met/Obs	18 Weather Detachment	Interned
						Sharon weather flight from Azores, interned in Portugal, crew repatriated.
22 November 1944	MacLeod, Malcolm J.	1/Lt.	Pilot	653 BS	Mosquito NS630	KIA
	Fitzgerald, Edward G.	2/Lt.	Navigator	653 BS		KIA
						Lost an engine on takeoff for weather mission and crashed.
22 November 1944	Harry, Russell E.	1/Lt.	Pilot	653 BS	Mosquito NS515	KIA
	Hopkins, Milford B.	12/Lt.	Navigator	653 BS		KIA
						Lost and engine on takeoff for weather mission and crashed.
23 December 1944	Spear, James D.	Flying Officer	Pilot	654 BS	Mosquito NS638	KIFA
	Bryan, Carroll	1/Lt	Navigator	654 BS		KIFA
						Crashed near Dursley, England, while on a training flight.
21 January 1945	Roberts, Jerry M.	2/Lt.	Pilot	653 BS	Mosquito NS509	POW
	Fisher, Ralph E.	1/Lt.	Navigator	653 BS		KIA
						Shot down over Germany during weather mission.
8 February 1945	Doroski, Victor	Captain	Pilot	654 BS	Mosquito NS583	KIA
	Hochman, Jacob	Captain	Navigator	654 BS		KIA
						Photo flash bomb ignited in bomb-bay during Joker Mission to Germany.
17 March 1945	Estrach, Calman T.	2/Lt.	Pilot	652 BS	B-17G 448601	KIA
	Clark, Robert N.	2/Lt.	Copilot			KIA
	Pelligrini, Anthony D.	2/Lt.	Navigator			KIA
	Eickhoff, Arthur G.	Cpl.	Radio/Gunner			KIA
	Allison, Robert C.	S/Sgt.	Engineer/Gunner			KIA
	Chressanthis, James G.	Cpl.	Gunner			KIA
	Gerlott, John C.	Sgt.	Gunner			KIA
	Sortomme, Charles L.	1/Lt.		Met/Obs	18 Weather Detachment	KIA
						Allah weather flight disappeared over the Atlantic Ocean.
19 March 1945	Emmel, Oliver N.	1/Lt.	Pilot	654 BS	A-26 43-22524	KIA
	Walch, John W.	Major	Navigator	HQ 25	BGH Rcn	KIA
	Tresemer, Edward C.	Major	Navigator	HQ 492	BG	KIA
	Brunner, Frederick J.	Pl. Sgt.	Gunner	USMC	OSS	KIA
	Grimm, Karl	Civilian OSS	Secret Agent			KIA
						Crashed at Schwege, Germany, during CHISEL insertion mission.

Date	Name	Rank	Duty	Unit	Aircraft	Notes
20 March 1945	Polovick, Joseph A.	2/Lt.	Pilot	654 BS	Mosquito RF996	POW
	Blaum, Bernard M.	1/Lt.	Navigator	654 BS		POW Suffered fuel vapor lock to both engines returning from Graypea mission.
24 March 1945	Stubblefield, Carrol, B.	1/Lt.	Pilot	653 BS	Mosquito NS711	KIA
	Richmond, James B.	1/Lt.	Navigator	654 BS		POW Shot down over Germany by Ninth AF P-47s.
25 March 1945	Boucher Bernard J.	1/Lt.	Pilot	653 BS	Mosquito NS752	POW
	Pessirilo, Louis	1/Lt.	Navigator	653 BS		KIA Shot down on weather mission over Germany.
3 April 1945	Podwojski, Alvin E.	Lt. Col.	Pilot	HQ 25 BG (Rcn)	Mosquito NS650	Interned
	Proulx, Lionel A.	Capt.	Navigator	HQ 25 BG (Rcn)		Interned Lost engine on Graypea to Kiel, Germany. Landed in Sweden.
4 April 1945	Smith, Theodore B.	1/Lt.	Pilot	653 BS	Mosquito NS635	POW
	Crawford, Troy	Colonel	Commander	445 BG		POW Shot down by B-24 gunners of CO's own 445 BG
9 April 1945	James, Robert W.	1/Lt.	Pilot	653 BS	Mosquito NS757	KIFA
	Hall, Terence S.	1/Lt.	Navigator	653 BS		KIFA Crashed at RAF Station Shepherds Grove due to unknown cause.
9 April 1945	Pruis, John A.	1/Lt.	Pilot	654 BS	Mosquito NS792	KIA
	Moore, Claude C.	1/Lt.	Navigator.	654 BS		Wounded Shot down by French Air Force P-51 Mustang on Graypea mission to Germany.

Best-case Scenario Synopsis of Mosquitos Assigned to 802 Rcn Grp (Sp)(P) and 25 BG (Rcn)(S) (Sp) Watton, England, W.W.II, with Assistance from Phil Butler.

MM308/G del. Alconbury 22 February 1944.
Mickey H2X, 482 BG conversion.
Accident 12 May 1944, H.M. Stewart, H.T. Gruis.
13 May 1944 Dismantled Alconbury.
7 December 1944 Reduced to Spares.

MM310 del. Langford Lodge 2 February 1944.
Cat. E FA Warton, bowed landing gear.
Repaired at Warton.
Accident 17 June 1944, C.W. Himes.
De Havilland Cat. B RW 2 July 1944.
Returned to USAAF (date not given).

MM311 delivered Alconbury 22 February 1944.
2nd Mickey H2X, 482 BG conversion.
13 May 1944 8 Special Squadron, A-Able.
7 August 1944, W.W. Thompson, C.C. Edgar, both KIA.

MM337 del. Burtonwood 11 February 1944.
29 February 1944 Nuthampstead, transition training.
Accident 3 March 1944, E.J. Maslow, O.H. Emmel.
RW Martin Hearn 7 March 1944.
USAAF 21 February 1945.
653 Squadron.
10 MU 16 June 1945.
SOC RTP 31 July 1946.

MM338 del. Burtonwood 9 February 1944.
Prototype WR aircraft 4 April 1944.
8 Light Weather Squadron, A-Able.
Accident 26 May 1944, H.H. Bozarth.
653 Squadron.
10 MU 16 June 1945.
Sold to France 15 October 1947.

MM340 del. Burtonwood 12 February 1944.
29 February 1944 Nuthampstead, transition training.
Accident 18 March 1944 Warton, C.W. Himes.
ROS DH 6 April 1944.
AW/CN 8 May 1944.
USAAF Watton 21 February 1945.
44 MU 10 August 1945.
Cat. E at DH 17 June 1946.

MM342 del. Burtonwood 14 February 1944.
2 March 1944 Cheddington training.
653 Squadron, P-Paul.
Accident 20 January1945, J.A Polovick.
ROS 31 January 1945.

MM344 del. Burtonwood 10 February 1944.
22 February 1944 Burtonwood for repairs.
Cheddington training.
Accident 22 March 1944, G.C. Hand.
RW DH Hatfield 13 April 1944.
USAAF Watton 20 September 1944.
653 Squadron.
654 Squadron, Chaff aircraft.
10 MU 16 June 1945.
SOC 25 July 1946 Broken down by HASA.

MM345 del. Burtonwood 10 February 1944.
2 March 1944 Cheddington training.
4 April 1944 Cheddington.
653 Squadron, Z-Zebra.
Accident 8 October 1944, R.C. Grimes.
Accident 7 December 1944, G.C. Hand, S.E .Horne.
Chaff aircraft.
Accident 20 February 1945, W.C. Davis.
ROS 71 MU 27 February 1945.
10 MU 10 June 1945.
SOC 25 July 1946 Broken down by HASA.

MM346 del. Burtonwood 15 February 1944.
4 April 1944 Cheddington training.
Accident 11 June 1944, R.D. Shoenhair.
653 Squadron, M-Mother.
44 MU 21 February 1946.
RNAS Fleet lands 17 December 1946.

MM364 del. Burtonwood 7 March 1944.
14 March 1944 Cheddington training.
8 Special Squadron, PRU aircraft.
654 Squadron, C-Charlie.
Accident 29 August 1944, C.P. Hackman.
Accident 14 January 1945, G.M. Vanderleest.
Ex-works civil repair organ.14 August 1945.
22 MU 22 August 1945.

MM368 del. Burtonwood 8 March 1944.
14 March 1944 Cheddington training.
25 April 1944 Burtonwood.
8 Light Weather Squadron, B-Baker.
Accident 7 August 1944, F.D .Porter. J.M. Davis.
ROS 21 August 1944.
USAAF Burtonwood 16 September 1944.
ROS 18 October 1944, Chaff aircraft.
USAAF Burtonwood 7 November 1944.
Accident 27 November 1944, S.G. Pilson.
ROS 7 MU 24 January 1945.
Accident 20 April 1945, L. Barnhard.
22 August 1945, 51 MU, Repair at Watton.
51 MU 23 October 1946.
RNAS Stretton 11 November 1946.

MM370 del. Burtonwood 11 March 1944.
4 April 1944 Cheddington training.
13 August 1944, D.H. Sanner POW, A. Kurjack KIA.

MM371 del. Burtonwood 11 March 1944.
4 April 1944 Cheddington training.
653 Squadron, Y-Yoke.
Prototype Chaff aircraft.
Accident 18 December 1944, J.M. Roberts.

MM382 (not listed)
Cheddington crashed in training.
Return to RAF.

MM384 del. Burtonwood 11 March 1944.
Accident 26 April 1944, R.E. Petersen, O.E. Kaellner.
ROS 4 May 1944.
USAAF Burtonwood 8 July 1944.
Accident 18 September 1944, C.O. Vinyard.

MM385 del. Burtonwood 11 March 1944.
Accident 24 March 1944, T.E. Hughes.
10 MU 4 April 1944.
RW DH Hatfield 11 April 1944.
Watton 19 January 1945, C-Charlie.
654 Squadron, Redtail.
10 MU 16 June 1945.
51 MU 30 November 1945.
Scrapped 23 July 1947.

MM386 del. Burtonwood 4 May 194.
8 Light Weather Squadron.
653 Squadron, O-Oboe.
Accident 6 November 1944
R.C. Grimes, C.W. Jodar, both KIA.

MM387 del. Burtonwood 18 March 1944.
25 April 1944 Burtonwood ready.
8 Light Weather Squadron.
653 Squadron, C-Charlie.
Accident 13 April 1945, R Gordon, J.W. Handren.
SOC 7 May 1945.

MM388 del. Burtonwood 19 March 1944.
8 Light Weather Squadron, D-Dog.
Cat. E FA 2 September 1944, V.J. Chipman.
AW/CN DH Hatfield 6 January 1945.
USAAF 9 February 1945, H-Harry.
Accident 21 February 1945, R.P. Houle.
ROS 71 MU 21 February 1945.
654 Squadron, BA-H (H-Harry).
Accident 18 June 1945, J.R. Hoover, E.R. Patterson.
SOC 19 October 1945.

MM391 del. Burtonwood 12 March 1944.
8 Light Weather Squadron.
653 Squadron, E-Easy.
Accident 9 September 1944, J.P. Noble, C.O. Gorham.
RW Martin Hearn 25 September 1944.
USAAF Watton 30 April 1945.
10 MU 16 June 1945.
51 MU 11 December 1945.
Sold to France 30 June 1946.

MM393 del. Burtonwood 14 March 1944.
8 Light Weather Squadron.
653 Squadron, F-Fox.
654 Squadron, Chaff aircraft.
10 MU 16 June 1945.
SOC 13 February 1947.

NS508 del. Burtonwood 4 April 1944.
8 Light Weather Squadron.
653 Squadron, G-George.
654 Squadron, Chaff aircraft.
Accident 20 February 1945, W.C. Davis.
Cat. E 20 February 1945.
654 Squadron, Chaff aircraft.

NS509 del. Burtonwood 4 April 1944.
8 Light Weather Squadron.
653 Squadron, H-Harry.
22 January 1945, J. Roberts POW, R. Fisher KIA.
Cat. E 10 April 1945.

NS510 (not listed)
8 Light Weather, I-Item.
Accident 25 July 1944, J.E. Goodbread, W.R. Barber.

NS512 del. Burtonwood 5 April 1944.
8 Light Weather Squadron.
653 Squadron, J-Jig.
654 Squadron, Chaff aircraft.
Accident 13 April 1945, J. Larkin, R. Gordon.
SOC 5 May 1945.

NS513 del. Burtonwood 4 April 1944.
653 Squadron, K-King.
Accident 14 September 1944, F.D. Porter, J. Davis.

NS514 del. Burtonwood 5 April 1944.
8 Light Weather Squadron.
653 Squadron, L-Love.
Accident 22 October 1944, M.J. MacLeod, E.J. Fitzgerald.
Cat. E FA 22 October 1944.

NS515 del. Burtonwood 6 April 1944.
8 Light Weather Squadron.
653 Squadron, M-Mike.
22 November 1944, R.E. Harry, M.B. Hopkins, both KIA.
Cat.E2, TT218 hrs, 120 hrs since overhaul.

NS516 del. Burtonwood 11 April 1944.
8 Light Weather Squadron, N-Nan.
653 Squadron, Chaff aircraft.
Accident 11 April 1945, E.F. Hodges, D.C. Bedard.
SOC 4 May 1945.

NS518 del. Burtonwood 12 April 1944.
8 Light Weather Squadron.
Accident 3 June 1944, H.L. James, W.E. Scharbach.

NS519 del. Burtonwood 12 April 1944.
8 Light Weather Squadron.
653 Squadron, P-Peter.
654 Squadron, Chaff aircraft.
Accident 27 December 1944, M.M. Hunt, R. Davies.
Cat. E 15 January 1945.

NS533 del. Burtonwood 5 April 1944.
Accident 15 June 44 Burtonwood, Lt. Hutchins.
8 Light Weather Squadron, Q-Queen.
12 August 1944, R.M. Nichols KIA, E.F. Harris evaded.

NS535 del. Burtonwood 12 May 1944.
8 Special Squadron, A-Able.
6 June 1944, 2nd Rooster Flight.
Accident 13 June 1944, O.E. Kaellner.
Cat. E FB 24 June 1944.

NS537 del. Burtonwood 8 May 1944.
8 Light Weather Squadron.
7 July 1944, J.J. Mann, W.L. Davis, both MIA.

NS538 del. Alconbury 4 July 1944.
8 Special Squadron, Mickey H2X.
654 Squadron, F-Fox.
Accident 8 September 1944,
R.C. Whitmer, R.G. Armstrong, both KIFA.

NS551 del. Burtonwood 12 May 1944.
10 MU 16 June 1945.
8 Special Squadron.
654 Squadron, Night Photo, K-Kilo.
5952M ground instructional airframe at RAF School.
of Photography 12 June 1946.

NS552 del. 31 May 1944.
654 Squadron, G-Golf.
Accident 21 September 1944, P.D. Dustman, C.A. Veldon.
Cat. A/C TT 140 hrs.
Nigh Photo.
Accident 23 March 1945, M.M. Hunt, A. Morrow.
71 MU 1 June 1945, Repair completed Watton.
10 MU 16 June 1945.
SOC 25 July 1946 Broken down by H&SA.

NS553 del. Burtonwood 15 May 1944.
Accident 26 May 1944, O.E. Kaellner, J.E. Matuska.
654 Squadron, P-Paul, Night Photo.
10 MU 18 June 1945.
SOC 25 July 1946 Broken down by H&SA.

NS554 del. Burtonwood 13 May 1944.
8 Special Squadron.
654 Squadron, Night Photo, V-Victor.
44 MU 22 June 1945.
27 July 1945 to Watton.
Cat. B RW Martin Hearn 18 March 1946.
Re-Cat. E 9 January 1947.

NS555 del. Watton 19 May 1944.
8 Special Squadron, A-Able.
8 June 1944, W..D. Gernand, E.C. Lynch, both KIA.

NS556 del. Watton 7 June 1944.
654 Squadron, Night Photo, J-Jig.
Accident 23 January 1945, R.N. Geary, P. Lovett.
SOC Cat.E2 27 February 1945.

NS557 del. Watton 7 June 1944.
8 Special Squadron, Night Photo aircraft.
Accident 6 August 1944, J.E. Matuska, R.A. Berge.
Cat.E1 MR 23 August 1944 SOC.

NS558 del. Alconbury 14 June 1944.
8 Special Squadron.
Accident 21 June 1944, R.F. Clounch, C.C. O'Connor, both KIFA.
Cat. E FB 21 June 1944.

NS559 del. Watton 9 June 1944.
8 Special Squadron.
4 August 1945, R. Tunnell, D. McCarthy, Cannon hole.
654 Squadron, Night Photo, I-Item.
Accident 1 February 1945, A.R. Bateman, H.F. Mann.
7 MU ROS 8 February 1945.
Watton 22 February 1945.
653 Squadron.
10 MU 16 June 1945.
SOC 25 July 1946. Broken down H&SA.

NS568 del. Watton 23 May 1944.
654 Squadron, PRU, L-Love.
Night Photo, B-Baker.
6 November 1944, accident,
O.E. Kaellner KIA, E.R. Cerruti injured.
SOC MR Cat. E 20 December 1944.

NS569 del. Watton 15 June 1944.
653 Squadron, N-Nan.
Accident 4 April 1945, J.F. Murphy.
SOC 18 April 1945.

NS570 del. Watton 14 June 1944.
653 Squadron.
25 September 1944, C.O. Vinyard, J.J. O`Mara,
both POWs.

NS574 del. Watton 5 July 1944.
Returned to RAF Benson 12 July 1944.
Assigned in error.

NS577 del. Watton 5 July 1944.
Probably returned to RAF.

NS581 del. Watton 17 June 1944.
654 Squadron, Night Photo, Q-Queen.
10 MU 16 June 1945.
22 MU 15 December 1945.
Scrapped 15 August 1947.

NS582 del. Watton 17 June 1944.
654 Squadron, Night Photo, R-Roger.
25 October 1944, G.M. Brooks KIA, R.C. Taylor returned.

NS583 del. Watton 17 June 1944.
654 Squadron, Night Photo.
8 February 1945, V.S. Doroski, J. Hochman, both KIA.

NS584 del. Alconbury 19 July 1944.
8 Special Squadron, Mickey H2X.
654 Squadron, D-Dog.
19 March 1945 492nd BG/OSS.
Accident 3 April 1945, R.L. Boone.

NS585 del. Watton 29 June 1944.
8 Light Weather Squadron.
653 Squadron, D-Dog.
Re-Cat. E 17 May 45 Martin Hearn Ltd.
22 MU 24 October 1945.
Sold to France 10 July 1946.

NS590 del. Watton 2 July 1944.
654 Squadron, PRU, B-Baker, S-B.
Accident 27 August 1944, O.E. Kaellner.
Accident 7 December 1944, J.M. Carter.
7 MU ROS 9 January 1945.
Watton 20 February 1945.
Accident 5 May 1945, L.F. Lakeman, Sgt G. Senseny.
1 June 1945, Repairs completed Watton.
10 MU 18 June 1945.
Sold 4 September 1947 (? to foreign air force)

NS591 del. Watton 23 June 1944.
653 Squadron, S-Sugar.
AW/CN Marshalls 30 July 1945.
9 MU 14 August 1945.
Sold France 17 July 1946.

NS592 del. Watton 29 June 1944.
653 Squadron, T-Tare.
654 Squadron, Chaff aircraft.
Marshalls 25 May 1945.
44 MU 26 September 1945.
Crashed en route, ROS DH.
RNAS Fleetlands 26 September 1946.

NS593 del. Watton 5 July 1944.
654 Squadron, U-Uncle, B-Baker.
18 September 1944, R.A. Tunnell, J.G. Cunney, both KIA.

NS594 del. Watton 28 June 1944.
653 Squadron, U-Uncle.
654 Squadron, Chaff aircraft.
Accident 28 January 1945, R.W. James, W.E. Loden.
Cat.E2 SOC 27 February 1945.

NS595 del. Watton 7 July 1944.
654 Squadron, PRU, A-Able.
Accident 29 January 1945, H.E. Peterson.
Accident 9 March 1945, R.B. Hastie.
10 MU 18 June 1945.
SOC 25 July 1946 Broken down by H&SA.

NS596 del. Watton 6 July 1944.
653 Squadron, M-Mother.
Accident 29 November 1944, W.B. Rouse, A. Morrow.

NS619 del. Watton 9 July 1944.
654 Squadron, PRU, U-Uncle.
Accident 22 March 1945, K. McGriffin.
Re-Cat.E1 3 May 1945.
25 May 1945, Cat. B, 43rd Air Group.

NS620 del. Watton 10 July 1944.
653 Squadron, W-William.
Accident 3 November 1944, M. Wayne, R. Wright.

NS625 del. Alconbury 2 August 1944.
Mickey H2X.
654 Squadron, K-King.
19 August 1944, H. Fordham POW, R.L. Musgrove KIA.

NS626 del. Watton 19 July 1944.
653 Squadron
654 Squadron, Chaff aircraft.
Re-Cat. E 17 May 1945.
22 MU Martin Hearn, 27 November 1945.
Sold to France 25 May 1946.

NS630 del. Watton 22 July 1944.
653 Squadron.
22 November 1944, M.S. MacLeod, E.G. Fitzgerald, both KIA.

NS632 del. Watton 28 July 1944.
653rd Squadron.
25 May 1945 Marshalls MI.
10 MU 14 January 1946.
Sold France 13 March 1946.

NS634 del. Watton 22 July 1944.
653 Squadron.
20 April 1945 Marshalls MI.
10 MU 10 January 1946.
Sold France 13 March 1946.

NS635 del. Watton 25 July 1944.
653 Squadron, I-Item.
Accident 16 September 1944, R.L Lee, W. Biggers.
Cat. A/C, TT 74 hours.
654th Squadron.
4 April 1945, T.B. Smith, T. Crawford, both POWs.

NS636 del. Watton 28 July 1944.
653 Squadron.
Marshalls 30 April 1945.
22 MU 23 December 1945.
Sold France 28 April 1946.

NS638 del. Watton 31 August 1944.
8 Special Squadron.
18 August 44 Tropical Mk 16, Mickey H2X.
654 Squadron.
23 December 1944, J.D. Spear, C.B. Bryan, both KIFA.
Cat. E FA Watton.

NS650 del. Watton 25 August 1944.
654 Squadron, Chaff aircraft H-Hotel.
3 April 1945 Interned Sweden, A.E. Podwojski, L.A. Proulx.
RAF Chalgrove (no date) Converted to instructional.
airframe '5710M' RAF Mount Farm, 9 October 1945.

NS651 del. Watton 25 August 1944.
653 Squadron, F-Fox.
654 Squadron, Chaff aircraft.
Accident 7 March 1945, J.C. Evans, R.G. Spaight.
1 May 1945 repair Watton.
44 MU 22 June 1945.
Scrapped 1 November 1945.

NS676 del. Watton 3 October 1944.
654 Squadron, Mickey H2X.
October 1944 OSS/JE.
18 March 1945 492 BG/OSS.
10 MU 18 June 1945.
57 MU 1 December 1945.
Scrapped 7 August 1947.

NS677 del. Alconbury 18 August 1944.
654 Squadron.
18 August 1944, Tropical Mk 16, Mickey H2X.
1 November 1944, V.J. Chipman, W.G. Cannon, both POWs.

NS686 Alconbury 9 August 1944.
654 Squadron, W-Wagon.
Tropical Mk 16, Mickey H2X.
18 March 1945, 492 BG/OSS.
44 MU 22 June 1945.
Scrapped 8 July 1948.

NS707 del. Alconbury 18 October 1944.
654 Squadron, Mickey H2X, E-Echo.
October 1944 JE/OSS.
18 March 1945 492 BG/OSS.
25 April 1945, J.G. Kuntz, R.L. Greene KIA, J.M. Nater.

NS708 10 MU 5 October 1944.
del. Watton 29 November 1944.
653 Squadron, M-Mother.
Accident 30 May 1945, R.M. Kenny, A.R. Earnest.
412 R&SU 7 June 1945, SOC.

NS709 del. Alconbury 24 October 1944.
654 Squadron, Mickey H2X.
10 MU 18 June 1945.
SOC 27 July 1946 Broken down by H&SA.

NS710 10 MU 5 October 1944.
del. Watton 30 November 1944.
653 Squadron, L-Love.
10 MU 4 October 1944.
10 MU 8 August 1945.
22 MU 23 May 1946.
Sold to France 25 July 1946.

NS711 10 MU 2 November 1944.
del. Watton 31 December 1944.
24 March 1945, C.B. Stubblefield KIA, J.B. Richmond POW.

NS712 10 MU 5 October 1944.
del. Watton 29 November 1944.
653 Squadron, O-Oboe.
10 MU 16 June 1945.
SOC 25 July 1949 Broken down by H&SA.

NS725 10 MU 6 October 1944.
del. Watton 24 November 1944.
654 Squadron, L-Love.
January 1945 OSS/JE.
15 March 1945 492 BG/OSS.
10 MU 16 June 1945.
SOC 4 September 1947 (Sold to foreign AF?).

NS730 10 MU 9 October 1944.
del. Watton.29 November 1944.
653 Squadron.
Accident 4 February 1945, W.B Rouse, J.M. Levy.
SOC 21 February 1945.

NS739 10 MU 19 October 1944.
del. Watton 7 December 1944.
653 Squadron, F-Fred.
44 MU 10 August 1945.
214 MU 16 January 1946.
SOC 30 September 1947.

NS740 10 MU 18 October 1944.
del. Watton 2 December 1944.
653 Squadron, M-Mike.
January 1945 OSS/JE.
18 March 1945 492 BG/OSS.
44 MU 21 March 1949.
Scrapped 8 July 1948.

NS742 10 MU 19 October 1944.
del. Watton 7 December 1944.
653 Squadron.
RNAS Stretton 14 May 1947.
44 MU 21 March 1948.
8 July 1948 registered G-AOCJ.
Israel Air Force 30 Oct. 1956, IDFAF 219.
SOC 1957.

NS743 10 MU 19 October 1944.
del. Watton 7 December 1944.
654 Squadron, Chaff aircraft, Z-Zebra.
10 MU 15 February 1946.
RNAS Fleetlands 7 November 1947 for spares.

NS744 10 MU 23 October 1944.
del. Watton 18 December 1944.
653 Squadron, E-Easy.
Accident 9 April 1944, J M. Carter, W.E. Loden.
SOC Re-Cat.E1 17 May 1945, Martin Hearn Ltd.

NS745 10 MU 28 October 1944.
del. Watton 18 December 1944.
653 Squadron, S-Sam.
654 Squadron, Chaff aircraft.
10 MU 16 June 1945.
SOC 4 September 1947 (Sold to foreign AF?).

NS748 10 MU 28 October 1944.
del. Watton 31 December 1944.
653 Squadron, A-Able.
Accident 19 April 1945, J.M. Carter.
SOC 16 May 1945.

NS752 10 MU 18 October 1944.
del. Watton 18 December 1946.
653 Squadron.
25 March 1945, B.J. Boucher POW, L. Pessirilo KIA.

NS753 10 MU 18 October 1944.
del. Watton 7 December 1944.
653 Squadron.
Martin Hearn RW 6 March 1945.
44 MU 21 March 1946.
RNAS Fleetlands 17 December 1946.

NS754 del. (not listed) 14 January 1945.
653 Squadron.
10 MU 28 October 1945.
10 MU 16 June 1945.
SOC 25 July 1946 Broken down by H&SA.

NS756 10 MU 2 November 1944.
del. Watton 16 December 1944.
653 Squadron.
10 MU 16 June 1945.
Sold to France 17 October 1946.

NS757 del. Watton 31 December 1944.
653 Squadron.
Accident 9 April 1945, R.W. James, T.S. Hall,
both KIFA.
Cat. E2, SOC 10 May 1945.

NS758 del. Watton 19 January 1945.
653 Squadron, T-Tom.
Accident 28 May 1945, H.F. Flook, W.F. McKinney.
1 August 1945 Repairs completed Watton.
27 MU 22 August 1945.
57 MU 10 April 1946.
Sold to France 14 August 1946.

NS772 del. Watton 19 January 1945.
653 Squadron.
44 MU 21 March 1946.
RNAS Fleetlands 27 November 1946.

NS773 del. Watton 17 January 1945.
653 Squadron.
44 MU 21 March 1946.
RNAS Fleetlands 27 November 1946.

NS774 del. Watton 14 January 1945.
653 Squadron, M-Mother.
Accident 25 March 1945, W. Borges, R. Gordon.
27 MU 1 August 1945.
22 MU 30 October 1945.
Scrapped 5 August 1947.

NS775 del. Watton 17 January 1945.
654 Squadron, Chaff aircraft, F-Fox.
44 MU 21 March 1946.
RNAS Fleetlands 16 December 1946.

NS782 del. Watton.17 January 1945.
653 Squadron, V-Victor.
Accident 8 March 1945, R.W. James, T.S. Hall.
Re-Cat. E 10 April 1945.

NS783 del. Watton 28 January 1945.
654 Squadron, Chaff aircraft.
10 MU 16 June 1945.
Sold to France 16 July 1946.

NS785 del. Watton 19 January 1945.
654 Squadron, Chaff aircraft, B-Baker.
10 MU 16 June 1945.
Sold to France 6 September 1946.
Israel Air Force 2 April 1952, IDFAF 2139.
SOC 1958.

NS792 del. Watton 22 March 1945.
654 Squadron, Chaff aircraft.
9 April 1945, J.A. Pruis KIA, C.C. Moore WIA.

NS793 del. Watton 28 January 1945.
654 Squadron, Chaff aircraft, X-X ray.
10 MU 18 June 1945.
Poolebrook 19 October 1945.
57 MU 24 October 1945.
RW DH 2 April 1946.
49 MU 11 June 1947.
Scrapped 2 March 1948.

NS794 del. Watton 19 March 1945.
654 Squadron, Chaff aircraft.
10 MU 16 June 1945.
Sold to France 22 September 1946.

NS796 del. (not listed) 19 April 1945.
44 MU 22 June 1945.
To be sold to BSAA.
Sold as scrap l6 July 1947.

NS802 del. Watton 20 April 1945.
44 MU 22 June 1945.
To be sold to BSAA.
Sold as scrap 16 June 1947.

NS804 del. Watton 20 April 1945.
44 MU 22 June 1945.
RNAS Stretton 16 May 1947.

NS805 del. Watton 20 April 1945.
44 MU 22 June 1945.
Sold to France 9 May 1946.

NS811 del. Watton 19 April 1945.
44 MU 22 June 1945.
Sold L. Cheshire 1 November 1946, G-AIRU.
Crashed, Corsica, en route to Israel, July 1948.

NS812 del. Watton 20 April 1945.
44 MU 22 June 1945.
Sold L. Cheshire 1 November 1946, G-AIRT.
Israel Air Force 5 July 1948, IDFAF serial 2101.

RF979 del. Watton 17 January 1945.
654 Squadron, Chaff aircraft, D-Dog.
57 MU 6 December 1945.
51 MU 30 January 1947.
SOC 11 March 1948.

RF983 del. Watton 14 January 1945.
654 Squadron, Chaff aircraft, O-Oscar.
10 MU 18 August 1945.
Sold To France 18 July 1948.

RF985 del. Watton 17 January 1945.
Abbots Ripton.
Accident 19 February 1945, R.W. Gilbert, J.K. Jackson.
44 MU 21 March 1946.
RNAS Stretton 9 April 1947.

RF986 del. Watton 11 January 1945.
654 Squadron, Chaff aircraft, W-Word.
44 MU 21 March 1946.
RNAS Fleetlands 26 November 1946.

RF988 del. Watton 17 January 1945.
654 Squadron, Chaff aircraft, J-Jig.
10 MU 18 June 1945.
SOC 13 February 1947.

RF989 del. Watton 1 February 1945.
Dispatched to Mediterranean.
Allied Air Force 5 March 1945.
SOC 6 April 1945.

RF992 del. Watton 17 January 1945.
653 Squadron, Chaff aircraft, R-Robert.
Battle damage 20 March 1945, R.W. Gilbert, R. Spoerl.
17 April 1945 Repaired Watton.
10 MU 16 June 1945.
Sold to France 4 July 1946.

RF996 del. Watton 17 January 1945.
654 Squadron, Chaff aircraft, Y-Yoke.
20 March 45, J.A. Polovick, B. Blaum, both POWs.

RF999 del. Watton 14 January 1945.
654 Squadron, Chaff aircraft, T-Tom.
44 MU 21 March 1946.
RNAS Fleetlands 16 December 1946.

RG113 del. Watton 14 January 1945.
654 Squadron, Chaff aircraft, S-Sugar.
Accident 7 April 1945, J.A. Pruis, C.C. Moore.
SOC 23 April 1945.

RG145 del. Watton 23 March 1945.
Accident 17 April 1945, W.C. Davis, W. Jorgensen.
27 MU 27 April 1945.
19 July 1945 Repairs completed Watton.
Scrapped 11 July 1947.

RG146 del. Watton 22 March 1945.
653 Squadron, J-Jack.
44 MU 28 April 1945.
Martin Hearn Cat. B RW 11 February 1946.
Re-Cat. 9 January 1947.

RG148 del. Watton 23 March 1945.
654 Squadron, Chaff aircraft.
10 MU 16 June 1945.
Sold to France 10 April 1948.

RG154 del. Watton 23 March 1945.
44 MU 21 March 1946.
27MU 16 December 1946.
Sold to France 18 August 1947.

RG155 del. Watton 23 March 1945.
654 Squadron, Chaff aircraft.
10 MU 16 June 1945.
57 MU 5 December 1945.
Sold to France 11 September 1946.
Israel Air Force 1952, IDFAF 2149.
W/O 19 March 1955 Crash on landing.

RG156 del. Watton 23 March 1945.
654 Squadron, Chaff aircraft.
44 MU 24 August 1945.
Martin Hearn Cat. B RW 30 January 1946.
Re-Cat. E 9 January 1947.

RG157 del. Watton 22 March 1945.
25 BG Q-Queen.
1 April 1945 492nd BG/OSS.
10 MU 18 June 1945.
57 MU 2 November 1945.
Scrapped 7 August 1947.

TA164 del. Watton 23 March 1945.
1 April 1945 492 BG/OSS.
44 MU 21 March 1946.
RNAS Stretton 14 May 1947.

British registered G-AOCN.
Israel Air Force as 4XFDL-92.

Mosquito TIII.
LR516
10 MU 30 December 1943.
Mount Farm 21 January 1944.
Accident Nuthampstead.
FA Cat. B 27 February 1944, G.C. Doherty, O.I. Emmel.
Re-Cat E1 31 March 1944.
Total flying hours 17:20.

LR530
10 MU 11 March 1944.
5 April 44 to Watton.
Cheddington 5 August 1944.
Accident 24 March 45, J.K. Werthman, E.F. Hodge.
Accident 10 April 45, A.C. Horvath.
10 MU 18 August 1945.
To ground instructional airframe 5910M 29 March 1946.

LR534
10 MU 14 April 1944.
Accident Cheddington 24 April 1944,
C.S. Satterlund, D.R. Scanlan.
10 MU 18 August 1945.
Marshall RW 11 October 1945.
37 MU 23 July 1947.
9 MU 26 October 1948.
64 Squadron 23 August 1949.
Brooklands Aviation 13 January 1950.
22 MU 1 August 1950.
204 AFS 24 July 1951.
231 CU 4 March 1952.
22 MU 12 May 1953.
FTU 2 September 1954.
15 MU 18 March 1956.
Scrapped 6 August 1957.

LR553
13 MU 7 April 1944.
10 MU 29 April 1944.
13 MU 7 April 1944.
0 MU 29 April 1944.
USAAF 25 May 1944.
38 MU 1 August 1946.
SOC 20 August 1947.
9 MU Cosford
10 MU Hulluvington
22 MU Siloth
44 MU Edzell
51 MU Lichfeld
57 MU Castle Kennedy

OC – Struck off charge.
RTP – Reduced to produce (spares material).
H&SA – Hants and Sussex Aviation (a contractor).
RNAS – Royal Naval Air Station, indicating a handover to the Royal Navy.
Martin Hearn, a contractor involved in Mosquito repairs located at Hooton Park near Liverpool.
Cat. E – Written off.
Cat. E followed by Cat. B indicates a re-assessment of the original damage.
Cat. B – repairs required are beyond the capacity of the unit to which the aircraft is assigned.
FA – Flying accident..
FB – Flying battle, indicating a loss in action.
RW – Repair in works.
ROS – Repair on site (of an incident).
AW/CN – Awaiting collection.
MR – Major repairs.
Marshalls of Cambridge, a major contractor involved in overhaul/repair/modification of Mosquito aircraft.
MI – Major inspection.
R&SU – Repair and Salvage Unit.
21/2/46 – Annual census date
BSAA – British South American Airline Corporation.
APDL – Awaiting Disposal.

Bibliography

Altner, Helmut. *Berlin. Dance of Death*. Havertown. PA: Casemate. 2002.

Anderson, Duncan. *Fall of the Reich*. Osceola, WI: MBI Pub. Co. 2000.

Anderton, David A. *Aggressors: Interceptor vs Heavy Bomber.* New York: Zokeisha Publications Ltd. 1991.

Andrews, Paul, & William H. Adams. *Mighty Eighth Combat Chronology & Supplement*. Savannah, GA: 8th AF Memorial Museum Foundation Publication. 1997.

—. *The Mighty Eighth Combat Chronology*. Savannah, GA: 8th AF Memorial Museum Foundation Publications. 1997.

—. *The Mighty Eighth Roll of Honor. Savannah, GA*: 8th AF Memorial Museum Foundation Publication. 1997.

—. *Heavy Bombers of the Mighty Eighth.* Savannah, GA: 8th AF Memorial Museum Foundation Publications. 1995.

Andrews, Paul; William H. Adams; & Jon H. Woolnough. *Bits and Pieces of the Mighty Eighth*. Savannah, GA: 8th AF Memorial Museum Foundation Publication. 1995.

Arnold, Henry H. *Global Mission.* Blue Ridge Summit. PA: Tab Books. 1989.

Beevor, Antony. *The Fall of Berlin 1945*. New York: Viking. 2002.

Boesen, Victor. *Storm: Irving Krick vs. the US Weather Bureaucracy*, New York: Putnam. 1978.

Brookes, Andrew J. *Photo Reconnaissance.* London: Ian Allan. 1975.

Brown, Anthony Cave. *Secret War Report of the OSS.* New York: Berkley Pub. Corp. 1976.

Brown, Ewing F. *The Weatherman Let Them Fly.* PA Furnace, PA: 8th AF Historical Society. 1993.

Carty, Pat. *Secret Squadrons of the Eighth*. London: Ian Allen. 1990.

Craven, Wesley F. & James L. Cate. *The AAF in WWII. Vol. III*: University Chicago Press. 1948.

Croman, G.P., & D. J. Smith. *DH Mosquito Crash Log.* Wigan, UK: Clarington Press Ltd. (not date)

Davis, Richard D. *Carl A. Spaatz and the Air War in Europe*, Center for Air Force History. Washington, DC: GPO. 1993.

Eisenhower, David. *Eisenhower At War 19431945*. New York: Random House. 1986.

Ensminger, Thomas L. *Spies. Supplies and Moonlit Skies. Vol. I, Vol. 2.* Fairborn, OH: Hillridge Wildlife Refuge. 2000-2004.

Freeman, Roger. *The Mighty Eighth.* London: Macdonald & Co. 1970.

—. *Mighty Eighth War Diary.* New York: Jane's. 1981.

—. *Mighty Eighth War Manual.* New York: Jane's. 1981.

—. *Mighty Eighth in Colour.* London: Cassell. 2001.

—. *Mighty Eighth. Warpaint and Heraldry.* London: Arms & Armour Press. 1997.

—. *Mighty Eighth*. London: Arms and Armour Press. 1986.

Fish, Robert W., *They Flew By Night,* 801/49 BG Association. 1992.

Giles, Brian D. *Meteorology and World War II.* University of Birmingham, UK: Royal Meteorological Society, Department of Geography. 1987.

Hagedorn, Dan. *Alae Supra A Canalem: Wings Over the Canal.* Paducah, Kentucky: Turner Publishing Co. 1995.

Halliday, Joe. *Mosquito.* Canada: Doubleday. 1970.

Henshall, Philip. *Vengeance.* Stroud. UK: Allan Sutton Pub. Ltd. 1995.

Henshall, Philip. *Hitler's V-Weapons Sites.* Stroud. UK: Sutton Publishing. 2002.

Hinsley, F.H. *British Intelligence in the Second World War.* Vol. 2. New York: Cambridge University Press. 1984.

Husted, H.H. *Thumb Nail History of WWII.* Boston: Humphries. 1948.

Infield, Glenn B. *Unarmed and Unafraid.* New York: Macmillan. 1970.

Ivie, Tom. *Aerial Reconnaissance: 10 Photo Recon Group in WWII.* Blue Ridge Summit, PA: Tab Books. 1981.

Jansen, Harold F. *History of the 446 BG (H).* The Hague, Netherlands. 1990.

Fussell-Keene, Pat. *Eyes of the Eighth.* Sun City, AZ: CAVU Pub. 1996.

Ketchum, Carolton G. *The Recollections of Colonel Retread.* Pittsburgh, PA: Hart Books. 1976.

Kreis, John F. *Piercing the Fog.* Bolling AFB. Washington, DC: Air Force History and Museums Program. 1996.

Leaf, Edward. *Above All Unseen. London:* Patrick Stephens. 1997.

Le Strange, Richard. *Century Bombers.* Aylsham, England: M.F. Barnwell. 1990.

Lewis, Brendan R. *Hitler Youth: the Hitlerjugend in War and Peace, 1933-45,* Osceola, WI: MBI Publishing Co. 2000.

Lloyd, Alwyn. *Liberator: Americas' Global Bomber.* Missoula, MT: Pictorial Histories Pub. Co. 1993.

Lucas, James. *Kommando.* New York: St. Martin's Press. 1985.

Powys-Lybbe, Ursula. T*he Eye of Intelligence.* London: W. Kimber. 1983.

Marshall, George C.; Henry H. Arnold: & Ernest .J. King, *The War Reports.* New York: J.B. Lipincott Co. 1947.

Maurer, Maurer. *Air Force Combat Units of World War Two.* Washington. DC: GPO. 1961.

MacCloseky, Monro. *Secret Air Missions.* New York: R. Rosen Press. 1966.

Mead, Peter. *Eye In The Air.* London: HMSO. 1983.

Merrick, K A. *Flights of the Forgotten.* London: Arms & Armour Press. 1989.

Middlebrook, Martin. *Arnhem 1944. The Airborne Battle.* Boulder. CO: Westview Press. 1994.

Middlebrook, Martin. *The Bomber Command War Diaries.* New York: Penguin Books. 1990.

Nesbitt, Roy Conyers. *Eyes of the RAF.* Washington, DC: Sutton Pub. 1996.

Olmsted, Merle C. *The Yoxford Boys.* Fallbrook. CA: Aero Publishers. 1971.

Olsen, Jack. *Aphrodite: Desperate Mission.* New York: G.P. Putnam's Sons. 1970.

Ostrom, Allen. *398th BG Remembrances.* Seattle, Washington: Vanguard Press. 1989.

Pape, Gary R. and R. C. Harrison. *Queen of the Midnight Skies.* Atglen. PA: Schiffer Publishing Ltd. 1992.

Parnell, Ben. *Carpetbaggers.* Austin. Texas: Eakin Press. 1987.

Parry, Simon. *Intruders Over Britain.* Surbition. England: Air Research. 1987.

Persico, Joseph E. *Piercing The Reich.* New York: Viking Press. 1979.

Polmar, Norman & Thomas B. Allen. *World War II: America at War 1941-45.* New York: Random House. 1991.

Poolman, Kenneth. *Fock-Wulfe Condor: Scourge of the Atlantic.* London: Macdonald and Janes's. 1978.

Powell, Geoffrey. *The Devil's Birthday: the Bridges to Arnhem 1944.* London: Buchan & Enright. 1984.

Powys-Lybbe, Ursula. *The Eye of Intelligence.* London: W. Kimber. 1983.

Roosevelt, Elliott, & James Brough. *A Rendezvous With Destiny.* New York: Putnam. 1975.

Roosevelt, Elliott. *As He Saw It.* New York: Dual, Sloan & Pearce. 1946.

Roosevelt, Kermit. *The Overseas Targets: War Report of the OSS. Vol. II.* New York: Walker and Company. 1976.

Schaffer, Ronald. *Wings of Judgment.* New York: Oxford University Press. 1985.

Sharp, Martin C. & Michael J. F. Bowyer. *Mosquito.* London: Faber. 1971.

Smith, Constance Babington. *Evidence In Camera.* London: Chatto and Windus. 1958.

Snyder, Louis L. *The War: A Concise History 1939-1945.* New York: J. Messner. 1960.

Stacey, P. *The Victory Campaign, Vol. III,* 1944-45. Ottawa: 1960.

Stagg, J. M. *Forecast For Overlord.* Ian Allen. London. 1971. London: Ian Allen. 1971.

Stanley, Roy M. *World War Two Photo Intelligence.* New York: Scribner. 1981.

Staerck, Chris. *Allied Photo Reconnaissance of World War Two.* San Diego: CA. Thunder Bay Press. 1998.

Swanberg, W.A. *Citizen Hearst.* New York: Galahad Books. 1996.

Trest, Warren A. *Air Force Roles and Missions: A History.* Washington, DC: USGPO. 1998.

Webster, Charles & Noble Franklin. *Strategic Air Offensive Against Germany 1939-1945. Vol. III.* London: HM Stationary Office. 1961.

Williams, Geoffrey. *Flying Through Fire.* London: Grange Books. 1996.

Whiting, Charles. *Ardennes. The Secret War.* New York: Stein and Day. 1985.

Magazine/Journal Articles

Bulletin Board. 25th BG Association Newsletter 11:2 (February 1997).

—. 25th BG Association Newsletter 11:3 (May 1997).

—. 25th BG Association Newsletters 18:3 (August 2004).

Clark, Ron. "Cloak and Dagger." *Air Enthusiast.* (July/August. 2007).

Evans, Stewart P. "Aphrodite Missions." *Fly Past.* (October-November 1988).

Ford, Daniel. "'Star N Bar' Mosquitoes." *Fly Past.* (January 2005).

Freeman, Roger A. "Project Aphrodite." *Fly Past.* (December 1998.

Journal of 7th Photo Recon Group. 10 (3 February 1991).

Newsletter. 3 Strategic Air Depot (SAD). (1 July 1989).

Roosevelt, Elliott. "If My Parents Could Hear Me." *Parade Magazine.* (10 October 1982).

Swanborough, Gordon. "British Aircraft at War. 1939-1945." *Air Pictorial.* (October/November 1994).

XBLCD, Newsletter of the OSS CommVets Amateur Radio Net.

III:12 (March-April) 1990.

XBLDC, Newsletter of the OSS CommVets Amateur Radio Net. IV:1 (May-June) 1990.

Internet

http://www.erfgoedleidschendam.nl/jaargang-2003/43-anton-schrader-en-de-werf-van-van-ravesteijn-in-19431

AF Historical Research Agency, Maxwell AFB, Alabama.

Aircraft Accident Reports.

168.7043-1. Charges made by E. Roosevelt 1942-43.

202.2-23. Night Photo.

202.2-24. Night Photography.

218.94-1. Pilotless Aircraft as Guided Missiles 1945-46. Reel No. A2242.

239.0512-997. Interview with E. Roosevelt. 1941-43 (13 Pages).

307.01. Caribbean Wing ATC. July 1943-June 1944.

486-4. CCS. 2 July 1944.

505.29-8. F-8 Mosquito Aircraft Record Cards. Reel No. B5093.

519.161-3. General C. Spaatz to Chief of Air Staff. letter dated 15 December 1944.

519.431-1. Third AD Tactical Mission Report. Reel No. A5687.

519.428-1. Aphrodite. Reel No. A5687.

519.600. History. Directorate of Intelligence.

519.809.90631-13. GCA in ETO.

519.9063. Radar Altimeter.

519.906-13. ASP Report No. 12. Survey of GCS (AN/MPN-1) in the ETO, Jan.-May 1945.

519.9071. Window ECM.

520.809.266. Aircraft Movements.

520.919. M-46 Photo Flash Bombs.

520.921. H2X.

520.923. Oct. 44-Sept. 45. Combat Camera Units.

527.3913. Aphrodite. Reel No. B5529.

527.431A-1 thru 29. Third Air Division. Final Aphrodite Project Report. Reel No. B5529.

527.476. Old, Archie J. Jr., Col. Report on Shuttle Mission to Russia 6 July 1944.

527.451. Frantic Mission.

1415 ABU Sweden. Reel No. A0173.

ABG 1410 ABU Fife. Metfield. Reel No. A0171.

CC-8-HI..

GP-Photo-3.

GP-FE-3-HI.

GP-25-SU-OP-S. Reel No. B0093 thru B0098.

GP-482-HI.

GP-482-SU-RE.

GP-482-SU-OP.

GP-492-SU-OPS.Reel No. B0654.

GP-RCN-802-SU-OP-S. Reel No. B0797 and B0798.

MACR. 19 March 1945.

SQ-Photo-5-HI.

SQ-RCN-8-HI.

SQ-FI-48-HI.50. Reel No. A0742.

SQ-SV-369-HI. Reel No. A9052.

SQ-Bomb-654-SU-RE.

SQ-Bomb-646-HI. 652 SQ. Reel No. A0650.

PW-90-HI.

WG-90-SW.

WG-325-HI. Reel No. C0164 and C0165.

WG-MAAF-HI.

Archives Canada, Ottawa.

RG 24. Volume 5187. File S-15-9-59. (Vol.5). Chiefs of Staff Memorandum. 5 April 1943.

NARA. College Park, Maryland.

RD2612. Airplanes-Mosquito-British-General.

RG 226. OSS files & Agent reports.

RG 226. Joan-Eleanor Log.

RG 331. SHAEF. Box 53.